Physical Climatology

Physical Climatology

Editor

Benja Rambeloson

scitus academics

Physical Climatology

Edited by **Benja Rambeloson**

ISBN: 978-1-68117-578-2
Library of Congress Control Number: 2016946432

SCITUS Academics LLC,
616, Corporate Way, Suite 2, 4766,
Valley Cottage, NY 10989

www.scitusacademics.com

Notice

Printed in United States of America on Acid Free Paper

Preface

The gaseous envelop surrounding the earth is called atmosphere while the science dealing with the study of the atmospheric components and characteristics is called meteorology and climatology. Climatology is the scientific study of climate and is a major branch of meteorology. Climatology is the tool that is used to develop long-range forecasts. There are three principal approaches to the study of climatology: physical, descriptive, and dynamic. The physical climatology approach seeks to explain the differences in climate in light of the physical processes influencing climate and the processes producing the various kinds of physical climates, such as marine, desert, and mountain. Physical climatology covenants with explanations of climate rather than with presentation.

Physical Climatology deals with the interpretation of factors responsible for the spatial and temporal variations of exchange of air circulations, heat and humidity. It studies various elements of weather namely insolation, temperature, precipitation, fogs, visibility etc. Different elements are formed due to combinations of these weather elements. The occurrences of different combinations of these weather elements are accomplished through different processes and mechanisms. Thus, these processes of exchange of heat, humidity, and momentum between atmosphere and earth's surface are also studied thoroughly. It is thus, evident that physical climatology studies the factors and processes of regional variations of climatic conditions.

Table of Contents

Chapter 1

Climatology Analysis of Aerosol Effect on Marine Water Cloud from Long-Term Satellite Climate Data Records

Xuepeng Zhao [1,*], Andrew K. Heidinger [2] and Andi Walther [3]

[1]National Centers for Environmental Information (NCEI), NOAA/NESDIS, Asheville, NC 28801, USA

[2]Center for Satellite Applications and Research (STAR), NOAA/NESDIS, Madison, WI 53706, USA

[3]Cooperative Institute for Meteorological Satellite Studies (CIMSS), UW-Madison, Madison, WI 53706, USA

ABSTRACT

Satellite aerosol and cloud climate data records (CDRs) have been used successfully to study the aerosol indirect effect (AIE). Data from the Advanced Very High Resolution Radiometer (AVHRR) now span more than 30 years and allow these studies to be conducted from a climatology perspective. In this paper, AVHRR data are used to study the AIE on water clouds over the global oceans. Correlation analysis between aerosol optical thickness (AOT) and cloud parameters, including cloud droplet effective radius (CDER), cloud optical depth (COD), cloud water path (CWP), and cloud cover fraction (CCF), is performed. For the first time from satellite observations, the long-term trend in AIE over the global oceans is also examined. Three regimes have been identified: (1) AOT < 0.08, where CDER increases with AOT; (2) 0.08 < AOT < 0.3, where CDER generally decreases when AOT increases; and (3) AOT > 0.3, where CDER first increases with

AOT and then levels off. AIE is easy to manifest in the CDER reduction in the second regime (named Regime 2), which is identified as the AIE sensitive/effective regime. The AIE manifested in the consistent changes of all four cloud variables (CDER, COD, CWP, and CCF) together is located only in limited areas and with evident seasonal variations. The long-term trend of CDER changes due to the AIE of AOT changes is detected and falls into three scenarios: Evident CDER decreasing (increasing) with significant AOT increasing (decreasing) and evident CDER decreasing with limited AOT increasing but AOT values fall in the AIE sensitive Regime 2.

KEYWORDS

aerosol; cloud; climate data record; aerosol indirect effect; satellite

1. INTRODUCTION

Cloud microphysical structures and properties play an important role in the cloud-radiative-precipitation feedbacks and provide a critical link between the energy and hydrological cycles of Earth's climate system. Atmospheric aerosol in the accumulation mode is the major source of cloud condensation nuclei and is therefore critical for the formation of cloud microphysical structures and properties [1,2,3]. Hygroscopic aerosol particles are called condensation nuclei (CN). A CN that can be activated by growth into the sub-micron size regime by absorption of water vapor in the cloud environment are called cloud condensation nuclei (CCN). Changes in atmospheric aerosols due to anthropogenic emissions will result in the modification of CCN and cloud microphysical properties, which will further change the cloud effect on the radiation budget and precipitation efficiency. This entire process is named as the aerosol indirect effect (AIE) [4,5,6,7,8,9,10].

Two types of aerosol indirect effect have been identified. The first type is called the first indirect effect (also named "Twomey effect" or "albedo effect"), which refers to the increase of cloud albedo due to cloud particle number increase and size reduction associated with the increase of aerosols under the assumption of constant cloud liquid water amount [4,5,6,9,11]. The second type is called the second indirect effect (also named "lifetime effect"), which refers to the prolonged cloud lifetime and cloud reflectance enhancement due to the reduction of precipitation efficiency resulted from cloud droplet number increase and size reduction induced from the increase of aerosols [7,10,12,13,14,15].

The first indirect effect has strong observational and modeling support. However, large ambiguities and uncertainties remain in the understanding of the second indirect effect. This is because the aerosol induced changes in cloud microphysical/optical properties (such as cloud liquid water amount, cloud optical depth, precipitation efficiency, *etc.*) are entangled with the changes in the cloud macrophysical properties (such as cloud height, cloud vertical motion, cloud lifetime, *etc.*) caused by cloud dynamics. These factors may suppress or amplify each other depending on the specific meteorological conditions and cloud regimes [10,16,17].

There have been many satellite observations, *in-situ* measurements and field campaigns of aerosol indirect effect in the past three decades trying to narrow the uncertainties of aerosol effect on cloud and precipitation. The *in-situ* and field measurements were mainly focused on process study of aerosol indirect effect for specific regions, seasons, and cloud types (e.g., [12,16,18,19,20,21,22,23,24]), while satellite observations were used either for studying the AIE for different aerosol loading scenarios in selected regions (e.g., [6,13, 15,25,26,27,28,29,30]) or for determining the correlation relationships between the aerosol loading and cloud microphysical properties, such as cloud droplet size, liquid water amount, and cloud optical depth, in broader areas (such as over the ocean, land, or globe) by using limit length of satellite data records (e.g., [28,31,32, 33,34,35,36,37]). Both field measurements and satellite observations of AIE have been used to develop databases for parameterization, initialization, and validation of regional and global climate models (e.g., [38,39, 40,41,42,43,44,45,46]), which makes it possible to more accurately simulate the AIE processes and perform climatological analysis of the AIE on both regional and global scales (e.g., [46,47, 48,49,50,51,52,53,54,55]). Due to the complexity of the AIE issues, remote sensing observation, *in-situ* measurement, and model simulation are often used together in the AIE investigation (e.g., [52,56,57]).

In this study, we analyze the correlation relationships between aerosol loading and cloud microphysical properties over global ocean for marine water clouds (mainly stratus and stratocumulus) from a climatology perspective by using more than 30-year climate data records (CDRs) of aerosols and clouds derived from the National Ocean and Atmospheric Administration (NOAA) operational satellite observations of the Advanced

Very High Resolution Radiometer (AVHRR). Identifying the aerosol indirect effect signatures contained in the long-term AVHRR satellite observation and examining the changing trend of the AIE signatures are the two major objectives of this study. The results will benefit the determination of global AIE climatology, which, so far, is mainly based on long-term model simulations. The content of the paper is arranged as follows: AVHRR satellite aerosol and cloud climate data record (CDR) datasets used in our analyses are introduced in the next section, the AIE climatology analysis and results are provided in Section 3, summary and conclusions are given in the closing section.

2. SATELLITE DATA

This study uses version 3 of NOAA operational satellite AVHRR aerosol optical thickness (AOT) CDR product and version 5.3 NOAA Pathfinder Atmospheres-Extended (PATMOS-x) AVHRR cloud CDR products. The AVHRR data comprise the longest operational global satellite observation of the Earth's atmosphere and surface. Observed radiances/reflectances have been widely used to derive long-term products of atmospheric variables, such as cloud and aerosol. AVHRR radiances have been carefully re-calibrated recently using more accurate Moderate Resolution Imaging Spectroradiometer (MODIS) radiances to achieve a quality suitable for long-term trend studies [58,59,60]. For example, an examination of cloud amount and cloud optical depth time series demonstrated that the resultant AVHRR inter-satellite biases are less than 1% after cross-calibration with MODIS and retrospectively inter-calibrated among AVHRR sensors from NOAA-6, -7, -8, -9, -10, -11, -12, -14, -15, -16, -17, -18, and -19 satellites [61]. The newly calibrated AVHRR global area coverage (GAC) radiances in 1 km × 4 km spatial resolution are subsampled rather than averaged to 0.1° × 0.1° degree equal angle orbital grid in the PATMOS-x Level-2b CDR data [61].

PATMOS-x cloud products used in our analysis are cloud droplet effective radius (CDER), cloud optical depth (COD), cloud water path (CWP), and cloud cover fraction (CCF) for marine water clouds (mainly stratus and stratocumulus). Different algorithms [61,62] were used to derive these cloud products for daytime and nighttime, respectively, by using the inter-calibrated AVHRR GAC radiance data. They are output together with the inter-calibrated radiances and selected ancillary data (e.g., surface types) as the Level-2b daily CDR products, which allow other applications and CDRs to be generated from the PATMOS-x CDR data. PATMOS-x CDR data span from 1978 to present and should extend past 2020. Only the data after 1980 will be used in the current study in view of many missing observations in the first three years (1978–1980) of AVHRR data records. PATMOS-x cloud products also include five cloud types: warm liquid water, supercooled water-mixed phase, opaque ice, nonopaque high ice (cirrus), and cloud overlap (multiple cloud layers). The cloud types are determined using a multispectral algorithm developed for AVHRR using a series of cloud phase threshold tests from the infrared and near-infrared channels. Specifically, three sequential tests are involved in the determination of warm liquid water type: (1) brightness temperature of 11 μm channel larger than 273.16 K (the melting point of pure water); (2) cloud overlap test; and (3) cirrus test. A detailed description of the algorithm and its performance can be found in Pavolonis *et al.* [63]. Both warm liquid water cloud type and ocean surface type contained in the PATMOS-x CDR products are used to identify the marine water cloud in the current analysis.

AVHRR AOT CDR is derived over global water (or ocean) surface from the 0.63 μm channel using a two-channel AVHRR aerosol retrieval algorithm [64] from AVHRR clear-sky reflectance, which is determined from the PATMOS-x AVHRR all-sky reflectance and cloud probability products contained in version 5.3 PATMOS-x AVHRR reflectance/cloud CDRs. Due to the limited channels of AVHRR instrument, only aerosol retrievals over water can be obtained and only the AOT retrieved from 0.63 μm channel achieves climate quality. The retrievals from the spectrally wider 0.86 μm channel are contaminated by water vapor absorption, which is difficult to be accurately quantified, especially in the context of AOT trend. The AVHRR AOT retrieval algorithm had been carefully validated by comparing its results with those from the AERONET ground AOT measurement [64,65] and the MODIS satellite AOT observations [66]. It had been applied to the previous versions of PATMOS-x reflectance to generate long-term AOT data for the study of AOT long-term trend over global and regional scales [67,68,69]. Residual cloud contamination on AOT has been greatly reduced in version 3 AVHRR AOT CDR product compared to the previous versions. It is more reliable for trend studies due to improvement in the accurate cloud detection algorithm. The cloud detection used for the AOT CDR comes from version 5.3 PATMOS-x AVHRR cloud CDR products [70].

Global AVHRR AOT (at 0.63 μm channel) CDR daily products along with the corresponding Level-2b PATMOS-x AVHRR cloud CDR products are used in the current study. The data period spans from 1981 to 2011 and the spatial resolution is 0.1° × 0.1° degree on an equal angle orbital grid. AVHRR AOT retrieval can be problematic at larger solar zenith angles [71,72] beyond 60°N and 60°S latitudes (especially for the winter months), which may result in poor data sampling in monthly AOT averages in these regions. At the same time, cloud detection is degraded over bright snow and ice surfaces at polar latitudes. Thus, we will confine our analysis to within 60°S and 60°N over the global oceans. For a trend analysis, two-and-a-half years of data after the major volcano eruptions of El Chichón in March 1982 and Mt. Pinatubo in June 1991 were discarded to avoid the abrupt AOT perturbations [67].

3. CLIMATOLOGY ANALYSIS AND RESULTS

3.1. Analysis Methodology

In order to evaluate AIE of water clouds over global ocean using the satellite observation, we analyze the correlations between AOT (or τ), which is considered as a proxy of column aerosol concentration [9,28,33,36,37,73], and cloud parameters, including CDER (or r_e), COD (or τ_c), CWP, and CCF. CDER and COD are retrieved directly from the satellite radiances while CWP is computed from CDER and COD. Their relationship can be expressed by the following equation [74,75]:

$$r_e = \frac{9}{5} \times \frac{CWP}{\rho \tau_c} \tag{1}$$

where ρ is the density of liquid water. When aerosol concentration increases (or AOT increases) and the cloud base is coupled with the marine boundary layer, more aerosols will be transported into the cloud layer to become activated CCN and grow to cloud droplets with smaller CDER given fixed cloud liquid water content (or constant CWP) [76,77]. Then COD will increase according to Equation (1). This chain of changes triggered by aerosol loading increase and described by Equation (1) with fixed CWP is the conventional first AIE mentioned in the introduction. Reduced CDER may further suppress precipitation efficiency and increase cloud lifetime and CCF along with the increase in CWP, which is the aforementioned conventional second AIE. Since CWP is derived from CDER and COD in our satellite observations, we hesitate to use CWP as an independent variable to detect the lifetime effect. However, CCF can be used as an independent variable to detect the cloud lifetime aerosol effect because it is derived independently from CDER and COD. A cloud detection algorithm based on the naïve Bayesian probabilistic methodology was used to determine PATMOS-x AVHRR cloud cover fraction. The algorithm was trained and validated by using CALIPSO (Cloud-Aerosol Lidar and Infrared Pathfinder Satellite Observations) lidar observations. For an AVHRR GAC pixel (~5 km in size), its cloud fraction was computed from five pixels of CALIPSO that were closest to the center of the AVHRR GAC pixel. Detailed description of the cloud detection algorithm can be found in Heidinger *et al.* [70].

Since the assumption of constant CWP in Equation (1) holds only occasionally for some scenarios in the real world due to the influence of cloud dynamics and meteorological circumstances, the conventional correlation changes of COD, CWP, and CCF with CDER (caused by increased aerosol loading) described in the first and second AIEs only hold for specific meteorological circumstances and cloud regimes [10]. Shao and Liu [73] found that the response of CDER, COD, CWP, and CCF to the increased aerosol loading for marine stratocumulus off the coast of California can be variable, which is determined by the coupled effects of aerosol loading and macro-scale cloud depth (or cloud dynamics) on cloud droplet size due to their coherent pattern. As a result, satellite derived correlation between AOT and CDER, COD, CWP, and CCF can be either negative, or positive, or neutral depending on the location and regime of the clouds [10,35,73]. Therefore, for the regions where AOT is negatively correlated with CDER from satellite observations, the AIE is manifested and detectable by the satellite observation. If positive correlations of AOT with COD, CWP, and CCF are also observed simultaneously in the same regions where AOT negatively correlated with CDER, the AIE detected by CDER reduction in the satellite observation is further manifested in the satellite observed changes in COD, CWP, and CCF. For the regions with positive or no correlation between AOT and CDER, the AIE may still happen but it is overridden by the cloud dynamic effect as suggested by Shao and Liu [73]. In this case, the AIE cannot be detected by the satellite observation.

3.2. Results

The analysis of the satellite AIE climatology, based on the long-term AVHRR cloud and aerosol climate data records, is presented in the following subsections.

3.2.1. Long-Term Averages

Figure 1 shows the distribution of long-term (1981–2011) averaged monthly mean AOT over the global ocean. The 21 red boxes (see Table 1 also) mark the locations selected for regional survey studies shown later. It is seen the AOT values over the coastal oceanic regions are generally higher than those over the remote oceans due to the abundant continental aerosols flowing out from their source regions over land. For example these continental aerosols can include dust aerosols over the west coast of North Africa, biomass burning aerosols over the west coast of Central/South Africa, and industrial pollution and dust over Arabian Sea and the northeast coast of China. Most of these aerosols with continental origin are hygroscopic and can be transported into the cloud layer, especially in the coupled boundary condition [76], to become activated CCN and modify the cloud microphysical properties. AOT values over the northern hemisphere (NH) are generally higher than over the southern hemisphere (SH) due to the greater abundance of continental aerosols in the NH. Thus, it is not hard to imagine that the AIE should be more active in the NH than in the SH, which will be confirmed in the analysis below.

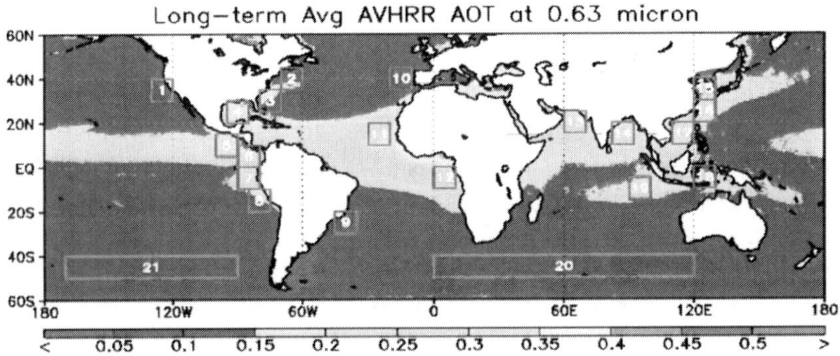

Figure 1. Global map for long-term (1981–2011) averaged monthly mean aerosol optical thickness (AOT). The 21 red boxes (see Table 1 also) are locations selected for regional survey study later.

Table 1. Latitude/longitude bounds of twenty-one regions selected for a regional survey study and marked by the red box in Figure 1.

Region	Longitude Bounds	Latitude Bounds	Region	Longitude Bounds	Latitude Bounds
1	−130.0°, −120.0°	30.0°, 40.0°	12	0.0°, 10.0°	−10.0°, 0.0°
2	−70.0°, −60.0°	35.0°, 45.0°	13	60.0°, 70.0°	15.0°, 25.0°
3	−80.0°, −70.0°	20.0°, 30.0°	14	82.0°, 92.0°	10.0°, 20.0°
4	−95.0°, −85.0°	20.0°, 30.0°	15	120.0°, 130.0°	30.0°, 40.0°
5	−100.0°, −90.0°	5.0°, 15.0°	16	120.0°, 130.0°	20.0°, 30.0°
6	−90.0°, −80.0°	0.0°, 10.0°	17	110.0°, 120.0°	10.0°, 20.0°
7	−90.0°, −80.0°	−10.0°, 0.0°	18	120.0°, 130.0°	−10.0°, 0.0°
8	−85.0°, −75.0°	−20.0°, −10.0°	19	90.0°, 100.0°	−15.0°, −5.0°
9	−45.0°, −35.0°	−30.0°, −20.0°	20	0.0°, 120.0°	−50.0°, −40.0°
10	−20.0°, −10.0°	35.0°, 45.0°	21	−170.0°, −90.0°	−50.0°, −40.0°
11	−30.0°, −20.0°	10.0°, 20.0°	-	-	-

Global maps for long-term (1981–2011) averaged monthly mean CDER, CWP, COD, and CCF of water clouds are shown in Figure 2. It is seen in Figure 2a that CDER values are generally below 10 μm over land and larger than 10 μm and 13 μm over the coastal and remote oceans, respectively. This is because aerosol loadings over land source regions are generally higher than over oceanic regions so that the AIE is more active over land than over ocean. The AIE is also more active over the coastal oceans than over the remote oceans due to abundant continental aerosols outflow from land (see Figure 1). This land and ocean contrast of CDER due to the AIE was also noticed by some previous global studies based on relatively short-length of satellite observations [9,31,52]. It was found by Sekiguchi *et al.* [35] that reduced CDER can only be identified along coastal regions where abundant continental aerosols flow out from land rather than in remote oceanic regions. This is corroborated by the gradual CDER increase from the coastal oceanic regions (with the values in the range of 10–13 μm) to the remote oceanic regions (with the values larger than 13 μm) seen in Figure 2a. Mean AOT over NH oceans (~0.16) is also higher than its SH counterpart (~0.10) due to more continental aerosols in the NH. Correspondingly, CDER also shows consistent hemispheric contrast with a mean value of 12.5 μm in the NH and 14.2 μm in the SH. AIE emerges not only in the reduction of CDER but also in the increase of CWP and COD (see Figure 2b,c) over the east coastal regions of China and North America, Bolivia, Argentina, and North Atlantic Ocean (>40°N). It is hard to see any aerosol indirect effects in the distribution of CCF displayed in Figure 2d. However, it is worth mentioning some interesting features of CCF distribution, which may facilitate the CCF related AIE discussions later. Regions with higher CCF of water clouds are located in several oceanic areas near the west coasts of the continents (including California/Mexico, central South America, South Europe/North Africa, South Africa, and Australia) along with Arabian Sea and the southeast coast of China.

The zonal distribution of CDER for long-term averaged monthly mean data is displayed in Figure 3. It is seen CDER values over land are systemically lower than over ocean due to the AIE, which is consistent with the finding by Han *et al.*[31], Bréon *et al.* [9], and Suzuki *et al.* [52]. However, land and ocean contrast is not manifested in zonal mean COD, CWP, and CCF as in CDER (not shown).

Figure 4 shows the correlation plot between AOT and the logarithm of CDER for long-term averaged global monthly mean data. Three regimes can be identified: (1) AOT < 0.08, where CDER increases with AOT; (2) 0.08 < AOT < 0.3, where CDER generally decreases when AOT increases; and (3)

AOT > 0.3, where CDER first increases with AOT and then levels off. A majority of the sample points falls in the second regime (named Regime 2 hereafter). When aerosol loading is low and falls in the first regime (named Regime 1 hereafter), the supersaturation is probably sufficient to maintain the activation of aerosols and their continuous growth to larger cloud droplets. Thus, CDER and AOT increase together. When aerosol loading reaches into Regime 2, more activated aerosols participate in the intense competition for the water vapor so that their growth to larger cloud droplets is suppressed. As a result, CDER decreases with AOT increases in Regime 2 (the first AIE), which becomes saturation in the negative response when AOT approaches to about 0.3. If aerosol loading further increases to the third regime (named Regime 3 hereafter), intense competition for water vapor evaporates the smaller droplets while the larger droplets will continue to grow at the expense of the smaller droplets until a new equilibrium is established. As a result, CDER increases first with AOT and levels off gradually in the Regime 3. Thus, the conventional AIE is only manifested in the form of CDER decreasing with AOT increasing in Regime 2.

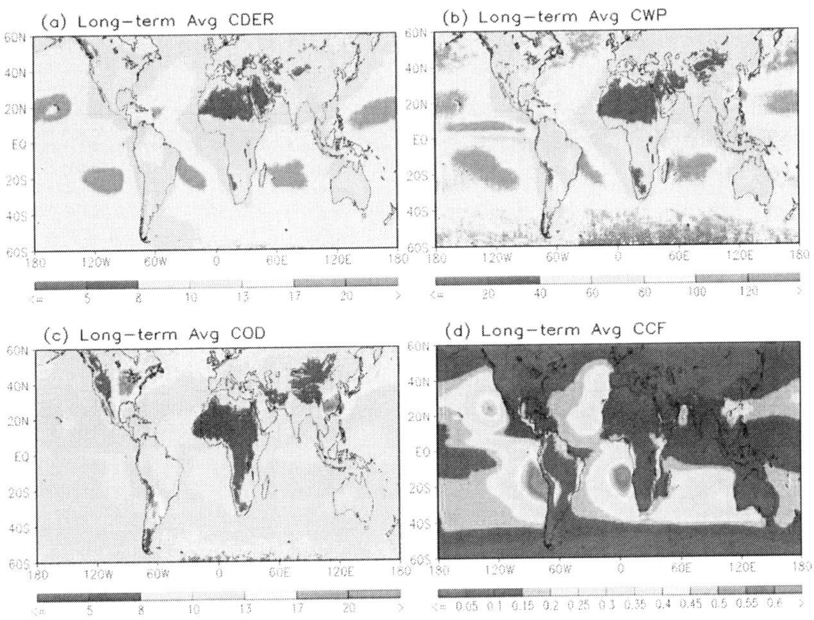

Figure 2. Global maps for long-term (1981–2011) averaged monthly mean: (**a**) cloud droplet effective radius (CDER); (**b**) cloud water path (CWP); (**c**) cloud optical depth (COD); and (**d**) cloud cover fraction (CCF) of water clouds.

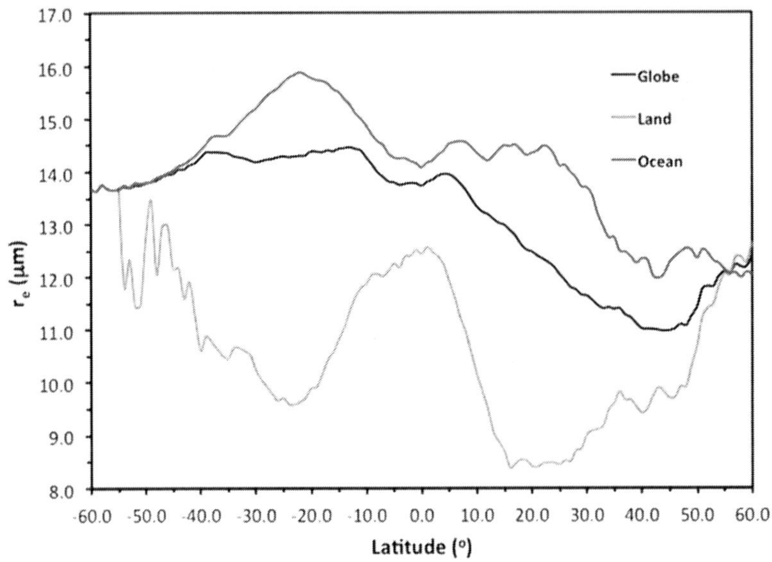

Figure 3. Zonal distribution of CDER for the long-term averaged monthly mean data.

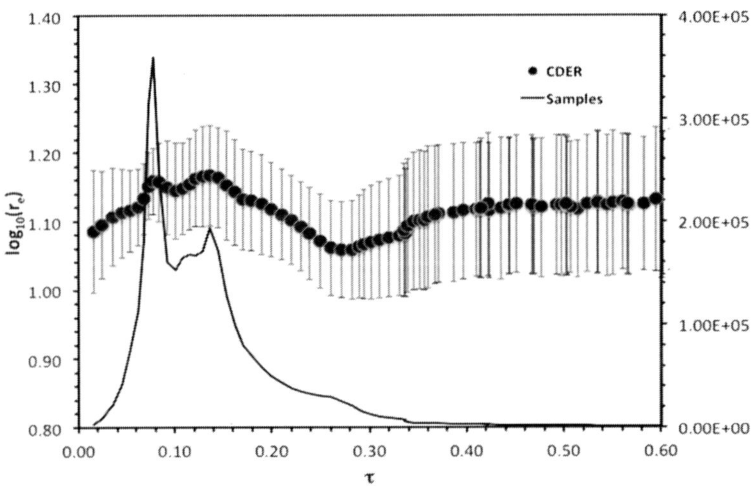

Figure 4. The correlation plot between AOT (τ) and the logarithm of CDER (r_e) for the long-term averaged global monthly mean data over ocean (dots) along with the number of data samples (line), which is binned according to AOT with a 0.01 increment denoted on the right-hand axis. The error bars indicate one standard deviation.

A similar correlation feature in Regime 2 along with the AOT saturation threshold (~0.3) has been noticed in some previous global correlation studies based on limited-length of satellite observations (e.g., [27,33,35]). Even though the correlation features in Regimes 1 and 3 haven't been reported in the previous observational studies, theoretic analysis and analytical model simulations (e.g., [3,27,78]) did find the correlation features similar to those identified in Regimes 1 and 3. Thus, it is worth to further exploring these two new regimes from both aspects of cloud microphysics and dynamics in future studies using model simulations, *in-situ* measurements and satellite observations.

The expected increase of COD associated with the decrease of CDER in Regime 2 due to the conventional AIE is not observed for the current long-term averaged monthly mean data. Sekiguchi *et al.* [35] pointed out that averaging procedures with shorter temporal periods (e.g., one day) in the satellite data analysis of aerosol–cloud interaction may produce different results from the averaging procedures with longer temporal periods (e.g., one month). The averaging process with longer temporal periods may obscure the cloud-aerosol interaction signature (such as the COD increase due to AIE). They suggested the daily averaging is more suitable for capturing the AIE signature than the monthly averaging for COD. By looking at the global maps of long-term averaged monthly mean CDER, COD, and CWP in Figure 2, we can see the AIE signature in the change of CDER is evident from land to coastal oceans and further to remote oceans while the AIE signature in the change of COD and CWP are only noticeable in limited locations. Thus, it is not hard to imagine the global statistics based on the long-term averaged monthly mean COD and CWP may damp or obscure the AIE signature contained in the COD (or CWP) change more than for CDER. At the same time, the averaging process may make some features associated with the cloud dynamic effect more distinct, which can also smear the AIE signature in the long-term averaged satellite observations.

To better identify the oceanic regions with clear manifestation of AIE in CDER, COD, CWP, and CCF, linear correlations between AOT and CDER, COD, CWP, and CCF have been calculated for 2.5° × 2.5° spatial grid over global oceans using their long-term averaged monthly mean values in 0.1° × 0.1° spatial resolution. Thus, the correlation coefficients reflect the relationship between AOT and each cloud variable in the spatial space.

Figure 5 shows global distribution of linear correlation slope for $\log_{10}(AOT)$ *vs.* $\log_{10}(CDER)$, $\log_{10}(COD)$, $\log_{10}(CWP)$, and CCF, respectively. The AIE manifested with distinct negative linear correlation slope of $\log_{10}(AOT)$ *vs.*

$\log_{10}(CDER)$ is observed in Figure 5a over several oceanic areas: (a) a swath extending from the east coast of China to the North Pacific Ocean at about 20°N–40°N latitudes; (b) a swath extending from the east coast of USA to the North Atlantic Ocean at about 25°N–40°N latitudes; (c) Arabian Sea; (d) tropical and subtropical Atlantic Ocean (extending from the west coast of North and Central Africa to the east coast of Central America); (e) a belt area at about ~10°S of tropical South Pacific Ocean extending from the west coast of South America to about 130°W longitude; and (f) the west coast of central South America. These six areas extending from the coastal oceans with offshore continental aerosol influence to the remote oceans will be mentioned several times in the following discussions so that they are named hereafter as area "CH-NPO", "US-NAO", "AS", "Tro-AO", "Tro-SPO", and "WcSAm", respectively.

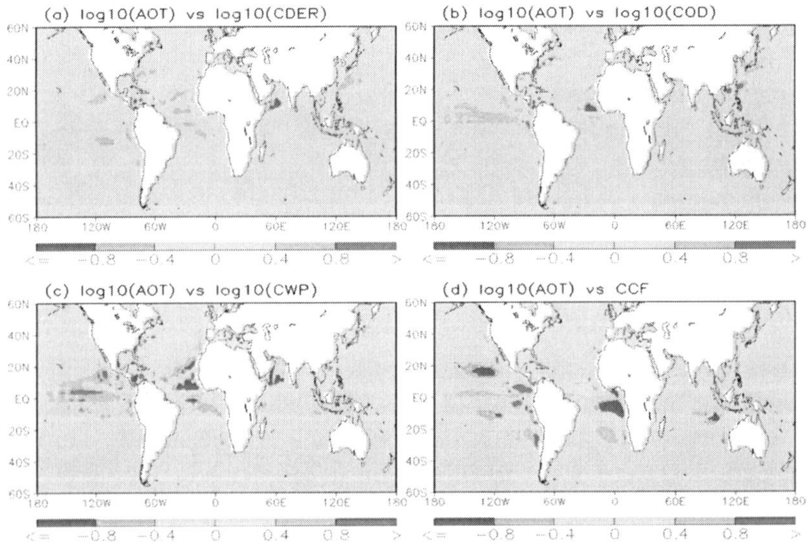

Figure 5. Global distributions of the linear correlation slope between: (**a**) $\log_{10}(AOT)$ and $\log_{10}(CDER)$; (**b**) $\log_{10}(AOT)$ and $\log_{10}(COD)$; (**c**) $\log_{10}(AOT)$ and $\log_{10}(CWP)$; and (**d**) $\log_{10}(AOT)$ and *CCF*. The linear correlation is calculated for 2.5° × 2.5° spatial grids over global oceans using the long-term averaged monthly mean values in 0.1° × 0.1° spatial resolution.

The AIE should manifest itself correspondingly as a positive linear correlation slope of $\log_{10}(AOT)$ *vs.* $\log_{10}(X)$ where *X* is any one of the cloud products used here (COD, CWP and CCF). The AIE signal in COD is seen only over three of the above six areas, CH-NPO, US-NAO, and WcSAm (see Figure 5b), with reduced spatial coverage. The AIE signal in CWP is only observed in CH-NPO and US-NAO areas but with much more reduced spatial coverage (see Figure 5c). The AIE signal in CCF is observed in CH-NPO and AS areas with much more reduced spatial coverage and magnitude (see Figure 5d).Distinct positive correlation between AOT and CCF is also observed in the west costal oceans of several continents (such as central South America, South Africa, and Australia), where higher cloud fraction of marine water clouds are expected (seeFigure 2d) but the corresponding AIE signature is only somewhat noticeable in the AOT and CDER correlation (see Figure 5a) and is not observed in the correlation of AOT *vs.* COD and CWP (see Figure 5b,c). This is probably a signature of the second AIE (lifetime effect) but needs to be confirmed by future research. The distinct positive (or negative) correlation feature observed in the tropical uplifting zone of the atmospheric circulation (or the corresponding subsidence zones over the south and north subtropical oceans) may be a signature of the cloud dynamic effect rather than the signature of AIE, which also warrants a future investigation.

3.2.2. Seasonal Features

Long-term averaged AOT and CDER for each month are shown in Figure 6. AOT values averaged for NH oceans are evidently higher than that averaged for SH oceans in MAM (March, April, May), JJA (June, July, August), and early SON (September, October, November) but close to each other in late SON and early DJF (December, January, February). CDER over ocean is about 4 μm larger than over land for all months and the NH values are systemically lower than their SH counterparts for nearly all the months due to the AIE.

Global distribution maps of the long-term averaged AOT for each season are shown in Figure 7. There are distinct seasonal variations in AOT, especially in the oceanic regions influenced by continental aerosols due to dynamic transport. The outflow of continental aerosols to downstream coastal oceanic regions is the strongest in MAM and JJA, weakest in SON, and moderate in DJF.

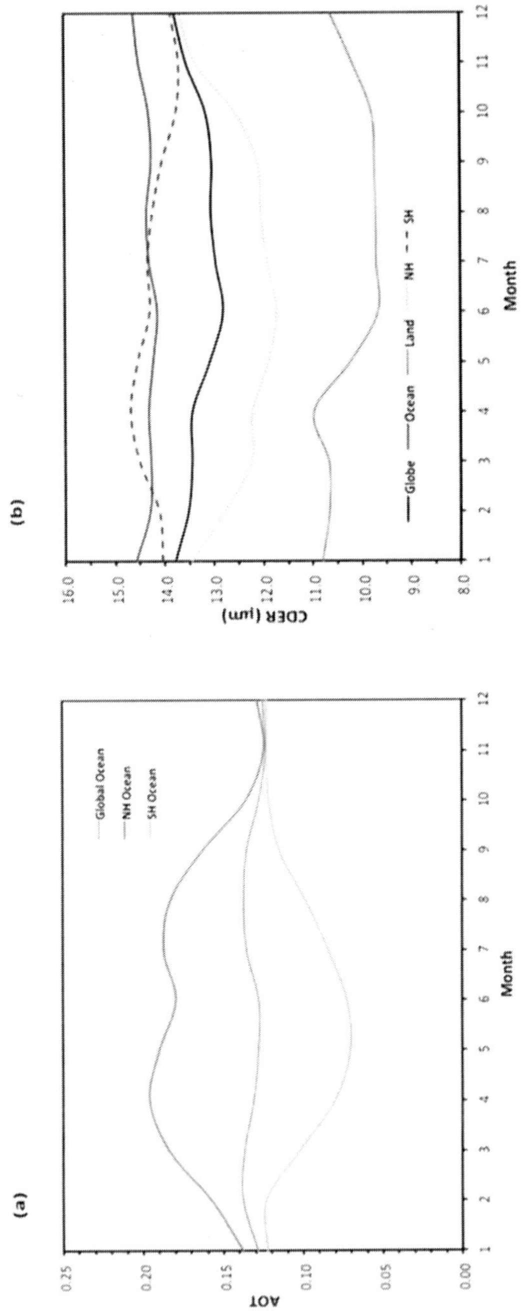

Figure 6. Long-term averaged (**a**) AOT and (**b**) CDER for each month in several geographical areas.

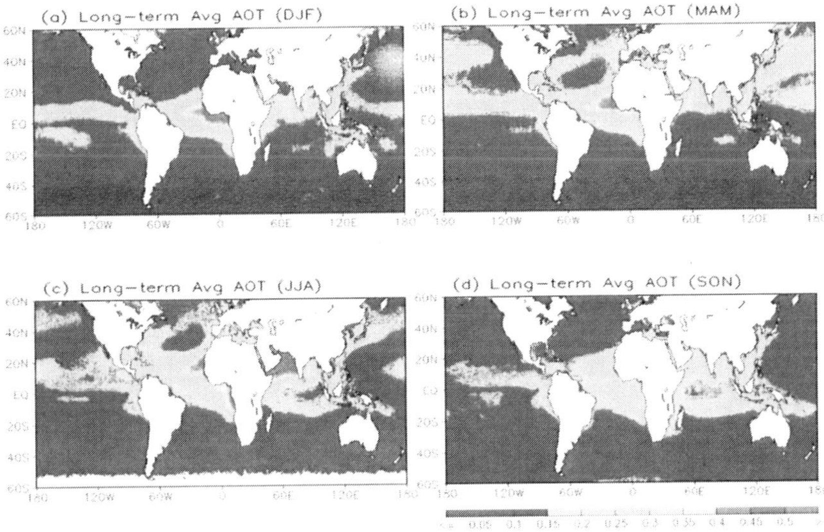

Figure 7. Global maps of the long-term averaged AOT for four seasons: (**a**) December, January, February (DJF); (**b**) March, April, May (MAM); (**c**) June, July, August (JJA); and (**d**) September, October, November (SON).

The global maps of the linear correlation slope for $\log_{10}(AOT)$ *vs.* $\log_{10}(CDER)$, $\log_{10}(COD)$, $\log_{10}(CWP)$, and *CCF* in each season are shown in Figure 8, Figure 9, Figure 10 and Figure 11, respectively. For CDER (see Figure 8), the AIE manifested in CH-NPO, US-NAO, and AS areas (defined above in Figure 5) is most evident in MAM. The AIE in Tro-AO area is a year round feature with some changes in the exact location and intensity. The AIE in Tro-SPO and WcSAm areas is moderate and with less seasonal variations. The corresponding AIE manifested in COD appears only in CH-NPO area (maximal in both intensity and coverage in MAM) and US-NAO area (maximal in both intensity and coverage in JJA) in Figure 9. The AIE manifested in CWP enhancement, when considered together with the AIE of CDER and COD, is only noticeable for CH-NPO area in MAM and somewhat noticeable for US-NAO area in JJA (see Figure 10).

The AIE manifested in CCF enhancement in CH-NPO and AS areas (identified in Figure 5d) is more evident (see Figure 11) in MAM and JJA, respectively. MAM and JJA are the season that AOTs are also in the highest values, respectively, over the two areas (see Figure 7). The AIE manifested in CCF enhancement over the west coast of central South America is more evident in SON, which is consistent with the seasonal variations of AOT values in the region. The AIE manifested in CCF enhancement over the west coast of South Africa is more evident in JJA and SON, which are also the seasons that AOTs are in the highest values in the region. Thus, the AIE manifested clearly for all four cloud variables (CDER, COD, CWP, and CCF) together are in limited locations and with evident seasonal variations, but are generally consistent with the AOT seasonal variations.

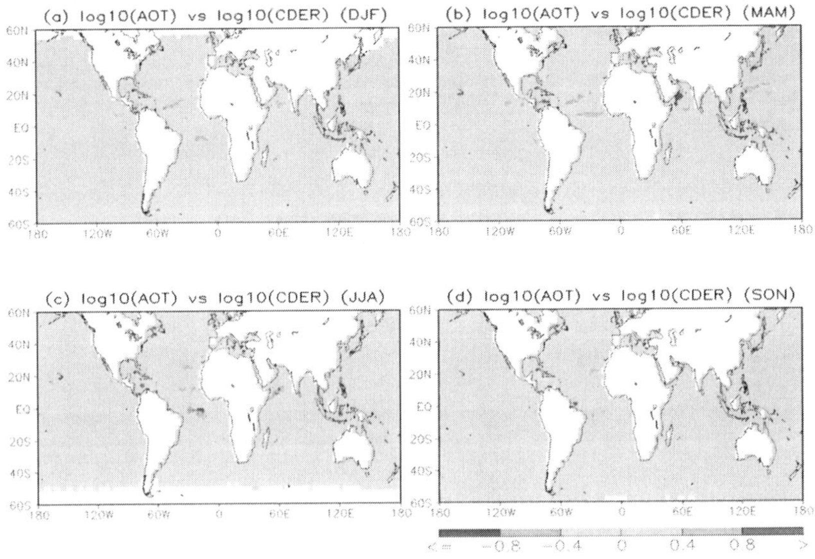

Figure 8. Similar to Figure 5a but shows the global maps of the linear correlation slope for $\log_{10}(AOT)$ *vs.* $\log_{10}(CDER)$ in four seasons: (**a**) DJF; (**b**) MAM; (**c**) JJA; and (**d**) SON. The linear correlation is calculated for 2.5° × 2.5° spatial grids over global oceans using long-term averaged seasonal mean values in 0.1° × 0.1° spatial resolution.

It is noticed that the AIE signatures observed in the correlation analysis for all months in Figure 5 seem to be more evident and consistent than for individual seasons in Figure 8, Figure 9, Figure 10 and Figure 11. As we pointed out above, the averaging processes involved in deriving long-term averaged data may damp AIE signatures as well as boost the effect of cloud dynamics. We suspect the cloud dynamic effect is more prominent in the seasonal scale than in the monthly scale while the AIE signature is reversed. As a result, the AIE signature is less clear and consistent in the correlation based on the seasonal mean values than that based on the monthly mean values. More in-depth studies in the future are needed to confirm this hypothesis.

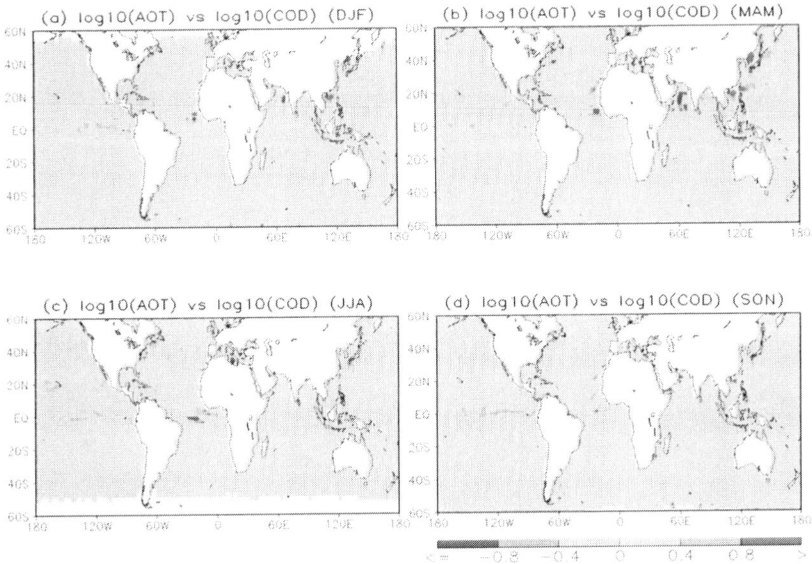

Figure 9. Similar to Figure 8 but shows the global maps of the linear correlation slope for $\log_{10}(AOT)$ *vs.* $\log_{10}(COD)$ in four seasons: (**a**) DJF; (**b**) MAM; (**c**) JJA; and (**d**) SON.

3.2.3. Regional Survey

Twenty-one regions in both coastal and remote oceans marked by the red boxes in Figure 1 are selected for a regional survey study of AIE. These regions are listed in Table 1 for their geographic bounds. Regions 1–19 (coastal oceans), 20 (remote ocean), and 21 (remote ocean) cover an area of 10° (latitude) × 10° (longitude), 10° (latitude) × 120° (longitude), and 10° (latitude) × 80° (longitude), respectively. Figure 12 shows the regional averages of CDER *vs.* AOT for long-term averaged monthly mean values in the 21 regions. It is seen CDER systematically decreases with AOT increases from the remote marine regions (19, 20, 21) with relatively clean air to the offshore regions under the strong influence of continental aerosols, such as the dust in Region 11 (west coast of north Africa), the biomass burning in

Region 12 (west coast of central Africa), and the industrial pollution and dust in Region 13 (Arabian Sea) and Region 15 (northeast coast of China). The AOT values in these regions also fall in the AIE sensitive Regime 2 identified above. Unlike long-term averaged monthly mean CDER, no monotonic changes of COD, CWP, and CCF with AOT are observed in these twenty-one regions.

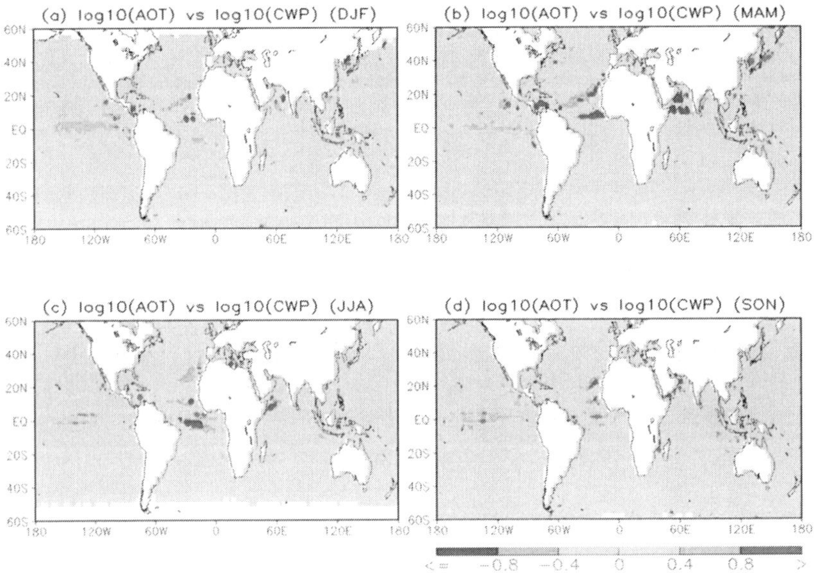

Figure 10. Similar to Figure 8 but shows the global maps of the linear correlation slope for $\log_{10}(AOT)$ *vs.* $\log_{10}(CWP)$ in four seasons: (**a**) DJF; (**b**) MAM; (**c**) JJA; and (**d**) SON.

Persistent marine stratus and stratus-cumulus cloud decks are frequently observed off the coasts of California (Region 1) and Southwest Europe (Region 10) and the intensive campaign studies of aerosol and cloud interaction have been performed in these two regions [16,23,28,79,80,81]. Figure 13 shows AOT *vs.* CDER for the monthly mean data averaged for Regions 1 and 10 over the time period from 1981 to 2011. The AOT values for the two regions indeed fall in AIE sensitive Regime 2 and there is clearly a decreasing trend of CDER along with AOT increase due to AIE. The negative correlation observed in Figure 13 is the AIE signature in the temporal space while the negative correlation observed in Figure 12 is the AIE signature in the spatial space.

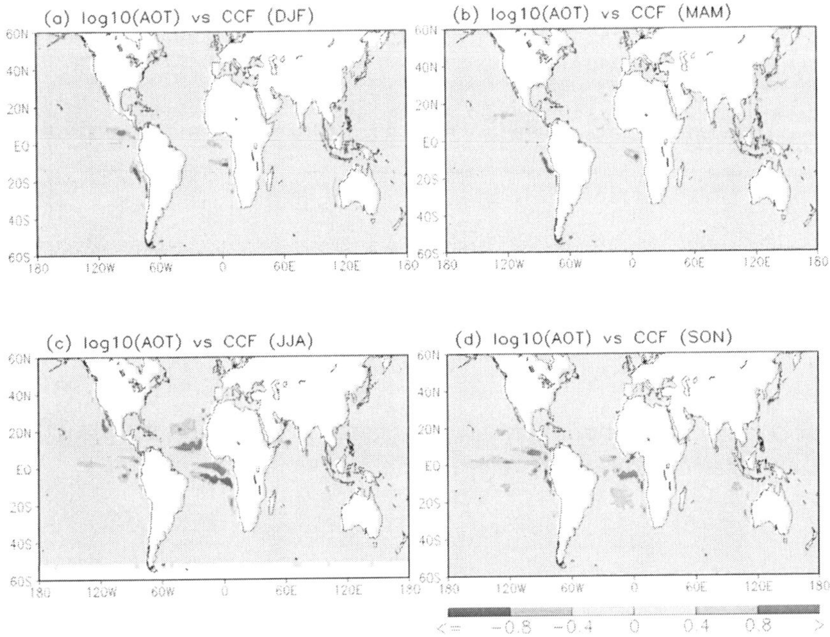

Figure 11. Similar to Figure 8 but shows the global maps of the linear correlation slope for $\log_{10}(AOT)$ *vs. CCF* in four seasons: (**a**) DJF; (**b**) MAM; (**c**) JJA; and (**d**) SON.

In addition to the two coastal oceanic regions shown in Figure 13, we only find another four coastal oceanic regions among the 21 selected regions that show evident negative correlation between AOT and CDER. There is no evident positive correlation between AOT and COD (CWP, CCF) being found in the 21 regions. The linear correlation slope and correlation coefficient between AOT and CDER for the six coastal oceanic regions with evident negative correlation between AOT and CDER along with the typical aerosol types in the six regions are listed in Table 2.

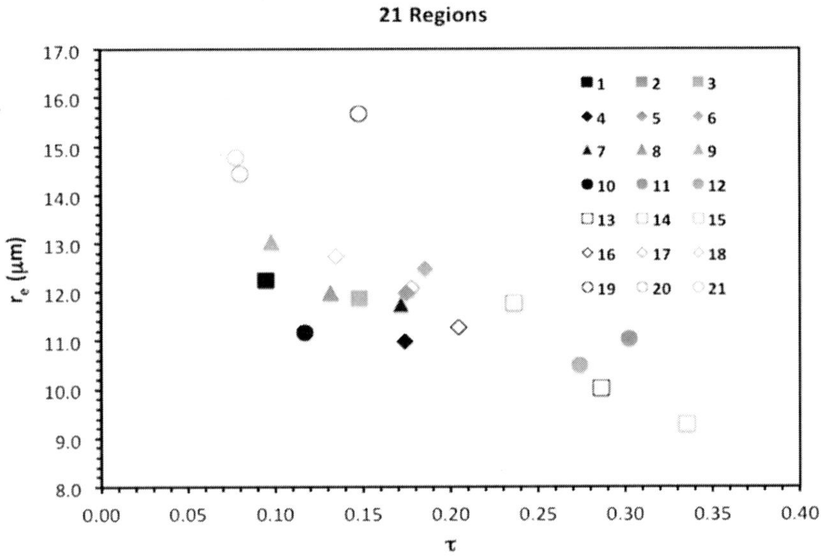

Figure 12. Scatter plot of regionally averaged AOT *vs.* CDER for their long-term averaged monthly mean values in the 21 regions listed in Table 1.

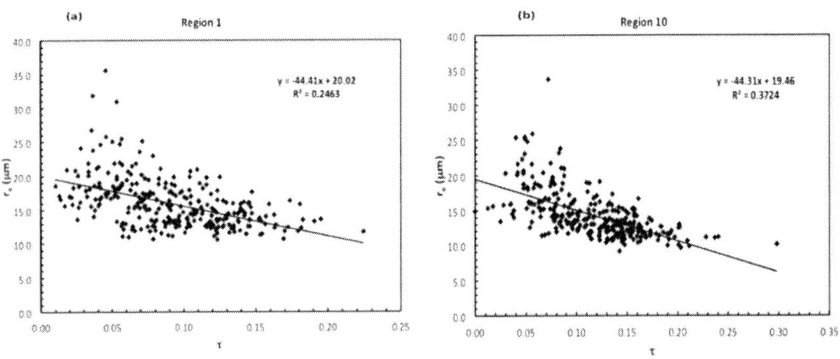

Figure 13. Scatter plot of AOT *vs.* CDER for the monthly mean data averaged for (**a**) Region 1 (offshore of California) and (**b**) Region 10 (offshore of Southwest Europe) over the time period from 1981 to 2011. The straight line is the linear regression line and its formula is also provided along with the square value of the correlation coefficient (R^2).

Table 2. The linear correlation slope and correlation coefficient for the six coastal oceanic regions with evident negative correlation between AOT and CDER among the 21 regions selected for AIE regional survey study. The typical aerosol types in the six regions are also listed in the last column.

Region	Linear Slope	Correlation Coefficient	Typical Aerosol Types
1	−44.43	0.50	Pollution
10	−44.31	0.61	Pollution
11	−14.21	0.34	Dust
13	−18.38	0.57	Dust and Pollution
14	−5.31	0.11	Pollution
16	−9.05	0.23	Pollution

3.2.4. Trend

The prominent advantage of using the long-term satellite CDR products for the correlation studies on the aerosol and cloud interaction is that we can examine the temporal trends of aerosol and cloud variables and investigate AIE in the context of long-term trend, which is not possible using satellite observations over shorter time periods. We will focus our trend study on the statistical linear tendency of CDER corresponding to AOT due to their evident correlation over global oceans identified in the above sections. The real tendencies of AOT and CDER may be much more complex than the statistical linear model. However, the linear trend allows for a simple approximation of the direction and magnitude of the changes in the data and may be adequate for many practical purposes. Results from the linear trend studies are not only commonly used by scientists but are also familiar to policy makers and the public. Thus, it is important to examine the climate change issues under such a linear model. We have successfully applied this linear trend analysis on the investigation of global and regional AOT changes and trends in some previous studies [67,68,69,82].

Here, we give a brief description on the linear trend analysis used in this study and a detailed description of linear trend analysis can be found in the literature (e.g., [83,84,85]). The linear change of a variable, such as CDER, is often measured with the linear trend (*LT*) in the unit of absolute (or percentage) changes per decade. CDER *LT* is the slope of linear regression line for the time series of monthly averaged CDER. We adopt the commonly used decision rule that a detectable trend is indicated at the 5% significance level or 95% confidence level, when $|LT/\sigma| > 2$ [85]. Here, σ is the standard deviation of CDER *LT*. It is difficult to determine σ from the observational data directly so that, as a first order approximation, we use the standard deviation (ε) of monthly averaged *LT* as σ after adjusting the unit of ε to match the unit of *LT*. We further name LT/σ ratio as the "significance" of CDER *LT* in this paper. A significance >+2 (or <−2) indicates the increasing (or decreasing) tendency detected is above 95% confidence level.

Figure 14 shows the global distribution of AOT linear trend (top panel) and the significance (bottom panel). It is seen that AOT increases evidently over the past three decades in the east coast of China, Arabian Sea, Bay of Bengal due to the industrial pollutions associated with fast economic growth in China and India. Evident AOT increases also observed over the west coast of central Africa and Southeast Asia due to the biomass burning. The AOT increase in these regions is above 95% confidence level. Distinct AOT reduction is observed over nearly the whole North Atlantic Ocean, especially over the east coast of USA and the coastal regions of Europe due to more rigid emission control in the developed countries. The AOT contributed from dust particles over the west coast of Sahara also tends to decrease in the past 30 years.

The corresponding global distribution of CDER linear trend (top panel) and the significance (bottom panel) is shown in Figure 15. Evident CDER decreasing trend is observed over the east coast of China due to enhanced AIE from AOT increase in the region. CDER increasing trend is observed over nearly the whole North Atlantic Ocean, especially over the east coast of USA, the coastal regions of Europe, and the west coast of Sahara. CDER increasing trend over these regions is due to the AIE compression since evident AOT decreasing is also observed over these regions. It is interesting to note that evident CDER decreasing trend is also observed over the broad oceanic regions extending from the west coasts of South/North America to the remote east oceanic regions of South/North Atlantic Ocean, which probably is also caused by the AIE enhancement since AOT increasing tendency (even though below 95% confidence level) is dominant in these regions (see Figure 14). Although AOT increasing trend is not significant in these regions but the AOT values (see Figure 1) are in the AIE sensitive Regime 2, which may still trigger active AIE and produce clear CDER decreasing trend. Likewise, evident CDER decreasing trend is also observed in the latitude belt between 20°S and 40°S of Indian and Atlantic Oceans, where AOT increasing trend is the dominant feature but below 95% confidence level. Since AOT values in these regions also fall in the AIE sensitive Regime 2, a minor AOT enhancement also produces intensive AIE and distinct CDER decreasing trend.

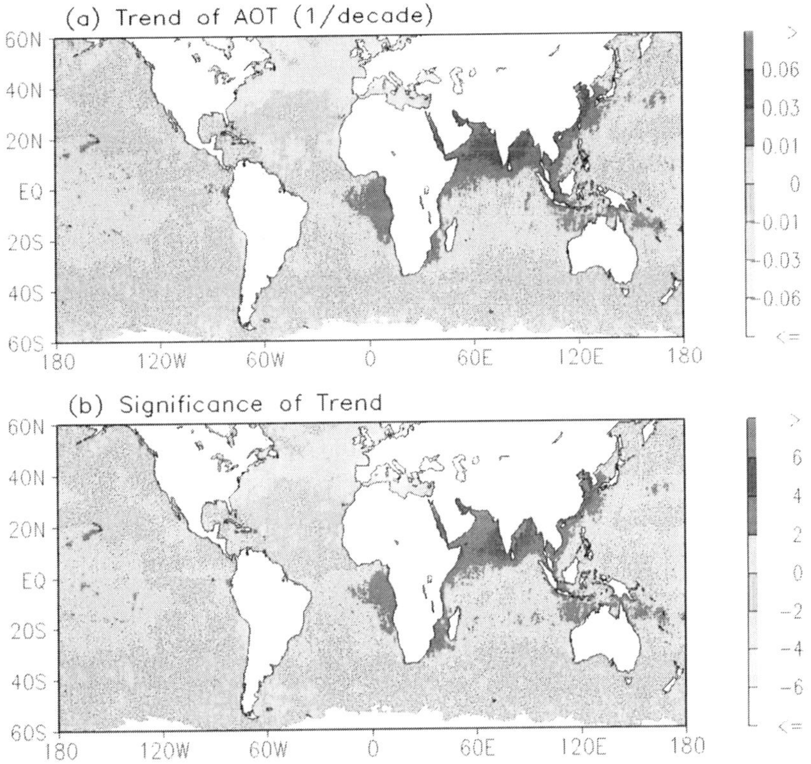

Figure 14. Global maps of (**a**) the linear trend and (**b**) the significance of the trend for AOT monthly mean values from 1981 to 2011. A significance >+2 (or <−2) indicates the increasing (or decreasing) tendency detected is above 95% confidence level.

4. SUMMARY AND CONCLUSIONS

More than 30-year of satellite aerosol and cloud climate data records have been used to study the aerosol indirect effect of marine water clouds (mainly stratus and stratocumulus) over the global ocean from a climatology perspective. Correlation studies between aerosol optical thickness and cloud parameters (CDER, COD, CWP, and CCF) are performed and the long-term trend of AIE over the global ocean is examined for the first time using satellite observations. Three regimes of global AOT have been determined to identify potential AIE manifestation in CDER: (1) AOT < 0.08, where CDER increases with AOT; (2) 0.08 < AOT < 0.3, where CDER generally decreases when AOT increases; and (3) AOT > 0.3, where CDER first increases with AOT and then levels off. The AIE is manifested as CDER decreasing with AOT increasing only in Regime 2, which is the most common regime and defined as the AIE sensitive/effective regime. Although the AIE manifested in CDER is a global phenomenon, it is more evident over land and coastal oceans heavily influenced by continental aerosols arising from wind-blown dust, biomass burning, and industrial emissions. The AIE clearly manifested together in CDER and COD is evident in seasonal variations and located in limited areas: including a swath extending from the east coast of China to the North Pacific Ocean at about 20°N–40°N latitudes (CH-NPO area) and a swath extending from the east coast of USA to the North Atlantic Ocean at about 25°N–40°N latitudes (US-NAO area). The AIE clearly manifested together in all four cloud variables (CDER, COD, CWP, and CCF) is also observed in CH-NPO area. The intensity and spatial coverage are greatly reduced for COD, CWP, and CCF.

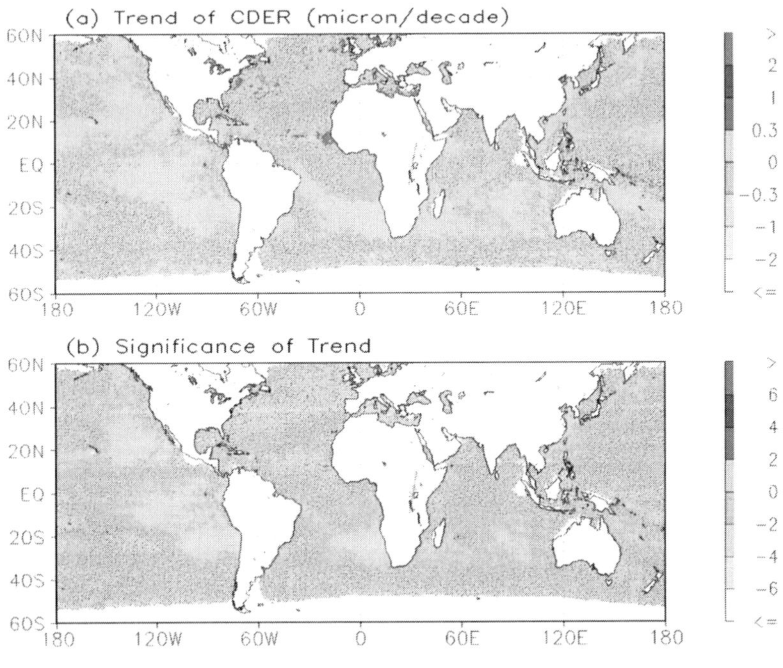

Figure 15. Global maps of (**a**) the linear trend and (**b**) the significance of the trend for CDER monthly mean values from 1981 to 2011. A significance >+2 (or <−2) indicates the increasing (or decreasing) tendency detected is above 95% confidence level.

For the coastal oceanic areas, such as offshore California and southwest Europe, with persistent marine stratus and stratus-cumulus cloud decks and continent aerosol outflows where AOT is in Regime 2 ($0.08 < AOT < 0.3$), the AIE manifested in CDER is evident. This finding over ocean is similar to that found by Feingold *et al.* [27] over land. They noticed the response of CDER to aerosol loading in Brazilian biomass burning regions is the greatest in the region where AOT is relatively small.

A trend analysis based on the 30-year data record is presented. A decreasing (increasing) trend of CDER (AOT) is observed over the east coast of China. An increasing (decreasing) trend of CDER (AOT) is observed over nearly the whole North Atlantic Ocean, especially over the east coast of USA, the coastal regions of Europe, and the west coast of Sahara. A distinct CDER decreasing trend is also observed over the broad oceanic regions of South and North American west coasts and in the latitude belt between 20°S and 40°S of the Indian and Atlantic Oceans even though only a minor AOT increase is identified in these regions, although their AOT values are in the AIE sensitive Regime 2.

There are two caveats in our global statistical analysis based on long-term averaged monthly mean data. The first is the averaging process whose longer temporal period may damp or obscure the cloud-aerosol interaction signatures as well as make some features associated with the cloud dynamic effect more distinct. This enhanced cloud dynamic effect may also smear the AIE signatures in the long-term averaged observations. Thus, the major objective of the current study is to identify AIE signatures retained in the long-term averages and to detect the long-term trend of these AIE signatures. The aim is not to provide the physical mechanisms of AIE from the current global statistical analysis. Even though the analysis of instantaneous observations or short-term averaged data is more suitable for examining the physical mechanisms of AIE, the global statistical analysis and trend detection of AIE signatures are also important elements of the overall climate impact of aerosol–cloud interactions, which depend on long-term global observations. The second is related to the collocation of aerosol and cloud observations. Since aerosol retrievals are performed only for the clear-sky conditions, the collocation error of aerosol and cloud may impact the correlation analysis based on the long-term averaged data. This is especially true for any specific geographic location. Thus, we are very cautious on the interpretation of the strong regional correlation features between AOT and the cloud variables. Only those correlation features consistent for multiple cloud variables are interpreted as possible AIE signatures. Thus, it is important to identify AIE signatures for multiple cloud variables (such as CDER, COD, CWP, and CCF) and check the consistency of these signatures in both global and regional scales as well as in both spatial and temporal spaces.

It is generally agreed that AIE is a local to regional scale process acting on a short time scale. Therefore the instantaneous observations (or the data averaged in a short time period) over a regional scale should retain more AIE signatures than the global data averaged over a long time period. However, the fact that AIE signatures are retained in the global long-term averaged data does suggest that the regional and short time scale aerosol–cloud interaction process does contribute to the long-term global climate signal. Investigations of the physical mechanisms of aerosol–cloud interactions using the instantaneous or short-term observations at regional scales are clearly necessary. However, this paper makes the case that the study of the climate impact of aerosol–cloud interaction using long-term averaged global observations is also necessary. Studying AIE radiative forcing from a climatology perspective also becomes possible by using the AIE climatology identified in the current correlation and trend analysis, which will be our future work. The observational results from the current study will also benefit the modeling studies of aerosol–cloud interaction, especially in the context of the long-term climate effect.

ACKNOWLEDGMENTS

This work is supported by the NOAA CDR program managed at the National Centers of Environment Information (NCEI) of NOAA/NESDIS. NCEI internal reviewers helped improve the paper. The constructive comments and suggestions provided by four anonymous reviewers were very helpful for improving the paper. The views, opinions, and findings contained in this paper are those of the author(s) and should not be construed as an official National Oceanic and Atmospheric Administration or U.S. Government position, policy, or decision.

AUTHOR CONTRIBUTIONS

X. Zhao produced AVHRR aerosol CDR product, performed data analysis, and drafted the paper. A. Heidinger and A. Walther produced PATMOS-x

AVHRR reflectance/cloud CDR products, provided help on the data analysis, improved and polished the paper.

REFERENCES

1. Twomey, S. The nuclei of natural cloud formation, part II, the supersaturation in natural clouds and the variation of cloud droplet concentration. *Pure Appl. Geophys.* **1959**, *43*, 243–249.

2. Verheggen, B.; Cozic, J.; Weingartner, E.; Bower, K.; Mertes, S.; Connolly, P.; Gallagher, M.; Flynn, M.; Choularton, T.; Baltensperger, U. Aerosol partitioning between the interstitial and the condensed phase in mixed-phase clouds. *J. Geophys. Res. Atmos.* **2007**, *112*.

3. Mason, B.J. *The Physics of Clouds*; Clarendon Press: Oxford, UK, 2010.

4. Twomey, S. Pollution and planetary albedo. *Atmos. Environ.* **1974**, *8*, 1251–1256.

5. Twomey, S.A.; Piepgrass, M.; Wolfe, T.L. An assessment of the impact of pollution on global cloud albedo. *Tellus B***1984**, *36*, 356–366.

6. Coakley, J.A.; Bernstein, R.L.; Durkee, P.A. Effect of ship-stack effluents on cloud reflectivity. *Science* **1987**, *237*, 1020–1022.

7. Charlson, R.J.; Schwartz, S.E.; Hales, J.M.; Cess, R.D.; Coakley, J.A.; Hansen, J.E.; Hofmann, D.J. Climate forcing by anthropogenic aerosols. *Science* **1992**, *255*, 423–430.

8. Ramanathan, V.; Crutzen, P.J.; Kiehl, J.T.; Rosenfeld, D. Atmosphere—Aerosols, climate, and the hydrological cycle.*Science* **2001**, *294*, 2119–2124.

9. Breon, F.M.; Tanre, D.; Generoso, S. Aerosol effect on cloud droplet size monitored from satellite. *Science* **2002**, *295*, 834–838.

10. Stevens, B.; Feingold, G. Untangling aerosol effects on clouds and precipitation in a buffered system. *Nature* **2009**, *461*, 607–613.

11. Radke, L.F.; Coakley, J.A.; King, M.D. Direct and remote-sensing observations of the effects of ships on clouds. *Science***1989**, *246*, 1146–1149.

12. Albrecht, B.A. Aerosols, cloud microphysics, and fractional cloudiness. *Science* **1989**, *245*, 1227–1230.

13. Kaufman, Y.J.; Fraser, R.S. The effect of smoke particles on clouds and climate forcing. *Science* **1997**, *277*, 1636–1639.

14. Kaufman, Y.J.; Koren, I. Smoke and pollution aerosol effect on cloud cover. *Science* **2006**, *313*, 655–658.

15. Rosenfeld, D.; Lohmann, U.; Raga, G.B.; O'Dowd, C.D.; Kulmala, M.; Fuzzi, S.; Reissell, A.; Andreae, M.O. Flood or drought: How do aerosols affect precipitation? *Science* **2008**, *321*, 1309–1313.

16. Lu, M.L.; Conant, W.C.; Jonsson, H.H.; Varutbangkul, V.; Flagan, R.C.; Seinfeld, J.H. The marine stratus/stratocumulus experiment (mase): Aerosol-cloud relationships in marine stratocumulus. *J. Geophys. Res. Atmos.* **2007**, *112*.

17. Stevens, B. Why are (precipitation mediated) aerosol effects on clouds so difficult to establish? *Geochim. Cosmochim. Acta* **2009**, *73*, A1274.

18. Twohy, C.H.; Hudson, J.G. Measurements of cloud condensation nuclei spectra within maritime cumulus cloud droplets—Implications for mixing processes. *J. Appl. Meteorol.* **1995**, *34*, 815–833.

19. Leaitch, W.R.; Banic, C.M.; Isaac, G.A.; Couture, M.D.; Liu, P.S.K.; Gultepe, I.; Li, S.M.; Kleinman, L.; Daum, P.H.; MacPherson, J.I. Physical and chemical observations in marine stratus during the 1993 North Atlantic regional experiment: Factors controlling cloud droplet number concentrations. *J. Geophys. Res. Atmos.* **1996**, *101*, 29123–29135.

20. Brenguier, J.L.; Pawlowska, H.; Schuller, L.; Preusker, R.; Fischer, J.; Fouquart, Y. Radiative properties of boundary layer clouds: Droplet effective radius *vs.* number concentration. *J. Atmos. Sci.* **2000**, *57*, 803–821.

21. Pawlowska, H.; Brenguier, J.L. Microphysical properties of stratocumulus clouds during ACE-2. *Tellus B* **2000**, *52*, 868–887.

22. Brenguier, J.L.; Bakan, S. Eucrex-1994: An experimental study of the radiative properties of marine boundary-layer clouds—Preface. *Atmos. Res.* **2000**, *55*, 1–2.

23. Durkee, P.A.; Noone, K.J.; Bluth, R.T. The monterey area ship track experiment. *J. Atmos. Sci.* **2000**, *57*, 2523–2541.

24. Li, Z.Q.; Niu, F.; Fan, J.W.; Liu, Y.G.; Rosenfeld, D.; Ding, Y.N. Long-term impacts of aerosols on the vertical development of clouds and precipitation. *Nat. Geosci.* **2011**, *4*, 888–894.

25. Kaufman, Y.J.; Nakajima, T. Effect of amazon smoke on cloud microphysics and albedo—Analysis from satellite imagery. *J. Appl. Meteorol.* **1993**, *32*, 729–744.

26. Rosenfeld, D. Suppression of rain and snow by urban and industrial air pollution. *Science* **2000**, *287*, 1793–1796.

27. Feingold, G.; Remer, L.A.; Ramaprasad, J.; Kaufman, Y.J. Analysis of smoke impact on clouds in Brazilian biomass burning regions: An extension of Twomey's approach. *J. Geophys. Res. Atmos.* **2001**, *106*, 22907–22922.

28. Matheson, M.A.; Coakley, J.A.; Tahnk, W.R. Aerosol and cloud property relationships for summertime stratiform clouds in the northeastern atlantic from advanced very high resolution radiometer observations. *J. Geophys. Res. Atmos.* **2005**, *110*.

29. Lin, J.C.; Matsui, T.; Pielke, R.A.; Kummerow, C. Effects of biomass-burning-derived aerosols on precipitation and clouds in the Amazon Basin: A satellite-based empirical study. *J. Geophys. Res. Atmos.* **2006**, *111*.

30. Koren, I.; Martins, J.V.; Remer, L.A.; Afargan, H. Smoke invigoration *vs.* inhibition of clouds over the amazon. *Science***2008**, *321*, 946–949.

31. Han, Q.Y.; Rossow, W.B.; Lacis, A.A. Near-global survey of effective droplet radii in liquid water clouds using ISCCP data. *J. Clim.* **1994**, *7*, 465–497.

32. Han, Q.Y.; Rossow, W.B.; Zeng, J.; Welch, R. Three different behaviors of liquid water path of water clouds in aerosol-cloud interactions. *J. Atmos. Sci.* **2002**, *59*, 726–735.

33. Wetzel, M.A.; Stowe, L.L. Satellite-observed patterns in stratus microphysics, aerosol optical thickness, and shortwave radiative forcing. *J. Geophys. Res. Atmos.* **1999**, *104*, 31287–31299.

34. Nakajima, T.; Higurashi, A.; Kawamoto, K.; Penner, J.E. A possible correlation between satellite-derived cloud and aerosol microphysical parameters. *Geophys. Res. Lett.* **2001**, *28*, 1171–1174.

35. Sekiguchi, M.; Nakajima, T.; Suzuki, K.; Kawamoto, K.; Higurashi, A.; Rosenfeld, D.; Sano, I.; Mukai, S. A study of the direct and indirect effects of aerosols using global satellite data sets of aerosol and cloud parameters. *J. Geophys. Res. Atmos.* **2003**, *108*.

36. Kaufman, Y.J.; Koren, I.; Remer, L.A.; Rosenfeld, D.; Rudich, Y. The effect of smoke, dust, and pollution aerosol on shallow cloud development over the Atlantic Ocean. *Proc. Natl. Acad. Sci. USA* **2005**, *102*, 11207–11212.

37. Loeb, N.G.; Schuster, G.L. An observational study of the relationship between cloud, aerosol and meteorology in broken low-level cloud conditions. *J. Geophys. Res. Atmos.* **2008**, *113*.

38. Ghan, S.J.; Chuang, C.C.; Penner, J.E. A parameterization of cloud droplet nucleation part I: Single aerosol type.*Atmos. Res.* **1993**, *30*, 198–221.

39. Menon, S.; Del Genio, A.D.; Koch, D.; Tselioudis, G. GCM simulations of the aerosol indirect effect: Sensitivity to cloud parameterization and aerosol burden. *J. Atmos. Sci.* **2002**, *59*, 692–713.

40. Menon, S.; Brenguier, J.L.; Boucher, O.; Davison, P.; Del Genio, A.D.; Feichter, J.; Ghan, S.; Guibert, S.; Liu, X.H.; Lohmann, U.; *et al.* Evaluating aerosol/cloud/radiation process parameterizations with single-column models and second aerosol characterization experiment (ACE-2) cloudy column observations. *J. Geophys. Res. Atmos.* **2003**, *108*.

41. Brenguier, J.L.; Pawlowska, H.; Schuller, L. Cloud microphysical and radiative properties for parameterization and satellite monitoring of the indirect effect of aerosol on climate. *J. Geophys. Res. Atmos.* **2003**, *108*.

42. Pawlowska, H.; Brenguier, J.L. An observational study of drizzle formation in stratocumulus clouds for general circulation model (GCM) parameterizations. *J. Geophys. Res. Atmos.* **2003**, *108*.

43. Morrison, H.; Curry, J.A.; Khvorostyanov, V.I. A new double-moment microphysics parameterization for application in cloud and climate models. Part I: Description. *J. Atmos. Sci.* **2005**, *62*, 1665–1677.

44. Morrison, H.; Curry, J.A.; Shupe, M.D.; Zuidema, P. A new double-moment microphysics parameterization for application in cloud and climate models. Part II: Single-column modeling of arctic clouds. *J. Atmos. Sci.* **2005**, *62*, 1678–1693.

45. Rotstayn, L.D.; Liu, Y.G. Sensitivity of the first indirect aerosol effect to an increase of cloud droplet spectral dispersion with droplet number concentration. *J. Clim.* **2003**, *16*, 3476–3481.

46. Rotstayn, L.D.; Liu, Y.G. Cloud droplet spectral dispersion and the indirect aerosol effect: Comparison of two treatments in a GCM. *Geophys. Res. Lett.* **2009**, *36*.

47. Rotstayn, L.D. Indirect forcing by anthropogenic aerosols: A global climate model calculation of the effective-radius and cloud-lifetime effects. *J. Geophys. Res. Atmos.* **1999**, *104*, 9369–9380.

48. Rotstayn, L.D.; Liu, Y.G. A smaller global estimate of the second indirect aerosol effect. *Geophys. Res. Lett.* **2005**, *32*.

49. Lohmann, U.; Feichter, J.; Penner, J.; Leaitch, R. Indirect effect of sulfate and carbonaceous aerosols: A mechanistic treatment. *J. Geophys. Res. Atmos.* **2000**, *105*, 12193–12206.

50. Lohmann, U.; Koren, I.; Kaufman, Y.J. Disentangling the role of microphysical and dynamical effects in determining cloud properties over the Atlantic. *Geophys. Res. Lett.* **2006**, *33*.

51. Lohmann, U.; Feichter, J. Global indirect aerosol effects: A review. *Atmos. Chem. Phys.* **2005**, *5*, 715–737.

52. Suzuki, K.; Nakajima, T.; Numaguti, A.; Takemura, T.; Kawamoto, K.; Higurashi, A. A study of the aerosol effect on a cloud field with simultaneous use of GCM modeling and satellite observation. *J. Atmos. Sci.* **2004**, *61*, 179–194.

53. Cheng, Y.J.; Lohmann, U.; Zhang, J.H. Contribution of changes in sea surface temperature and aerosol loading to the decreasing precipitation trend in southern China. *J. Clim.* **2005**, *18*, 1381–1390.

54. Chou, C.; Neelin, J.D.; Lohmann, U.; Feichter, J. Local and remote impacts of aerosol climate forcing on tropical precipitation. *J. Clim.* **2005**, *18*, 4621–4636.

55. Morrison, H.; Pinto, J.O. Mesoscale modeling of springtime arctic mixed-phase stratiform clouds using a new two-moment bulk microphysics scheme. *J. Atmos. Sci.* **2005**, *62*, 3683–3704.

56. Qian, Y.; Gong, D.Y.; Fan, J.W.; Leung, L.R.; Bennartz, R.; Chen, D.L.; Wang, W.G. Heavy pollution suppresses light rain in China: Observations and modeling. *J. Geophys. Res. Atmos.* **2009**, *114*.

57. Bennartz, R.; Fan, J.W.; Rausch, J.; Leung, L.R.; Heidinger, A.K. Pollution from china increases cloud droplet number, suppresses rain over the east China sea. *Geophys. Res. Lett.* **2011**, *38*.

58. Heidinger, A.K.; Cao, C.Y.; Sullivan, J.T. Using moderate resolution imaging spectrometer (MODIS) to calibrate advanced very high resolution radiometer reflectance channels. *J. Geophys. Res. Atmos.* **2002**, *107*.

59. Cao, C.; Weinreb, M.; Xu, H. Predicting simultaneous nadir overpasses among polar-orbiting meteorological satellites for the intersatellite calibration of radiometers. *J. Atmos. Ocean. Technol.* **2004**, *21*, 537–542.

60. Heidinger, A.K.; Straka, W.C.; Molling, C.C.; Sullivan, J.T.; Wu, X.Q. Deriving an inter-sensor consistent calibration for the AVHRR solar reflectance data record. *Int. J. Remote Sens.* **2010**, *31*, 6493–6517.

61. Heidinger, A.K.; Foster, M.J.; Walther, A.; Zhao, X.P. The pathfinder atmospheres-extended AVHRR climate dataset.*Bull. Am. Meteorol. Soc.* **2014**, *95*.

62. Walther, A.; Heidinger, A.K. Implementation of the daytime cloud optical and microphysical properties algorithm (DCOMP) in PATMOS-x. *J. Appl. Meteorol. Clim.* **2012**, *51*, 1371–1390.

63. Pavolonis, M.J.; Heidinger, A.K.; Uttal, T. Daytime global cloud typing from AVHRR and VIIRS: Algorithm description, validation, and comparisons. *J. Appl. Meteorol.* **2005**, *44*, 804–826.

64. Zhao, X.-P.; Dubovik, O.; Smirnov, A.; Holben, B.N.; Sapper, J.; Pietras, C.; Voss, K.J.; Frouin, R. Regional evaluation of an advanced very high resolution radiometer (AVHRR) two-channel aerosol retrieval algorithm. *J. Geophys. Res.***2004**, *109*, D02204.

65. Zhao, X.-P.; Laszlo, I.; Holben, B.N.; Pietras, C.; Voss, K.J. Validation of two-channel VIRS retrievals of aerosol optical thickness over ocean and quantitative evaluation of the impact from potential subpixel cloud contamination and surface wind effect. *J. Geophys. Res. Atmos.* **2003**, *108*.

66. Zhao, X.-P.; Laszlo, I.; Minnis, P.; Remer, L. Comparison and analysis of two aerosol retrievals over the ocean in the terra/clouds and the earth's radiant energy system—Moderate resolution imaging spectroradiometer single scanner footprint data: 1. Global evaluation. *J. Geophys. Res. Atmos.* **2005**, *110*.

67. Zhao, X.-P.; Laszlo, I.; Guo, W.; Heidinger, A.; Cao, C.; Jelenak, A.; Tarpley, D.; Sullivan, J. Study of long-term trend in aerosol optical thickness observed from operational AVHRR satellite instrument. *J. Geophys. Res.* **2008**, *113*, D07201. Zhao, X.-P.; Heidinger, A.K.; Knapp, K.R. Long-term trends of zonally averaged aerosol optical thickness observed from operational satellite AVHRR instrument. *Meteorol. Appl.* **2011**, *18*, 440–445.

68. Zhao, X.P. Satellite observed aerosol optical thickness and trend around megacities in the coastal zone. *Adv. Meteorol.***2015**, *2015*, 170672-7.

69. Heidinger, A.K.; Evan, A.T.; Foster, M.J. A naive Bayesian cloud-detection scheme derived from Calipso and applied withing PATMOS-x. *J. Appl. Meteorol. Clim.* **2012**.

70. Stowe, L.L.; Ignatov, A.M.; Singh, R.R. Development, validation, and potential enhancements to the second-generation operational aerosol product at the national environmental satellite, data, and information service of the national oceanic and atmospheric administration. *J. Geophys. Res. Atmos.* **1997**, *102*, 16923–16934.

71. Ignatov, A.; Stowe, L. Aerosol retrievals from individual AVHRR channels. Part I: Retrieval algorithm and transition from dave to 6 s radiative transfer model. *J. Atmos. Sci.* **2002**, *59*, 313–334.

72. Shao, H.F.; Liu, G.S. Why is the satellite observed aerosol's indirect effect so variable? *Geophys. Res. Lett.* **2005**, *32.*

73. Stephens, G.L. Radiation profiles in extended water clouds. 1. Theory. *J. Atmos. Sci.* **1978**, *35*, 2111–2122.

74. Bennartz, R. Global assessment of marine boundary layer cloud droplet number concentration from satellite. *J. Geophys. Res. Atmos.* **2007**, *112*, D02201.

75. Dong, X.Q.; Schwantes, A.C.; Xi, B.K.; Wu, P. Investigation of the marine boundary layer cloud and CCN properties under coupled and decoupled conditions over the azores. *J. Geophys. Res. Atmos.* **2015**, *120*, 6179–6191.

76. Logan, T.; Xi, B.K.; Dong, X.Q. Aerosol properties and their influences on marine boundary layer cloud condensation nuclei at the arm mobile facility over the Azores. *J. Geophys. Res. Atmos.* **2014**, *119*, 4859–4872.

77. Pruppacher, H.R.; Klett, J.D. *Microphysics of Clouds and Precipitation*; Kluwer Academy: Dordrecht, the Netherlands, 1997.

78. Johnson, D.W.; Osborne, S.; Wood, R.; Suhre, K.; Johnson, R.; Businger, S.; Quinn, P.K.; Wiedensohler, A.; Durkee, P.A.; Russell, L.M.; *et al.* An overview of the Lagrangian experiments undertaken during the North Atlantic regional aerosol characterisation experiment (ACE-2). *Tellus B* **2000**, *52*, 290–320.

79. Brenguier, J.L.; Chuang, P.Y.; Fouquart, Y.; Johnson, D.W.; Parol, F.; Pawlowska, H.; Pelon, J.; Schuller, L.; Schroder, F.; Snider, J. An overview of the ACE-2 cloudycolumn closure experiment. *Tellus B* **2000**, *52*, 815–827.

80. Raes, F.; Bates, T.; McGovern, F.; Van Liedekerke, M. The 2nd aerosol characterization experiment (ACE-2): General overview and main results. *Tellus B* **2000**, *52*, 111–125.

81. Zhao, T.X.P.; Chan, P.K.; Heidinger, A.K. A global survey of the effect of cloud contamination on the aerosol optical thickness and its long-term

trend derived from operational AVHRR satellite observations. *J. Geophys. Res. Atmos.* **2013**,*118*, 2849–2857.

82. Tiao, G.C.; Reinsel, G.C.; Xu, D.M.; Pedrick, J.H.; Zhu, X.D.; Miller, A.J.; Deluisi, J.J.; Mateer, C.L.; Wuebbles, D.J. Effects of autocorrelation and temporal sampling schemes on estimates of trend and spatial correlation. *J. Geophys. Res. Atmos.* **1990**, *95*, 20507–20517.

83. Wilks, D.S. *Statistical Methods in the Atmosphere Sciences*; Academic Press: San Diego, CA, USA, 1995.

84. Weatherhead, E.C.; Reinsel, G.C.; Tiao, G.C.; Meng, X.L.; Choi, D.S.; Cheang, W.K.; Keller, T.; DeLuisi, J.; Wuebbles, D.J.; Kerr, J.B.; *et al.* Factors affecting the detection of trends: Statistical considerations and applications to environmental data. *J. Geophys. Res. Atmos.* **1998**, *103*, 17149–17161.

Chapter 2

Changes in Climate Extremes over North Thailand, 1960–2099

Mohammad Badrul Masud,[1] Peeyush Soni,[2] Sangam Shrestha,[3] and Nitin K. Tripathi[3]

[1]Global Institute for Water Security, School of Environment and Sustainability, University of Saskatchewan, 11 Innovation Boulevard, Saskatoon, SK, Canada S7N 3H5

[2]Agricultural Systems and Engineering, School of Environment, Resources and Development, Asian Institute of Technology, P.O. Box 4, Klong Luang, Pathumthani 12120, Thailand

[3]School of Engineering and Technology, Asian Institute of Technology, P.O. Box 4, Klong Luang, Pathumthani 12120, Thailand

ABSTRACT

This study analyzes 24 climate extreme indices over North Thailand using observed data for daily maximum and minimum temperatures and total daily rainfall for the 1960–2010 period, and HadCM3 Global Climate Model (GCM) and PRECIS Regional Climate Model simulated data for the 1960–2100 period. A statistical downscaling tool is employed to downscale GCM outputs. Variations in and trends of historical and future climates are identified using the nonparametric Mann-Kendall trend test and Sen's slope. Temperature extreme indices showed a significant rising trend during the observed period and are expected to increase significantly with an increase in summer days and tropical nights in the future. A notable decline in the number of cool days and nights is also expected in the study area while the number of warm days and nights is expected to increase. There was an insignificant decrease in total annual rainfall, number of days with rainfall more than 10 and 20 mm. However, the annual rainfall is projected to increase by 9.65% in the future 2011–2099 period compared to the observed 1960–2010 period.

1. INTRODUCTION

Analyses of climate change, climate variability, and their extremes are becoming more important as they clearly affect the human society and are essential for exploration of ecological and societal changes. One of the motivations for such analyses is to establish if and how global warming [1] might have influenced the climate on a global as well as on regional scale. In the environment and on the economic activities, the impact of climate change is reflected by the change of frequency, duration, and intensity of climate extreme events. After the third assessment report of the Intergovernmental Panel on Climate Change (IPCC), projections and detection of climate change are no longer relegated to global averages and have expanded to include extremes [2]. Various climate extreme events such as more recurrent and more intense droughts, floods, cyclones, and extreme rainfall events pose increasing and, oftentimes, limitless threats to environment, water resources, and agricultural production [3, 4].

General Circulation Models (GCMs) provide a reasonable basis for estimating future climates; however, outputs from GCM cannot be applied directly in many environmental and water resources protection studies because of some limitations, which include poor representation and simulation of local-scale processes and coarse spatial resolution. Therefore, these issues can be addressed by downscaling approaches. First approach of downscaling (statistical downscaling) climate data is through the use of statistical regressions that establish the relationship between large-scale variables, such as the driving factors derived from GCMs, to local level climate conditions. In the second approach (dynamical downscaling) Regional Climate Models (RCMs) are embedded into GCMs. The RCM is essentially a numerical model in which GCMs are used to fix boundary conditions. So, applying a downscaling technique is recommended for regional-scale analysis in IPCC-AR4 [5]. More recently, downscaling is widely applied in climatology for scenario construction and simulation/prediction of mean, minimum, and maximum air temperature and rainfall [6–8].

A number of investigations have attempted to identify observed and projected future changes in climate extremes on a global scale (e.g., [9–12]). A study on the variability in occurrence of extreme climate events at a global scale during the second half of the twentieth century [9] showed general increase in warm nights, decrease in the number of frost days, more frequent heavy rainfall events, less frequent cold temperature extremes, and decrease in the intra-annual extreme temperature range. Alexander et al. [10] showed that over 70% of global land area observed significant increase in the annual occurrence of warm nights and significant decrease in occurrence of cold nights. Projected changes in precipitation and dryness extremes are more ambiguous than that in temperature extremes at the end of 21st century [11]. Another study by Donat et al. [12], a research to improve global gridded dataset of extreme indices (HadEX to HadEX2, [10]), showed widespread significant changes in temperature extremes with warming especially for those indices derived from minimum temperature. Precipitation also showed similar changes but the changes were much more spatially heterogeneous compared to temperature changes.

A few studies have examined trends in climate extremes, other than changes in mean values across the Southeast Asia. Hence, little is known about trends in extremes of temperature and rainfall in this region. Manton et al. [13] analysed climate extreme indices for historical period in the Southeast Asia and South Pacific. Regional studies across the Asia–Pacific region have shown significant increases in occurrences of annual number of hot days and warm nights and decreases in occurrences of annual number of cool days and cold nights over the past few decades (e.g., [13, 14]). Results from Manton et al. [13] revealed significant decrease in the number of rainy days, significant increase in the proportion of total rainfall from extreme rainfall, significant increase in the minimum and maximum temperature, increase in the number of warm nights, and decrease in the number of cold nights and cool days at Nan Province of North Thailand. Chinvanno et al. [15] developed high-resolution climate scenarios using PRECIS (Providing Regional Climates for Impact Studies) RCM and showed that, during the observed period, most part of the North and Northeastern Thailand had 3-4 hot months over the year, while the Central Plain and the Southern Region had slightly longer summertime. Preliminary finding on MM5-RCM simulations of future projection by Kreasuwan [16] during 2010–2029 revealed that the average temperature in Thailand will increase slightly, and this trend will continue further in the future (during 2040–2059). In a study of Southeast Asia START (System for Analysis, Research and Training) Regional Centre (SEA START RC) [17], it was found that the night time temperature will get warmer than the daytime temperature by the 2045–2065 period. For the Northern mountainous and valley zone, annual average maximum and minimum temperature are expected to rise from 32.41 to 35.82°C and 20.43 to 24.82°C, respectively, when compared to the median value among the results from 8 GCMs.

North Thailand is a critical area for a number of regional sustainable developmental issues. The mountainous landscape has relatively high forest cover. It contains the upper reaches of most of the watersheds feeding into the Chao Phraya River system, including the largest one, the Ping River Basin (PRB, catchment area of about 35,000 km^2). This basin is an important wet-season source of water stored in the downstream Bhumibol Reservoir (located in Tak Province of North Thailand), which serves irrigation and commercial users during a lengthy dry season when drought may be problematic [18]. The basin area was previously covered by subtropical forests that have been slowly converted to agricultural lands in the recent past due to a host of economic, social, and political drivers that have commonly been driving land-cover/land-use changes through montane mainland Southeast Asia [19]. Greater Chiang Mai city area experienced a farm-to-city transition where built-up areas increased from 9% to over 33% between 1989 and 2009 [20–22]. The effects of urbanization that include the construction of major ring roads and changes in the Ping River bank and floodplains are blamed for causing more serious floods in recent years [21, 23–25]. Flooding and mudslides affected a large area in North Thailand, including Chiang Mai, Chiang Rai, Phayao, and Mae Hong Song Provinces [26]. The 2011 Chao Phraya River flood that caused $45 billion damage in Thailand highlights the difficulty of water management [27, 28]. Despite today's popularization of irrigation, rainfed crop fields remain in a large part of North Thailand. Since this type of cultivation largely depends on weather conditions, it has potential vulnerability against future climate changes accompanied by global warming. Changes in climate extremes have relations and significant impact on all of the anthropogenic activities and natural hazards. These issues constituted the motivation for this research. The purpose of the study is not to find out the best climate model rather than identifying the past and future changes of climate extremes in the regional scale through different emission scenarios. Therefore, in this study, observed trends in temperature and rainfall extremes in North Thailand were investigated over the last 51 years. Also, to detect the possible changes of extremes up to the year 2100, assessments of regional climate projections were made, derived from PRECIS regional model and HadCM3 (Hadley Centre Coupled Model version 3) global model.

The paper is organized as follows. Section 2 presents an overview of the study area and data used in this research, followed by methodology in Section 3. In Section 4, results of various analyses are presented and finally conclusions of the study are given in Section 5.

2. STUDY AREA AND DATA USED

2.1. Geography

Administratively as well as geographically Thailand has four regions, namely, the Northeastern, North, Central Plain, and Southern. This study focused on four provinces of North Thailand, namely, Chiang Mai, Chiang Rai, Lamphun, and Tak (Figure 1). These provinces cover an important drainage area of the PRB and have a significant impact on country's agricultural activities. The Ping River, a major tributary of the Chao Praya River, plays a vital role in the country's economy. The Bumibol Dam on the Ping River adds additional importance to this region. Geographically, this region is mountainous and hilly with narrow valleys and occasional intermountain depressions forming a highly complex relief that merges across the international borders into Myanmar, Laos, and Yunnan (China), situated between 14.56°–20.27° North and 97.20°–101.47° East. These north-south oriented hill ridges are parallel from west to east and intersected by some major valeeys, particularly those near Chiang Mai and Chiang Rai provinces. It experiences a strong monsoon climate with heavy summer rains between mid-May or June and October but experiences dry weather in winter because of the northeast monsoon which is a main factor that controls the climate of this region. The northern part of the country usually experiences an extended period of warm weather because of its inland nature and tropical latitude zone [29]. In this region, the average annual temperature ranges from 20 to 34°C and the average annual rainfall varies between 1000 and 1800 mm depending on the location.

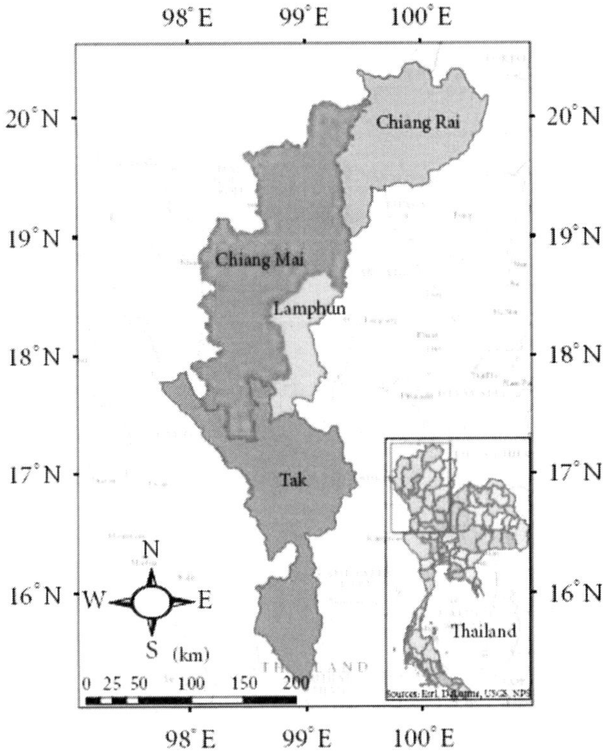

Figure 1. Location map of the study area in Thailand.

2.2. Data Used

Observed daily minimum temperature (°C), maximum temperature (°C), and total daily rainfall (mm) data were collected from Thailand Meteorological Department for the period of 1960 to 2010 for Chiang Rai, Chiang Mai, and Tak provinces (Table 1). In Lamphun Province recording of climate data started from 1981. Two sets of model data, each of which has two SRES scenarios (A2 and B2), were used in this study to the end of 21st century (1960–2099). The first set of data was collected from SEA START RC, where the climate data were downscaled using PRECIS RCM, which used ECHAM4 GCM data as an initial condition. Another set of climate data was produced by using a Statistical Downscaling tool (Statistical Downscaling Model (SDSM)) for the coarse resolution climate variables obtained from HadCM3 GCM.

Table 1. Detail information for meteorological station.

Station name	Station code	Data availability	Elevation (m)	Latitude (°N)	Longitude (°E)
Chiang Rai	303201	1960–2010	393	19.92	99.83
Chiang Mai	327501	1960–2010	311	18.77	98.97
Lamphun	329201	1981–2010	296	18.57	99.03
Tak	376201	1960–2010	98	16.88	99.15

3. METHODOLOGY

3.1. Downscaling Climate Data

The SDSM [30] calculates statistical relationships, based on multiple linear regression techniques, between large-scale (the predictors) and local (the predictand) climates. There are three implicit assumptions involved in statistical downscaling [31]. Firstly, the predictors are variables of relevance and are realistically modeled by the host GCM. Secondly, the empirical relationship is also valid under altered climatic conditions. Thirdly, the employed predictors fully represent the climate change signal. These relationships are developed using observed weather data, assuming that these relationships will remain valid in the future, that can be used to obtain downscaled local information for some future time period by driving the relationships with GCM-derived predictors. The observed large-scale predictors have been derived from the National Centers for Environmental Prediction (NCEP) reanalysis dataset [32]. This model becomes more accepted in recent years due to its applicability in a wide region and simplicity of establishing a relationship between predictor and predictand variables for future time zone [33].

Substantial differences may exist between projections from climate models; use of a single model does not provide results that adequately reflect the uncertainty inherent in regional climate scenario generation [34]. This study incorporates output of two models for a reliable projection of future climate. The SDSM is used to downscale the scenarios (A2 and B2) of HadCM3 by using the predictor variables developed by NCEP. HadCM3 was chosen for this analysis because it is one of four sets of climate model (HadCM2, HadCM3, CGCM2, and CSIRO) output provided with SDSM. The SDSM has successfully been used in several other downscaling studies [35–37]. Before forecasting the future climate variables, SDSM has been validated for the 1980–1989 period. The SEA START RC used ECHAM4 GCM (resolution 280 × 280 km^2; Max Plank Institute, Germany) for climate scenario generation and PRECIS RCM (developed at the Hadley Centre at the Meteorological Office, spatial resolution of 25 × 25 km^2 and emission scenarios B1, B2, A2, and A1FI) for downscaling the GCM data for specific region. The main reason to select PRECIS RCM in this research is that SEA START RC specially handles Southeast Asian region to predict more precise climate and freely provide data to researchers. Hence, A2 and B2 scenarios of PRECIS RCM data were collected from Southeast Asian START Regional Centre website (http://www.start.or.th/), where data are readily available to download.

A2 and B2 scenarios are the storylines which describe the development of different social, economic, technological, environmental, and policy dimensions. A2 is the future of the differentiated world, with emphasis on human wealth. Major underlying themes include strengthening regional cultural identities, with an emphasis on family values and local traditions, high population growth, and less concern for rapid economic development. B2 scenario emphasizes on sustainability and equity (Mixed Green Bag) and local solutions to economic, social, and environmental sustainability in the world [38].

3.2. Extreme Indices

Extreme climatic events, through the prolonged hot and cold spells, have numerous impacts on a nation which include significant loss of life, agriculture, production, water resources, forestry, energy, and infrastructure. It is essential to observe the climate extremes and continue to look for identification of changes in extremes as our resources and infrastructure are becoming more vulnerable to ruthless and extreme weather [39]. All of the extreme indices (13 for temperature and 11 for rainfall, Table 2) are calculated from the daily observation of temperature and rainfall by using RClimDex software [40]. These indices are recommended by the Climate Variability and Predictability (CLIVAR) Expert Team on Climate Change Detection, Monitoring, and Indices (ETCCDMI), adopted by the IPCC AR4. Data quality control is a necessary step before the analysis of climate variation, because erroneous outliers can impact seriously on trends [41]. For this, each data point was examined based on neighboring data, as well as on the understanding of the climate of the region. The data were edited if the problem is obvious (e.g., negative precipitation, 195 degrees changed to 19.5, and maximum temperature less than minimum temperature), set to missing if it is clearly a problem with unknown solution, or kept if deemed probably valid.

3.3. Trend Analysis

The trend of climate extremes presented in this paper is calculated by using the nonparametric Mann-Kendall (MK) test [39, 40] that has been recommended by the World Meteorological Organization in assessing trends in environmental time series data [42]. The test has been widely used in trend detection analysis for hydrological and meteorological data [12, 43–47]. It is nearly as powerful as its parametric competitors. This method was chosen because it does not consider any distribution of the variable and some of the climate indices do not consider Gaussian distribution [12]. To estimate the true slope of an existing trend (as change per year), Sen's nonparametric method [48] is used. Sen's slope method gives a robust estimation of the trend and is used in various trend analyses of climate data [10, 12, 42]. Different significant levels (10%, 5%, 1%, and 0.1%) were used to consider a trend to be statistically significant.

4. RESULTS AND DISCUSSION

In this section of the paper, results of various analyses ranging from trend analysis for observed period are presented first, followed by seasonal variation of climate parameters, model performance evaluation, and, finally, projected changes of climate parameters for the future period. Where necessary, link to other related studies is established in different subsections. For brevity, analyses of climate extremes results are presented for Chiang Rai Province and selected results for other provinces.

4.1. Trend Analysis for Observed Period (1960–2010)

Trends and statistical results of temperature extremes are presented in Table 3, where increasing trend of temperature is experienced in the observed period. Both maximum and minimum temperature indices (TXx, TXn, TNx, and TNn) increased with significant trend, with the fact that minimum temperature was increased more than maximum temperature. The increasing trend of summer days (SU35) and tropical nights (TR25) favored the increment of temperature in the study area except in Chiang Rai Province. Warm days and warm nights (TX90p and TN90p) also increased with very high significant trend whereas cool days (TX10p) and cool nights (TN10p) decreased [13, 49] with the same trend along with mild to steep slope. Similar results were found by another study of Sharma and Babel [50] for Western Thailand. In the observed period, significant increases in warm extremes (TX90p, TN90p, SU35, TR25, and WSDI) were experienced in North Thailand where reverse trend was observed in case of cold extremes (TX10p, TN10p, and CSDI) which is also favored by the negative trend of diurnal temperature range (DTR) except in Lamphun

Province (Figure 2). Zhang et al. [51] and Liu et al. [52] reported decreasing trend of the diurnal temperature range (DTR) because of rapid increase of the minimum temperature in recent decades. Based on Mann-Kendall normalized test statistic and Sen's slope value, Tak Province experienced relatively hot climate compared to any other provinces whereas Lamphun observed relatively cool climate.

All rainfall extremes showed insignificant trend except consecutive dry days (CDD) which was increasing with significant trend during the observed period of 1960–2010 (Table 3) in Chiang Rai and Tak provinces. Similar results were observed in another study of Sharma and Babel [50]. Except Lamphun Province, negative trends of PRCPTOT and R10 indicate that the rainfall was decreasing (Figure 2) over the study area. This less rainfall distribution supports the negative trend of consecutive wet days (CWD).

4.2. Seasonal Variation in Observed Period

During observed period, maximum temperature () showed a significant increasing trend in all seasons except spring which has an insignificant negative trend in all provinces. Minimum temperature () increased significantly in Chiang Rai Province in all seasons, which is followed by Tak, Chiang Mai, and Lamphun provinces. The Mann-Kendall test statistic showed that has increasing trend more than . In mountainous provinces like Chiang Rai and Chiang Mai, the temperature is observed to be higher than other provinces. A similar trend was identified by Zongxing et al. [41]. They also noticed the influence of topography on the regional pattern of temperature indices in Southwestern China; significant trends were on the Eastern Xizang Plateau and the Hengduan Mountains, while stations with insignificant trends were in the Sichuan Basin and on the Yunnan-Guizhou Plateau. There was no significant trend of rainfall detected during the observed period except winter (downward trend) and spring (upward trend) in Chiang Rai and Chiang Mai provinces, respectively (Table 4).

4.3. Model Performance Evaluation

Climate parameters can theoretically be projected by using appropriate GCM and RCM for intended place, but this does not imply that all the data will perfectly correspond to the real situation. Some data may be consistent with a particular time of month or year while other may not. To evaluate model performance, after screening all the 26 GCM variables in the SDSM model for significant correlation, 12 out of 26 variables were found significantly correlated () to the observed temperature and rainfall data. Later on, these variables were used for the calibration of the model to develop a statistical relationship with the observed data and GCM output. For checking the consistency of observed and simulated maximum and minimum temperature and cumulative rainfall data, 1980s was taken as the base period. Results (Figure 3) revealed that none of the SRES of two models is fully unswerving with the observed data. However, based on the coefficient of determination, both PRECIS RCM and HadCM3 GCM have consistency with the base period. Similar performance was observed for HadCM3 GCM, which was downscaled using SDSM model in another study by Deb and Babel [53]. The maximum temperature difference between the observed and simulated values was about ±0.5°C, which imply that two models are reliable to simulate future data. Both models showed higher accuracy in case of predicting the rainfall compared to the base period. The average difference was ±60 mm for all provinces except Tak which shows the maximum difference for HadCM3 GCM and minimum for PRECIS RCM. This performance evaluation is not to select the best model to simulate future scenarios rather than to provide more alternatives to know about the future climate and their extremes.

4.4. Future Temperature and Rainfall Scenarios

In this research, two IPCC scenarios are developed by using two climate models for the 1960–2099 period, which includes the observed 1960–2010 period and simulated 2011–2099 period for four provinces of North Thailand.

Table 2. Definition of extreme temperature and rainfall indices used in this study.

ID	Indicator name	Definition	Unit
		Temperature indices	
SU	Summer days	Annual count when daily max temperature (TX) > 35°C	Days
TR	Tropical nights	Annual count when daily minitemperature (TN) > 25°C	Days
TXx	Hottest day	Monthly maximum value of daily max temperature	°C
TNx	Warmest night	Monthly maximum value of daily min temperature	°C
TXn	Coldest day	Monthly minimum value of daily max temperature	°C
TNn	Coldest night	Monthly minimum value of daily min temperature	°C
TN10p	Cool nights	Percentage of days when TN < 10th percentile	Days
TX10p	Cool days	Percentage of days when TX < 10th percentile	Days
TN90p	Warm nights	Percentage of days when TN > 90th percentile	Days
TX90p	Warm days	Percentage of days when TX > 90th percentile	Days
DTR	Diurnal temp range	Monthly mean difference between TX and TN	°C
WSDI	Warm spell duration index	Annual count of days with at least 6 consecutive days when TX > 90th percentile	Days
CSDI	Cold spell duration index	Annual count of days with at least 6 consecutive days when TN < 10th percentile	Days

		Rainfall indices	
CDD	Consecutive dry days	Maximum number of consecutive days with RR < 1 mm	Days
CWD	Consecutive wet days	Maximum number of consecutive days with RR ≥ 1 mm	Days
PRCPTOT	Annual total rainfall	Annual total rainfall (RR ≥ 1 mm)	mm
R10	Moderate rainy days	Annual count of days when PRCP ≥ 10 mm	Days
R20	Heavy rainy days	Annual count of days when PRCP ≥ 20 mm	Days
R100	Very heavy rainy days	Annual count of days when PRCP ≥ 100 mm, 100 is user defined threshold	Days
R95p	Very wet days	Annual total PRCP when RR > 95th percentile	mm
R99p	Extremely wet days	Annual total PRCP when RR > 99th percentile	mm
RX1day	Max 1-day rainfall amount	Monthly maximum 1-day rainfall	mm
Rx5day	Max 5-day rainfall amount	Monthly maximum consecutive 5-day rainfall	mm
SDII	Simple daily intensity index	The ratio of annual total rainfall to the number of wet days (≥1 mm)	mm/day

Table 3. Observed trend of temperature and rainfall indices.

Annual series	Provinces							
	Chiang Rai		Chiang Mai		Lamphun		Tak	
	MK Z	Sen's slope	MK Z	Sen's slope	MK Z	Sen's slope	MK Z	Sen's slope
	Temperature indices							
SU35	−0.12	0.000	0.87	0.143	1.46	0.500	2.47^{*}	0.385
TR25	−0.32		2.54^{*}	0.161	1.56	0.364	4.11^{***}	0.857
TXx	2.32^{*}	0.011	3.28^{**}	0.019	2.44^{*}	0.029	2.76^{**}	0.017
TNx	5.71^{***}	0.034	3.10^{**}	0.019	2.07^{*}	0.021	3.53^{***}	0.028
TXn	2.71^{**}	0.025	1.80^{+}	0.018	−0.32	−0.005	2.12^{*}	0.017
TNn	5.56^{***}	0.038	5.50^{***}	0.041	2.04^{*}	0.018	5.20^{***}	0.047
TN10p	-6.20^{***}	−0.220	-4.82^{***}	−0.185	—	—	-4.21^{***}	−0.180
TX10p	-3.35^{***}	−0.080	-1.99^{*}	−0.064	—	—	-3.17^{**}	−0.093
TN90p	6.64^{***}	0.684	3.66^{***}	0.224	—	—	4.34^{***}	0.328
TX90p	4.42^{***}	0.290	3.66^{***}	0.246	—	—	3.90^{***}	0.234
DTR	-3.16^{**}	−0.017	-2.75^{**}	−0.017	1.09	0.013	−1.44	−0.008
WSDI	5.23^{***}	0.545	3.81^{***}	—	—	—	3.65^{***}	—
CSDI	−1.00	0.000	−0.644	—	—	—	−1.06	—

	Rainfall indices							
CDD	1.67^{+}	0.462	1.36	0.368	0.04	0.059	2.39^{*}	0.676
CWD	−1.27	−0.034	0.17	0.000	1.15	0.000	−0.34	0.000
PRCPTOT	−0.02	−0.192	−0.84	−1.747	0.57	4.092	−0.34	−0.594
R10	−0.48	−0.029	−0.63	−0.036	0.21	0.000	−0.38	0.000
R20	0.33	0.000	−1.29	−0.056	0.04	0.000	0.26	0.000
R100	−0.74	—	0.27	—	1.27		1.32	—
R95P	0.02	0.024	0.02	0.057	1.28	3.450	0.36	0.476
R99P	−1.54	−0.362	1.20	0.000	2.55^{*}	3.666	1.29	0.000
RX1day	−0.98	−0.241	1.92^{+}	0.429	1.23	0.495	1.02	0.260
RX5day	−1.29	−0.428	0.65	0.221	2.12^{*}	1.333	1.47	0.726
SDII	0.55	0.012	−0.46	−0.006	0.46	0.008	1.90^{+}	0.031
V. Heavy	−1.49	—	−1.46	—	1.27	—	0.30	—
Heavy	0.00	—	0.74	—	1.62	—	1.57	—
Moderate	1.21	−0.069	−0.38	−0.024	0.80	0.182	−1.13	−0.111

$^{+}$Moderately significant, *significant, **highly significant, and ***very highly significant.
Moderately significant = 90% level of confidence, significant = 95% level of confidence, highly significant = 99.9% level of confidence, and very highly significant = 99.99% level of confidence.

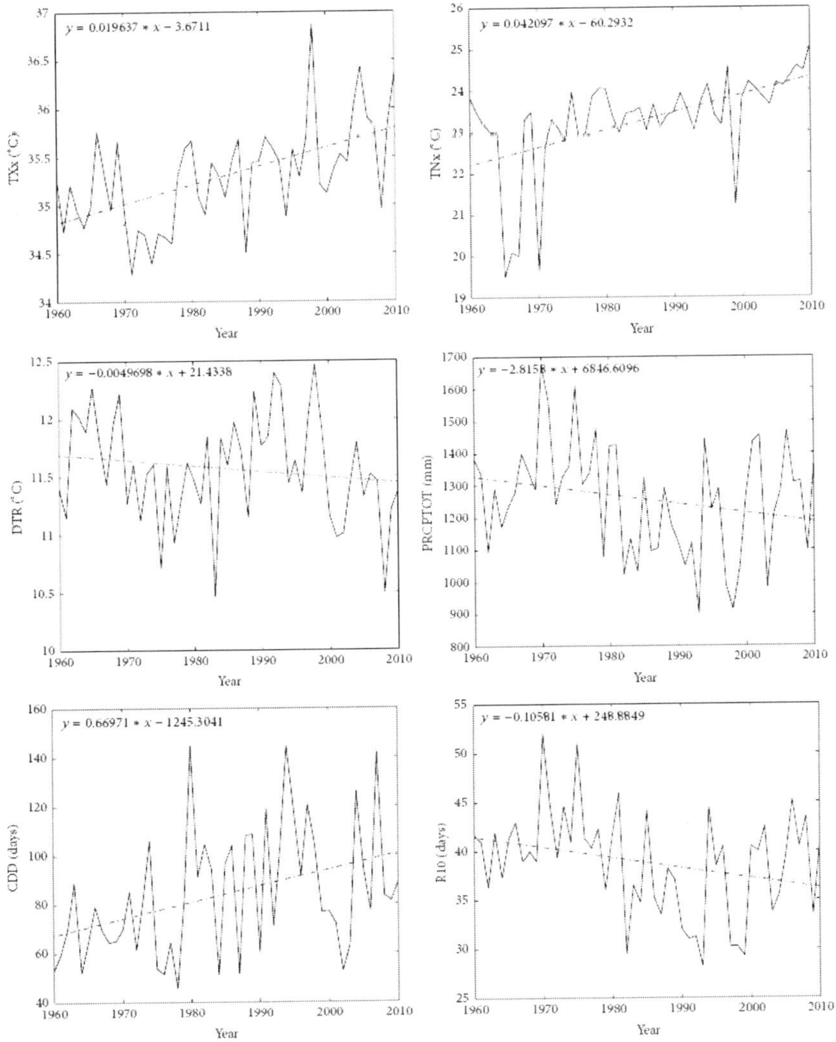

Figure 2. Monthly indices of temperature and rainfall extremes for the observed period 1960 to 2010 averaged across the study area. The dotted line is a trend-line computed by least square fit and corresponding regression equation is showed for each index.

Table 4. Observed seasonal trend of temperature and rainfall.

Climatic parameter	Season	Chiang Rai		Chiang Mai		Lamphun		Tak	
		MK Z	Sen's slope	MK Z	Sen's slope	MK Z	Sen's slope	MK Z	Sen's slope
T_{max}	Winter	3.80***	0.034	3.28**	0.028	2.68**	0.057	3.20**	0.032
	Spring	−0.80	−0.006	0.31	0.003	0.00	0	−0.08	−0.001
	Summer	2.08*	0.011	2.36*	0.016	1.82+	0.02	2.47*	0.016
	Autumn	3.78***	0.026	3.01**	0.019	2.07*	0.027	3.33***	0.021
T_{min}	Winter	3.62***	0.041	4.26***	0.053	2.25*	0.053	3.78***	0.05
	Spring	6.22***	0.042	4.42***	0.044	2.43*	0.037	2.73**	0.015
	Summer	7.02***	0.033	3.48***	0.014	0.86	0.007	4.00***	0.019
	Autumn	3.40***	0.025	2.45*	0.014	−0.93	−0.014	3.43***	0.022
Rainfall	Winter	−1.65+	−0.004	−0.23	0.000	0.20	0.000	−0.45	0.000
	Spring	1.62	0.020	1.66+	0.015	0.14	0.005	0.36	0.007
	Summer	−1.41	−0.027	−1.45	−0.020	0.71	0.027	0.52	0.007
	Autumn	−0.15	−0.002	−0.93	−0.012	−0.82	−0.023	−0.34	−0.007

+Moderately significant, * significant, ** highly significant, and *** very highly significant.

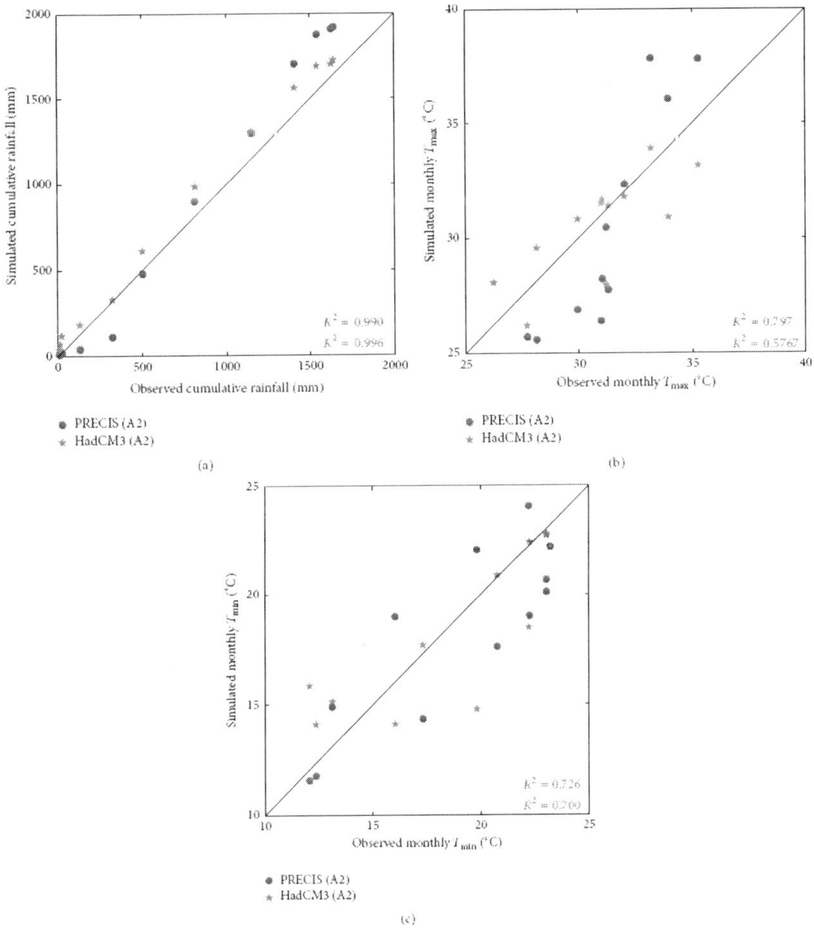

Figure 3. Comparison of observed and simulated cumulative rainfall (a), monthly maximum temperature (b), and monthly minimum temperature (c) for 1980–1989 of Chiang Rai. Results for the case of PRECIS RCM are shown in blue color and those for the case of HadCM3 GCM in red.

Average maximum and minimum temperature variations in the future period relative to the observed period are plotted in Figure 4 for all provinces. None of the SRES is considered to have an outcome more likely than another but it is obvious that all demonstrate the continuation of the trends demonstrated by the historical data, which indicates increasing rates of temperature change for North Thailand. There are some exceptions like when HadCM3 GCM predicts a flat trend of maximum temperature, which is below the observed average maximum temperature. Similarly, PRECIS RCM did the same for Talk Province. These exceptions are similar for minimum temperature. The ensemble averaged maximum temperature for the future period (2011–2099) in Chiang Rai, Chiang Mai, Lamphun, and Tak is 31.29, 33.45, 33.29, and 34.96°C, whereas the observed maximum temperature is 30.89, 31.91, 32.91, and 33.32°C, respectively. The ensemble averaged minimum temperature in the future period will increase by 0.87, 1.42, 0.93, and 0.49°C relative to the observed period in Chiang Rai, Chiang Mai, Lamphun, and Tak, respectively. For both models and scenarios considered herein, the average monthly maximum and minimum temperature for all provinces average increases projected for the 2011–2099 period are 0.99 and 0.93°C above the observed period (1960–2010) mean of 32.26 and 20.66°C, respectively. Erda et al. [54] reported that the temperature increase in China by the end of the twenty-first century may be between 3 and 4°C, simulated by A2 scenario of PRECIS model. In Southeast Asia, mean warming of temperature will be 2.5°C in 2080–2099 relative to 1980–1999 [55]. According to IPCC ensemble-mean predictions, results estimated a net annual increase in temperature between 1.32°C (lowest emissions scenario B1) and 2.01°C (highest emissions scenario A1F1) for the 2040–2069 periods relative to the baseline 1961–1990 period [56]. It is clear from the above discussion that all scenarios indicate higher future temperature for North Thailand than what would result from a continuation of the trend from the historical record. Multimodel and multiensembles temperature anomaly, regionally averaged, showed that the trend of increasing monthly minimum temperature is higher than that of monthly maximum temperature (Figure 5).

As a result of higher interannual variability, prediction and identification of trends are less apparent for rainfall than temperature. However, each model and scenario evaluated project an increase in average annual rainfall of 9.65% for 2011–2099 compared to the observed average annual rainfall 1235 mm. Total annual rainfall projections for the range of SRES suggest small differences (Figure 6) for Chiang Rai. For the period 2011–2099, the HadCM3 GCM and PRECIS RCM models project increase in total annual rainfall by 20.95% and 18.46%, respectively, above the observed period mean of 1725 mm in Chiang Rai Province. PRECIS RCM projects increases in total rainfall but HadCM3 GCM projects decreases for Chiang Mai and Lamphun. This state is opposite for Tak Province where HadCM3 GCM projects increases in total rainfall. Future projection of monsoon rainfall is quite complicated due to the presence of aerosols (carbon aerosols) in the

Asian monsoon [55]. However, the rainfall anomaly, regionally averaged, showed a continuous increasing trend of annual rainfall (Figure 7).

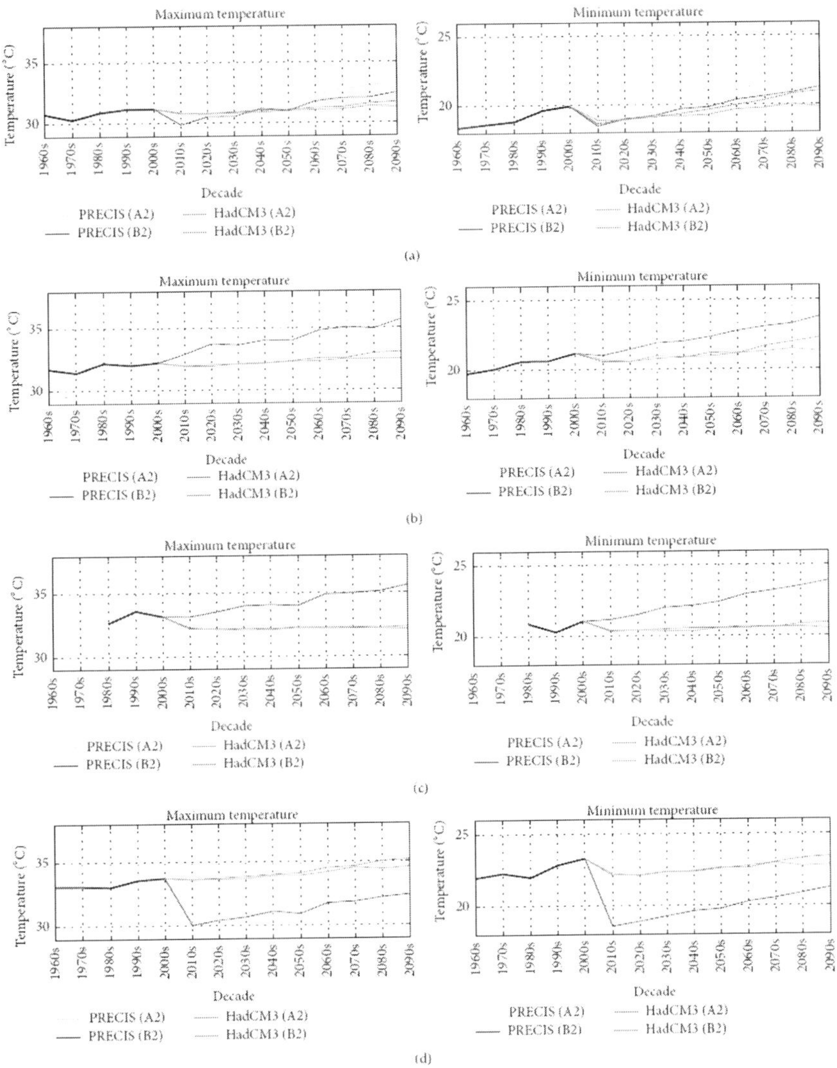

Figure 4. Average monthly maximum and minimum temperature variation in the future for (a) Chiang Rai, (b) Chiang Mai, (c) Lamphun, and (d) Tak. Black solid thick line up to 2000s indicates the observed data.

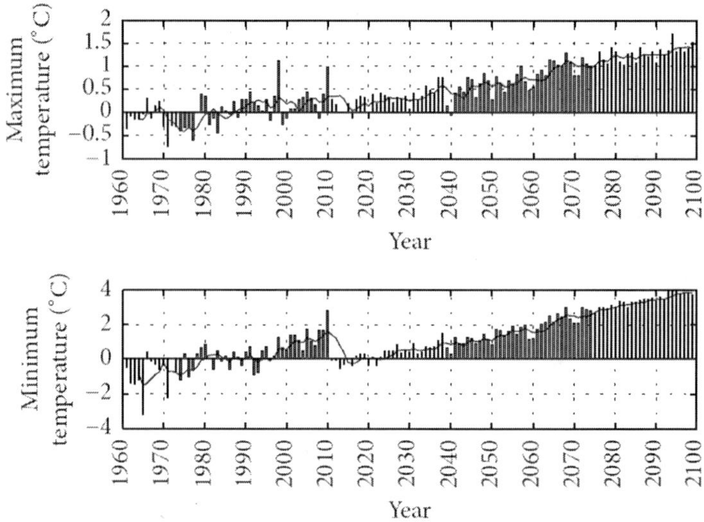

Figure 5. Multimodel and multiensembles temperature anomaly with respect to the control period 1970 to 2000. Results for the case of observed period (observed data) are shown in blue and those for the case of future period (simulated data) in red. The black solid is representing the 5-year moving average of anomalies.

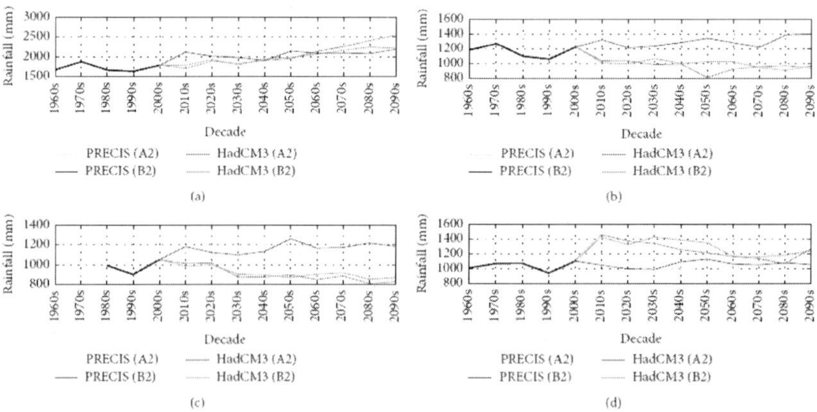

Figure 6. Annual rainfall variation in the future for (a) Chiang Rai, (b) Chiang Mai, (c) Lamphun, and (d) Tak. Black solid thick line up to 2000s indicates the observed data.

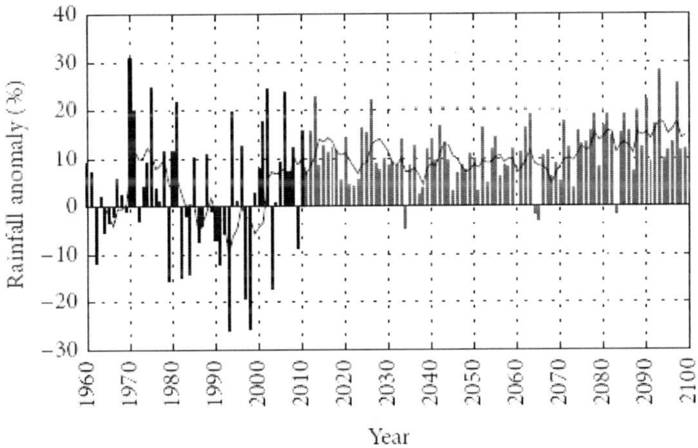

Figure 7. Multimodel and multiensembles rainfall anomaly with respect to the control period 1970 to 2000. Results for the case of observed period (observed data) are shown in black and those for the case of future period (simulated data) in blue. The black solid is representing the 5-year moving average of anomalies.

4.5. Projected Seasonal Variation

This section explains the ensemble average seasonal temperature and rainfall variation for 2020s, 2050s, and 2080s compared to the observed base period (1980s) (Tables 5 and 6). Daily maximum temperature in winter is projected to decrease during 2020s with respect to the base period. The daily maximum temperature projections of A2 and B2 scenarios of PRECIS RCM increase at higher rates in winter than that of the HadCM3 GCM and for the 2080s both scenarios of PRECIS RCM project daily maximum temperature increases by 2.33 and 1.40°C, respectively. Similarly, both models project an increasing daily maximum temperature for spring, summer, and autumn seasons except 2020s, where only PRECIS RCM showed a decline of daily maximum temperature. Daily minimum winter temperature is going to increase in 2020s followed by 2050s and 2080s compared to the base period 1980s. The maximum increment (4.99°C) is found for A2 scenario of PRECIS RCM in 2080s. In spring, PRECIS RCM projects an increase in daily minimum temperature in 2020s, 2050s, and 2080s regardless of scenarios but HadCM3 GCM projects the opposite. In summer and autumn, this minimum temperature is projected to increase in

2050s and 2080s by both scenarios of PRECIS and HadCM3 models. It is noteworthy that the increasing rate of daily minimum temperature is higher during winter and summer than that of daily maximum temperature, whereas this fact is opposite in case of spring and autumn.

Seasonal variations of annual rainfall in the future decades with respect to the base period 1980s are tabulated in Table 6. In winter, A2 and B2 scenarios of HadCM3 GCM project an increase in annual rainfall by 80% in 2080s, whereas PRECIS RCM's predictions are not consistent with this result. However, each scenario and model evaluated project a decrease in annual rainfall ranging from 13.84 to 61.10% in spring. However, in summer, the projection is opposite to spring where annual rainfall is increasing in 2020s, 2050s, and 2080s ranging from 19.83 to 46.55%. For autumn, the annual rainfall projections for the range of SRES suggest big differences, because PRECIS RCM projects an increase in rainfall whereas HadCM3 GCM projects the opposite. Although both temperature and rainfall are rising in the future, there is no significant correlation found for temperature and rainfall in this region [57].

4.6. Trend Analysis for the Future Period

Figure 8 illustrates the two IPCC SRES for Chiang Rai Province as predicted by PRECIS RCM and HadCM3 GCM to see the trend of different warm and cold temperature extremes, developed by RClimDex, for the period of 2011–2099. Mann-Kendall test statistics with Sen's slope are provided in Table 7 for the same province.

Almost all the temperature rising extreme indices (TX90p, TN90p, TXx, TNx, SU35, TR25, and WSDI) are increasing (Table 7 and Figure 8) with highly (very highly) significant trend, and the temperature falling indices (TX10p, TN10p, TXn, TNn, and CSDI) are decreasing with same trend for both SRES A2 and B2. Also, the values of Mann-Kendall test statistic and Sen's slope indicate that minimum temperature extremes (TNx and TNn) have stronger trend and steeper slope than maximum temperature extremes (TXx and TXn) and that the cold extremes (TX10p and TN10p) have weaker trend and milder slope than warm extremes (TX90p and TN90p). On an average, maximum and minimum temperature will increase by 1.56 and 2.24°C, respectively, in 2090s as compared to 2020s (Figure 8). In contrast to summer days (SU35) and tropical nights (TR25), tropical nights will increase far beyond summer nights with the same level of significance. Hence, minimum temperature will increase more than the maximum temperature which is supported by the very high significant negative trend of diurnal temperature range (DTR); that is, day-night temperature variation will be smaller in future. Daily minimum temperatures are projected to increase faster than daily maximum temperatures, leading to a decrease in diurnal temperature range in the world [55].

The summary of the statistical analysis of rainfall extreme indices with their level of significance is tabulated in Table 8 and their temporal trends are illustrated in Figure 9. The consecutive dry days (CDD) are indicating less rainfall in the study area. This extreme index is projected to decrease by an ensemble average of 1.38 days/year. The trend of annual total rainfall (PRCPTOT) reveals a very high significant increasing trend except B2 scenario of PRECIS RCM. This indicates that the study area will receive more rainfall in the future. Number of days with rainfall ≥10 mm (R10) and ≥20mm (R20) are indicators of moderate and heavy rainfall. Both of these indices are projected to increase significantly. However, R20 will increase with higher trend than that of R10. Two important measures of extreme rainfall are the percentage of total rainfall due to events above the 95th and 99th percentiles (R95p and R99p). Results revealed that both indices increase with the significant trend. The sign of the trend for the annual total rainfall is compared with the sign of the trends in the R99p index. At a station where the annual amount decreases, negative R99p trends are indicative of a large contribution of the extremes and vice versa [50]. The maximum event intensity indices (RX1day and RX5day) show a significant increasing trend, whereas all scenarios of both models show a very high significant trend for RX5day compared to that of RX1day. The simple daily intensity index (SDII) is the combined effect of the trend of PRCPTOT and consecutive wet days (CWD). All scenarios of both models predict an increase of SDII with nonsignificant trend except A2 scenario of HadCM3 GCM. Table 9 provides the summary of the statistical analyses of rainfall and temperature indices with their significance for Chiang Mai, Lamphun, and Tak provinces. With some exception especially for rainfall, the future projections for all temperature and rainfall indices are similar to the projection for Chiang Rai Province.

Table 5. Average seasonal variation of daily maximum and minimum temperature in the future relative to 1980s of observed period.

Decade	Winter T_{max} (T_{min}) (°C)				Spring T_{max} (T_{min}) (°C)			
	PRECIS		HadCM3		PRECIS		HadCM3	
	A2	B2	A2	B2	A2	B2	A2	B2
2020s	−0.18 (1.71)	−0.23 (1.41)	−0.40 (1.97)	−0.33 (1.86)	2.67 (1.03)	2.98 (1.18)	0.39 (−1.41)	0.33 (−1.46)
2050s	0.37 (2.52)	0.09 (2.10)	−0.01 (2.61)	−0.15 (2.43)	3.93 (2.38)	3.36 (2.06)	0.76 (−0.76)	0.57 (−0.94)
2080s	2.33 (4.99)	1.40 (3.42)	0.86 (4.02)	0.30 (3.15)	4.62 (4.07)	4.52 (3.24)	1.02 (0.13)	0.85 (−0.38)
	Summer T_{max} (T_{min}) (°C)				Autumn T_{max} (T_{min}) (°C)			
	PRECIS		HadCM3		PRECIS		HadCM3	
	A2	B2	A2	B2	A2	B2	A2	B2
2020s	0.88 (−0.38)	0.55 (−0.07)	0.74 (0.32)	0.64 (0.40)	−0.55 (−0.69)	−0.30 (−0.67)	0.46 (−0.04)	0.65 (0.02)
2050s	1.99 (1.09)	1.20 (0.70)	1.40 (1.01)	0.94 (0.79)	0.94 (0.81)	0.52 (0.47)	1.13 (0.50)	0.85 (0.39)
2080s	3.56 (2.70)	2.77 (1.69)	2.00 (1.92)	1.47 (1.29)	2.20 (2.79)	1.62 (1.62)	1.89 (1.26)	1.27 (0.75)

Table 6. Average seasonal variation of annual rainfall in the future relative to 1980s of observed period.

Decade	Winter rainfall (%)				Spring rainfall (%)			
	PRECIS		HadCM3		PRECIS		HadCM3	
	A2	B2	A2	B2	A2	B2	A2	B2
2020s	25.97	−6.45	40.83	32.29	−44.35	−54.20	−15.20	−21.64
2050s	−42.03	6.34	52.45	53.08	−61.10	−39.75	−21.51	−15.68
2080s	−29.89	3.44	70.89	79.38	−32.08	−64.30	−16.01	−13.84
	Summer rainfall (%)				Autumn rainfall (%)			
	PRECIS		HadCM3		PRECIS		HadCM3	
	A2	B2	A2	B2	A2	B2	A2	B2
2020s	26.00	37.31	46.16	43.86	28.77	3.09	−11.50	−15.32
2050s	25.48	46.55	32.35	35.33	31.48	14.34	−32.24	−21.16
2080s	34.63	19.83	26.57	42.70	52.74	18.76	−28.12	−36.76

Table 7. Trend of temperature indices for 2011–2099 period in Chiang Rai.

Annual temperature series	PRECIS (A2)	PRECIS (B2)	HadCM3 (A2)	HadCM3 (B2)
	MK test (Sen's slope)			
SU35	7.42*** (0.600)	5.99*** (0.416)	6.79*** (0.263)	5.81*** (0.188)
TR25	9.97*** (1.055)	7.63*** (0.589)	11.03*** (0.813)	9.07*** (0.400)
TXx	10.11*** (0.052)	9.07*** (0.035)	5.01*** (0.010)	5.84*** (0.011)
TNx	11.40*** (0.057)	10.08*** (0.037)	9.76*** (0.031)	6.42*** (0.016)
TXn	6.57**** (0.032)	4.53*** (0.021)	6.25*** (0.011)	2.39* (0.004)
TNn	10.27*** (0.050)	7.89*** (0.029)	8.69*** (0.028)	4.68*** (0.012)
TN10p	−7.56*** (−0.054)	−5.84*** (−0.053)	−9.15*** (−0.073)	−5.57*** (−0.043)
TX10p	−6.52*** (−0.068)	−5.63*** (−0.063)	−5.69*** (−0.037)	−2.97** (−0.019)
TN90p	11.12*** (0.738)	9.81*** (0.560)	11.34*** (0.428)	9.15** (0.211)
TX90p	9.69*** (0.534)	7.91*** (0.373)	8.39*** (0.117)	7.10** (0.066)
DTR	−6.11*** (−0.011)	−2.32* (−0.003)	−9.32*** (−0.018)	−3.68*** (−0.006)
WSDI	9.68*** (1.750)	7.71*** (0.416)	4.71*** (—)	3.12* (—)
CSDI	−3.11** (—)	−4.39*** (—)	−0.22 (—)	0.00 (—)

+Moderately significant, *significant, **highly significant, ***very highly significant, and ( ) = Sen's slope.

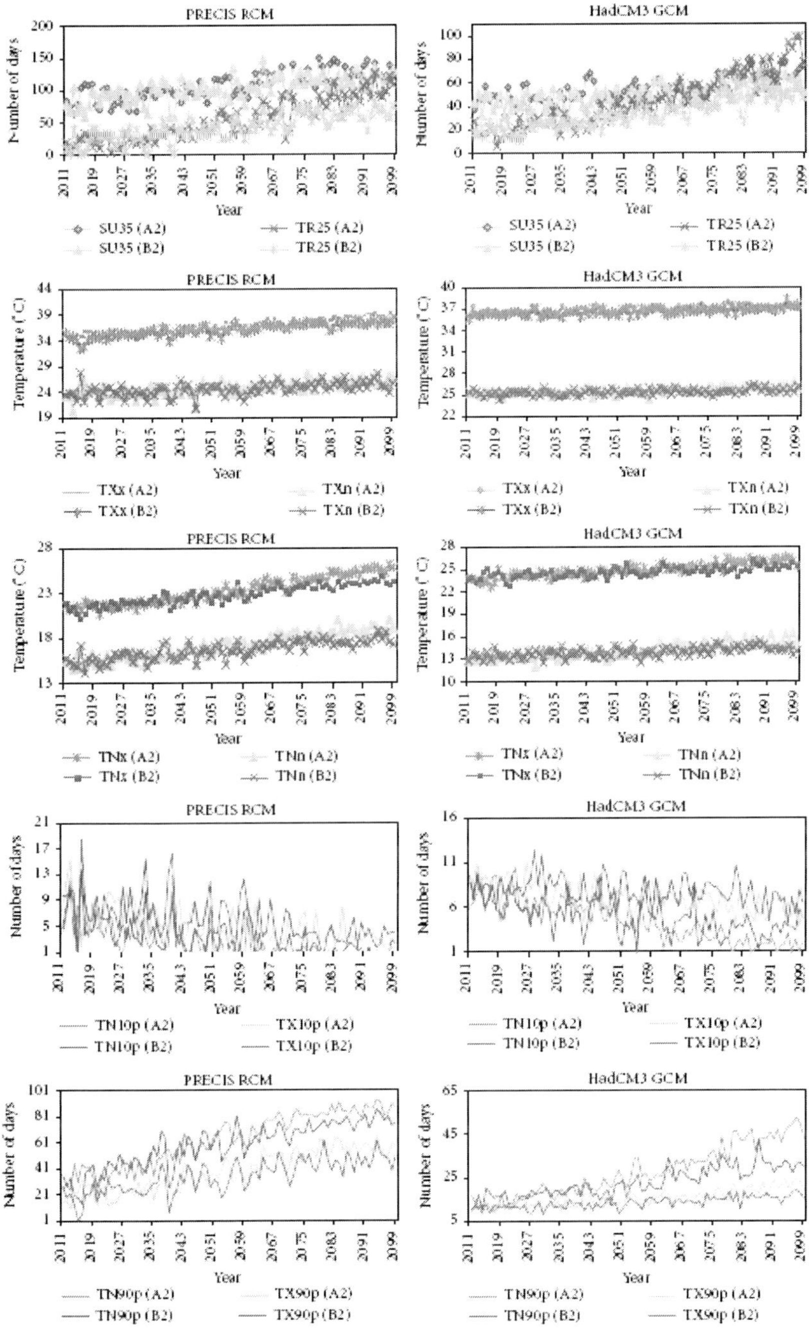
PRECIS RCM
Number of days
Year
SU35 (A2)
TR25 (A2)
SU35 (B2)
TR25 (B2)
HadCM3 GCM
Number of days
Year
SU35 (A2)
TR25 (A2)
SU35 (B2)
TR25 (B2)
PRECIS RCM
Temperature (°C)
Year
TXx (A2)
TXn (A2)
TXx (B2)
TXn (B2)
HadCM3 GCM
Temperature (°C)
Year
TXx (A2)
TXn (A2)
TXx (B2)
TXn (B2)
PRECIS RCM
Temperature (°C)
Year
TNx (A2)
TNn (A2)
TNx (B2)
TNn (B2)
HadCM3 GCM
Temperature (°C)
Year
TNx (A2)
TNn (A2)
TNx (B2)
TNn (B2)
PRECIS RCM
Number of days
Year
TN10p (A2)
TX10p (A2)
TN10p (B2)
TX10p (B2)
HadCM3 GCM
Number of days
Year
TN10p (A2)
TX10p (A2)
TN10p (B2)
TX10p (B2)
PRECIS RCM
Number of days
Year
TN90p (A2)
TX90p (A2)
TN90p (B2)
TX90p (B2)
HadCM3 GCM
Number of days
Year
TN90p (A2)
TX90p (A2)
TN90p (B2)
TX90p (B2)

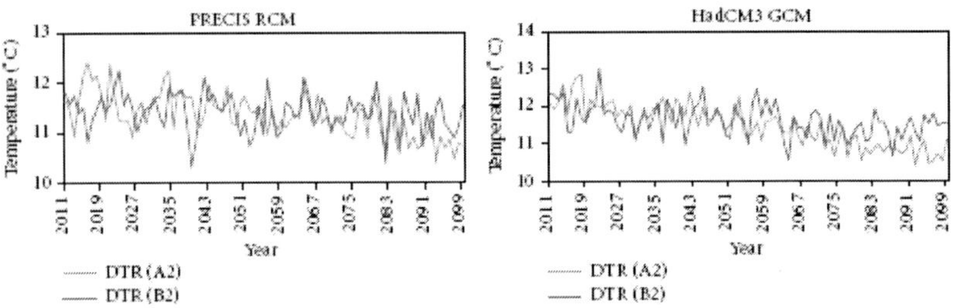

Figure 8. Annual series of temperature extremes for Chiang Rai province as predicted by PRECIS RCM and HadCM3 GCM.

Table 8. Trend of rainfall indices for 2011–2099 period in Chiang Rai.

Annual rainfall series	PRECIS (A2)	PRECIS (B2)	HadCM3 (A2)	HadCM3 (B2)
		MK test (Sen's slope)		
CDD	−0.03 (0.000)	0.01 (0.000)	-5.10^{***} (−0.378)	-4.39^{***} (−0.340)
CWD	-2.21^{*} (−0.153)	−0.67 (−0.055)	3.60^{***} (0.056)	0.28 (0.000)
PRCPTOT	3.54^{***} (3.834)	0.89 (0.958)	8.43^{***} (10.278)	5.89^{***} (5.801)
R10	1.95^{+} (0.073)	0.17 (0.000)	8.11^{***} (0.360)	5.82^{***} (0.190)
R20	5.99^{***} (0.163)	3.50^{***} (0.091)	7.17^{***} (0.255)	5.21^{***} (0.167)
R100	1.28 (—)	—	0.00 (—)	0.00 (—)
R95P	3.76^{***} (3.366)	1.64 (1.396)	4.49^{***} (2.361)	2.91^{**} (1.444)
R99P	1.39 (—)	0.31 (0.000)	3.15^{***} (—)	1.31 (—)
RX1day	1.68^{+} (0.155)	0.22 (0.031)	2.17^{*} (0.058)	0.86 (0.024)
RX5day	5.04^{***} (0.315)	0.61 (0.128)	4.79^{***} (0.388)	3.59^{***} (—)
SDII	1.82 (0.027)	1.61 (0.009)	3.13^{**} (0.017)	1.85 (0.311)

$^{+}$Moderately significant, *significant, **highly significant, and ***very highly significant.

Table 9. Trend of temperature and rainfall indices for 2011–2099 period.

Annual series	Chiang Mai				Lamphun				Tak			
	PRECIS		HadCM3		PRECIS		HadCM3		PRECIS		HadCM3	
	A2	B2	A2	B2	A2	B2	A2	B2	A2	B2	A2	B2
	Temperature indices											
SU35	++	++	++	++	++	++	−	−	++	++	++	++
TR25	++	++	++	++	++	++	++	++	++	++	++	++
TXx	++	++	++	++	++	++	++	−	++	++	++	++
TNx	++	++	++	++	++	++	++	+	++	++	++	++
TXn	++	++	++	++	++	++	++	+	++	++	++	++
TNn	++	++	++	++	++	++	++	+	++	++	++	++
TN10p	−−	−−	−−	−−	−−	−−	−	−	−−	−−	−	−
TX10p	−−	−−	−−	−−	−−	−−	−	−	−−	−−	−−	−−
TN90p	++	++	++	++	++	++	++	++	++	++	++	++
TX90p	++	++	++	++	++	++	++	++	++	++	++	++
DTR	−−	−	−−	−	−−	−−	−−	−−	−−	−−	−	−
	Rainfall indices											
CDD	+	+	++	−	+	−	+	+	+	−	++	++
CWD	−	+	−	−−	−	−	−	−	−	−	+	+
PRCPTOT	+	+	−	−	+	−	−	−	+	+	+	+
R10	−	−	−−	−	+	+	−−	−	+	+	−−	−−
R20	++	++	+	+	++	++	−	−	++	++	−	−
R95P	++	++	++	+	++	++	−	−	++	+	+	−
SDII	+	++	++	++	++	++	−	−	+	++	++	++

+ positive trend, − negative trend, and statistical significant trend (++, −−) signs.

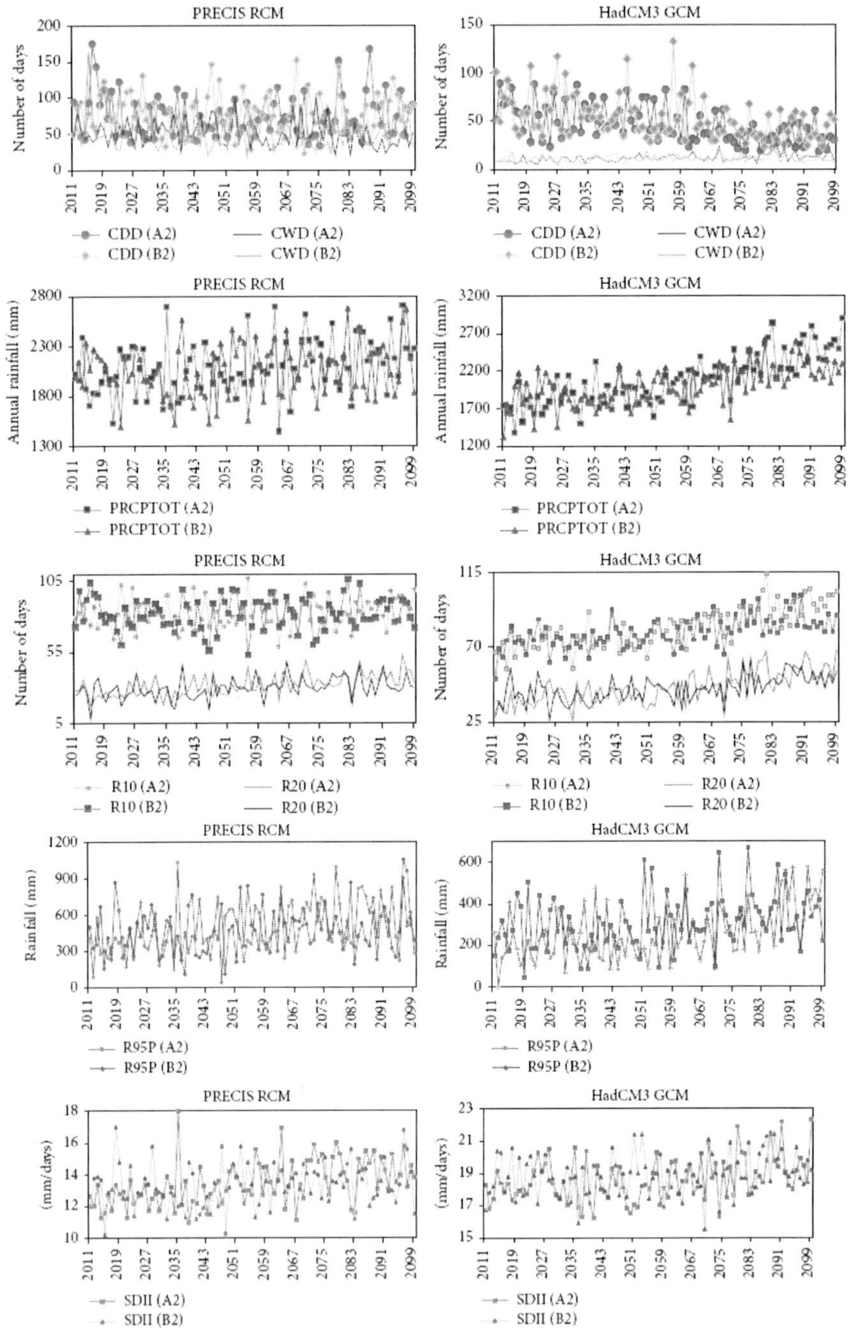

Figure 9. Annual series of rainfall extremes for Chiang Rai province as predicted by PRECIS RCM and HadCM3 GCM.

5. CONCLUSIONS

Our analysis enables a number of conclusions to be drawn as follows:(1)During observed 1960–2010 period, the entire temperature extremes showed a warming trend which is also supported by the statistical analyses. A warming trend is indicated by the summer days, tropical nights, Max, Min, Max, Min, cool nights, cool days, warm nights, warm days, diurnal temperature range, warm spell duration indicator, and cold spell duration indicator. The warming trend is higher in the southern part of the study area, that is, in Tak. Trend and slope of night extremes (TNx, TNn, TN10, and TN90) appear to have been higher than those in day extremes (TXx, TXn, TX90, and TX10) resulting in the decrease of day-night temperature variation which is also indicated by diurnal temperature range (DTR).(2)Considering all seasons (winter, spring, summer, and autumn), maximum temperature () and minimum temperature () have increased significantly in all seasons except spring which is insignificant over the whole study area. Statistical analyses indicated that the increasing trend of was higher than that of. (3)In future, the maximum temperature beyond 35°C (summer days) and minimum temperature beyond 25°C (tropical nights) will increase with the very high significant trend. For all SRES (A2 and B2) of both models considered in this study, the maximum and minimum temperature increases projected for the period of 2011–2099 were 0.99 and 0.93°C above the observed period. However, monthly maximum and minimum temperature will boost up by 1.56 and 2.24°C at the end of this century as compared to the 2020s. In future, all the day, night, cool, and warm extremes will change same as observed period. Diurnal temperature range will go down (0.64°C) notably over the study area.(4)During the observed period, annual total rainfall (PRCPTOT) and moderate rainfall (R10) indices were decreased which is supported by the negative trend of consecutive wet days (CDD). However, none of the indices showed significant trend in the observed period and there was no significant trend found for seasonal rainfall also. Each model and scenario evaluated project an increase in average annual rainfall of 9.65% for 2011–2099 compared to the observed annual rainfall 1235 mm.Topographic changes of the earth greatly influence the variation of climatic parameters and their extremes and anomalies. Hence, it is recommended to explore the changes of climate extremes and their anomalies over topography using a large number of stations.

ACKNOWLEDGMENTS

The authors would like to acknowledge the financial support from Asian Development Bank and Japan Scholarship Program (ADB-JSP) to finish this research work. Thanks to Jannatul Ferdous for her help in data analysis which helped the authors to finish this work.

REFERENCES

1. IPCC, Climate Change 2007. The Physical Science Basis, Contribution of Working Group I to the Fourth Assessment Report of the Intergovernmental Panel on Climate Change, Cambridge University Press, Cambridge, UK, 2007.

2. C. Tebaldi, K. Hayhoe, J. M. Arblaster, and G. A. Meehl, "Going to the extremes—an intercomparison of model-simulated historical and future changes in extreme events," Climatic Change, vol. 79, no. 3-4, pp. 185–211, 2006.

3. J. Alcamo, N. Dronin, M. Endejan, G. Golubev, and A. Kirilenko, "A new assessment of climate change impacts on food production shortfalls and water availability in Russia," Global Environmental Change, vol. 17, no. 3-4, pp. 429–444, 2007.

4. R. P. Allan and B. J. Soden, "Atmospheric warming and the amplification of precipitation extremes,"Science, vol. 321, no. 5895, pp. 1481–1484, 2008.

5. J. H. Christensen, B. Hewitson, A. Busuioc, et al., "Regional climate projections," in Climate Change 2007: The Physical Science Basis. Contribution of Working Group I to the Fourth Assessment Report of the Intergovernmental Panel on Climate Change, S. Solomon, D. Qin, M. Manning, et al., Eds., Cambridge University Press, Cambridge, UK, 2007.

6. H. Kettle and R. Thompson, "Statistical downscaling in European mountains. Verification of reconstructed air temperature," Climate Research, vol. 26, no. 2, pp. 97–112, 2004.

7. R. L. Wilby, T. M. L. Wigley, D. Conway et al., "Statistical downscaling of general circulation model output: a comparison of methods," Water Resources Research, vol. 34, no. 11, pp. 2995–3008, 1998.

8. F. Bouraoui, G. Vachaud, L. Z. X. Li, H. Le Treut, and T. Chen, "Evaluation of the impact of climate changes on water storage and groundwater recharge at the watershed scale," Climate Dynamics, vol. 15, no. 2, pp. 153–161, 1999.

9. P. Frich, L. V. Alexander, P. Della-Marta et al., "Observed coherent changes in climatic extremes during the second half of the twentieth century," Climate Research, vol. 19, no. 3, pp. 193–212, 2002.

10. L. V. Alexander, X. Zhang, T. C. Peterson, et al., "Global observed changes in daily climate extremes of temperature and precipitation," Journal of Geophysical Research: Atmospheres, vol. 111, no. 5, Article ID D05109, 2006.

11. B. Orlowsky and S. I. Seneviratne, "Global changes in extreme events: regional and seasonal dimension,"Climatic Change, vol. 110, no. 3-4, pp. 669–696, 2012.

12. M. G. Donat, L. V. Alexander, H. Yang et al., "Updated analyses of temperature and precipitation extreme indices since the beginning of the twentieth century: the HadEX2 dataset," Journal of Geophysical Research: Atmospheres, vol. 118, no. 5, pp. 2098–2118, 2013.

13. M. J. Manton, P. M. Della-Marta, M. R. Haylock et al., "Trends in extreme daily rainfall and temperature in southeast Asia and the south Pacific: 1961–1998," International Journal of Climatology, vol. 21, no. 3, pp. 269–284, 2001.

14. G. M. Griffiths, L. E. Chambers, M. R. Haylock et al., "Change in mean temperature as a predictor of extreme temperature change in the Asia-Pacific region," International Journal of Climatology, vol. 25, no. 10, pp. 1301–1330, 2005.

15. S. Chinvanno, V. Luang-Aram, C. Saengmanee, and J. Thanakitmetavut, Developing High Resolution Climate Scenarios for Southeast Asia Region and Mekong River Basin for 1960–2100, SEA START Regional Centre, Bangkok, Thailand, 2011,

16. J. Kreasuwan, "Simulation of future climate scenario for Thailand," Tech. Rep., Thailand Research Fund, Bangkok, Thailand, 2009,

17. Southeast Asia START Regional Center, Preparation of Climate Change Scenarios for Climate Change Impact Assessment in Thailand, Southeast Asia START Regional Center, 2011

18. H. S. Lim, K. Boochabun, and A. D. Ziegler, "Modifiers and amplifiers of high and low flows on the Ping river in Northern Thailand (1921–2009): the roles of climatic events and anthropogenic activity," Water Resources Management, vol. 26, no. 14, pp. 4203–4224, 2012.

19. J. Fox, J. B. Vogler, O. L. Sen, T. W. Giambelluca, and A. D. Ziegler, "Simulating land-cover change in Montane Mainland Southeast Asia," Environmental Management, vol. 49, no. 5, pp. 968–979, 2012.

20. S. Sangawongse, "Land-use/land-cover dynamics in Chiang Mai: appraisal from remote sensing, GIS and modeling approaches," CMU Journal, vol. 5, no. 2, pp. 243–254, 2006.

21. L. Lebel, P. Garden, N. Subsin, and S. Na Nan, "Averted crises, contested transitions: water management in the Upper Ping River basin, Northern Thailand," in Water Policy Entrepeneurs. A Research Companion to Water Transitions Around the Globe, D. Huitema and S. Meijerink, Eds., pp. 137–157, Edward Elgar, Cheltenham, UK, 2009.

22. S. Sangawongse, S. Prabudhanitisarn, and E. Karjagthimaporn, "Agricultural land use change and urbanization in Thailand," in Proceedings of the UM Conference, 2011.

23. J. Rigg and M. Ritchie, "Production, consumption and imagination in rural Thailand," Journal of Rural Studies, vol. 18, no. 4, pp. 359–371, 2002.

24. W. Jompakdee, "Rivers in jeopardy and the role of civil society in river restoration: Thai experiences,"CMU Journal, vol. 3, no. 1, pp. 59–71, 2004.

25. EPSEA, "Adapting to floods in Chiang Mai: confronting the gaps," EEPSEA Climate Change Policy Brief no. 2010CC-PB6, 2010.

26. S. H. Wood and A. D. Ziegler, "Floodplain sediment from a 100-year-recurrence flood in 2005 of the Ping River in northern Thailand," Hydrology and Earth System Sciences, vol. 12, no. 4, pp. 959–973, 2008.

27. A. D. Ziegler, "Water management: reduce urban flood vulnerability," Nature, vol. 481, no. 7380, p. 145, 2012.

28. A. D. Ziegler, L. H. She, C. Tantasarin, N. R. Jachowski, and R. Wasson, "Floods, false hope, and the future," Hydrological Processes, vol. 26, no. 11, pp. 1748–1750, 2012.

29. Thai Meteorological Department, The Climate of Thailand, 2015,

30. R. L. Wilby, C. W. Dawson, and E. M. Barrow, "SDSM—a decision support tool for the assessment of regional climate change impacts," Environmental Modelling and Software, vol. 17, no. 2, pp. 147–157, 2002.

31. B. C. Hewitson and R. G. Crane, "Climate downscaling. Techniques and application," Climate Research, vol. 7, no. 2, pp. 85–95, 1996.

32. E. Kalnay, M. Kanamitsu, R. Kistler et al., "The NCEP/NCAR 40-year reanalysis project," Bulletin of the American Meteorological Society, vol. 77, no. 3, pp. 437–471, 1996.

33. R. L. Wilby, C. W. Dawson, and E. M. Barrow, "SDSM—a decision support tool for the assessment of regional climate change impacts," Environmental Modelling and Software, vol. 17, no. 2, pp. 147–159, 2002.

34. U. Cubasch, G. A. Meehl, G. J. Boer, et al., "Projections of future climate change," in Climate Change 2001. The Scientific Basis, J. T. Houghton, Y. Ding, D. J. Griggs, et al., Eds., pp. 525–582, Cambridge University Press, New York, NY, USA, 2001.

35. R. Fealy and J. Sweeney, "Statistical downscaling of precipitation for a selection of sites in Ireland employing a generalised linear modelling approach," International Journal of Climatology, vol. 27, no. 15, pp. 2083–2094, 2007.

36. C. J. R. Williams, D. R. Kniveton, and R. Layberry, "Assessment of a climate model to reproduce rainfall variability and extremes over

Southern Africa," Theoretical and Applied Climatology, vol. 99, no. 1-2, pp. 9–27, 2010.

37. L. Tryhorn and A. DeGaetano, "A comparison of techniques for downscaling extreme precipitation over the Northeastern United States," International Journal of Climatology, vol. 31, no. 13, pp. 1975–1989, 2011.

38. IPCC, Special Report on Emissions Scenarios: A Special Report of Working Group III of the Intergovernmental Panel on Climate Change, Cambridge University Press, Cambridge, UK, 2000.

39. H. B. Mann, "Nonparametric tests against trend," Econometrica, vol. 13, pp. 245–259, 1945.

40. M. G. Kendall, Rank Correlation Methods, Hafner Publishing, New York, NY, USA, 1948.

41. L. Zongxing, Y. He, P. Wang et al., "Changes of daily climate extremes in southwestern China during 1961–2008," Global and Planetary Change, vol. 80-81, pp. 255–272, 2012.

42. S. Yue, P. Pilon, and G. Cavadias, "Power of the Mann–Kendall and Spearman's rho tests for detecting monotonic trends in hydrological series," Journal of Hydrology, vol. 259, no. 1–4, pp. 254–271, 2002.

43. V. D. P. R. daSilva, "On climate variability in Northeast of Brazil," Journal of Arid Environments, vol. 58, no. 4, pp. 575–596, 2004.

44. Q. Liu, Z. Yang, and B. Cui, "Spatial and temporal variability of annual precipitation during 1961–2006 in Yellow River Basin, China," Journal of Hydrology, vol. 361, no. 3-4, pp. 330–338, 2008.

45. L. V. Alexander and J. M. Arblaster, "Assessing trends in observed and modelled climate extremes over Australia in relation to future projections," International Journal of Climatology, vol. 29, no. 3, pp. 417–435, 2009.

46. L. Liang, L. Li, and Q. Liu, "Temporal variation of reference evapotranspiration during 1961–2005 in the Taoer River basin of Northeast China," Agricultural and Forest Meteorology, vol. 150, no. 2, pp. 298–306, 2010.

47. J. A. Marengo, L. M. Alves, W. R. Soares et al., "Two contrasting seasonal extremes in tropical South America in 2012: flood in Amazonia and drought in Northeast Brazil," Journal of Climate, vol. 26, pp. 9137–9154, 2013.

48. P. K. Sen, "Estimates of the regression coefficient based on Kendall's tau," Journal of the American Statistical Association, vol. 63, pp. 1379–1389, 1968.

49. V. L. Tran, D. C. Hoang, and T. T. Tran, "Building of climate change scenario for Red River catchments for sustainable development and environmental protection," in Proceedings of the Science Workshop on Hydrometeorological Change in Vietnam and Sustainable Development, pp. 70–82, Ministry of Natural Resource and Environment, Hanoi, Vietnam, 2005.

50. D. Sharma and M. S. Babel, "Trends in extreme rainfall and temperature indices in the western Thailand,"International Journal of Climatology, vol. 34, no. 7, pp. 2393–2407, 2014.

51. X. B. Zhang, E. Aguilar, S. Sensoy, et al., "Trends in Middle East climate extreme indices from 1950 to 2003," Journal of Geophysical Research: Atmospheres, vol. 110, no. 22, Article ID D22104, 2005.

52. X. Liu, Z. Cheng, L. Yan, and Z.-Y. Yin, "Elevation dependency of recent and future minimum surface air temperature trends in the Tibetan Plateau and its surroundings," Global and Planetary Change, vol. 68, no. 3, pp. 164–174, 2009.

53. P. Deb and M. S. Babel, "Assessment of impact of climate change and adaptation measures for maize production in East Sikkim, India," Journal of Hydrology and Meteorology, vol. 9, no. 1, pp. 15–27, 2015.

54. L. Erda, X. Wei, J. Hui et al., "Climate change impacts on crop yield and quality with CO_2 fertilization in China," Philosophical Transactions of the Royal Society B: Biological Sciences, vol. 360, no. 1463, pp. 2149–2154, 2005.

55. G. A. Meehl, T. F. Stocker, W. D. Collins et al., "Global climate projections," in Climate Change 2007. The Physical Science Basis, S. Solomon, D. Qin, M. Manning et al., Eds., Contribution of Working Group I to the Fourth Assessment Report of the Intergovernmental

Panel on Climate Change, Cambridge University Press, Cambridge, UK, 2007.

56. V. Cruz Rex, H. Hideo, L. Murari et al., "In climate change 2007. Impacts, adaptation and vulnerability," in Working Group II to the 4th Assessment Report of the Intergovernmental Panel on Climate Change, M. L. Parry, O. F. Canziani, J. P. Palutikof, P. J. van der Linden, and C. E. Hanson, Eds., pp. 469–506, Cambridge University Press, Cambridge, UK, 2007.

57. A. G. Reda, N. K. Tripathi, P. Soni, T. Tipdecho, and A. Phalke, "Temporal climate trend of Ping basin of Thailand and implications for Mekong Region," Journal of Earth Science and Climatic Change, vol. 4, article 146, 2013.

Chapter 3

A New Global Climatology of Annual Land Surface Temperature

Benjamin Bechtel

Institute of Geography, University of Hamburg, Bundesstraße 55, 20146 Hamburg, Germany

ABSTRACT

Land surface temperature (LST) is an important parameter in various fields including hydrology, climatology, and geophysics. Its derivation by thermal infrared remote sensing has long tradition but despite substantial progress there remain limited data availability and challenges like emissivity estimation, atmospheric correction, and cloud contamination. The annual temperature cycle (ATC) is a promising approach to ease some of them. The basic idea to fit a model to the ATC and derive annual cycle parameters (ACP) has been proposed before but so far not been tested on larger scale. In this study, a new global climatology of annual LST based on daily 1 km MODIS/Terra observations was processed and evaluated. The derived global parameters were robust and free of missing data due to clouds. They allow estimating LST patterns under largely cloud-free conditions at different scales for every day of year and further deliver a measure for its accuracy respectively variability. The parameters generally showed low redundancy and mostly reflected real surface conditions. Important influencing factors included climate, land cover, vegetation phenology, anthropogenic effects, and geology which enable numerous potential applications. The datasets will be available at the CliSAP Integrated Climate Data Center pending additional processing.

KEYWORDS

land surface temperature; annual temperature cycle; annual cycle parameters; cloud gaps; thermal infrared; MODIS

1. INTRODUCTION

Land surface temperature (LST) is the temperature of the Earth's surface or "skin" [1]. It is an important parameter in the Earth system at the land-atmosphere interface. LST determines the long-wave radiation as well as the turbulent heat flux into the atmosphere and hence is related to both the surface radiation and energy balance. Therefore, LST is frequently used in a number of fields including hydrology, climatology, and geophysics. It shows great spatiotemporal variability due to the heterogeneity of influencing factors such as land cover, vegetation, soil moisture, geology and topography [2]. For instance, it can be used to estimate surface fluxes [3,4], evaporation [5,6], and to analyze the surface urban heat island [7,8,9,10].

The LST is directly linked to thermal infrared (TIR) radiation by Plank's law describing the radiation emitted by a black body in thermal equilibrium. Therefore, derivation of LST by TIR remote sensing from satellites has a long history that can be traced back to the Television and InfraRed Observation Satellite 2 satellite in the early 1960s [11]. Today, numerous satellite LST retrieval algorithms are available [1,2], which can be classified into single-channel algorithms [12,13,14], multi-channel algorithms [15,16,17] and multi-angle algorithms [15]. While the first class is largely dependent on detailed knowledge of the atmospheric influences e.g., by radiative transfer modelling [18], multi-channel or split-window algorithms take advantage of the different atmospheric absorptions in different TIR bands to directly empirically correct the atmospheric influence from the measurements.

Besides substantial progress in the processing of TIR data, significant challenges remain with respect to LST retrieval. Particularly, the atmospheric influence depends on atmospheric composition and meteorology. Further, most real world objects are not perfect black bodies and emit less radiation, which is described by a wavelength-dependent factor, called emissivity. Since the emissivity is usually not known *a priori*, several methods for its estimation have been developed. Among the most frequently used approaches, are classification based methods [19] and normalized difference vegetation index (NDVI) based methods [20,21]. While the former assume constant emissivity within a particular land cover class, the latter assume that the vegetation fraction is the dominant influence on the emissivity. However, both methods do not work well in regions with very heterogeneous land cover which are not dominated by vegetation, such as urban areas for instance. More sophisticated approaches attempt to solve the Planck's equation. Day-night approaches assume constant emissivity at two subsequent acquisitions and hence reduce the number of unknown variables. Likewise, the temperature emissivity separation (TES) approach [22] uses an empirical relationship between the spectral contrast and the minimum emissivity to increase the number of equations and make the retrieval problem solvable. Further challenges include cloud masking and anisotropy. A comprehensive review of all available methods and their advantages and disadvantages are given in [2].

Another important limitation to the application of remote sensing LST is data availability. Currently available instruments [23] either have low temporal or coarse spatial resolution. Instruments on geostationary satellites like the Spinning Enhanced Visible Infra-Red Imager (SEVIRI) onboard Meteosat Second Generation (MSG) satellites have a high temporal resolution (15 min) but a poor spatial resolution (3.1 km at nadir), which is inadequate for many applications, e.g., in urban areas. Instruments in polar orbits either have a wide swath width with a revisit time of approximately one day and a spatial resolution of about 1000 m (e.g., the Moderate-resolution Imaging Spectroradiometer, MODIS) or a low swath width with a spatial resolution of about 100 m but a long revisit time of about 1–2 weeks. Unfortunately, all sensors of this class (*i.e.*, ASTER and Landsat 5, 7 and 8) have very similar equator passing times. Hence, the spatiotemporal variation is not sufficiently covered. The availability is further reduced by clouds, which systematically inhibit simultaneous acquisitions over large areas in humid and moderate climates.

Promising methods to enhance the spatiotemporal resolution and availability of LST data include downscaling (also referred as disaggregation or fusion) [24,25,26,27,28] or estimation of sub-cloud temperature [29]. A robust and already established method is diurnal temperature cycle (DTC) modelling [30,31,32,33] which allows interpolation of missing values due to single cloudy acquisitions. For instance in [30] a Levenberg-Marquardt minimization scheme was utilized to fit a DTC model to the time series of cloud-screened brightness temperatures and describe the thermal behavior of the land surface by five determined model parameters per day, which were found to be related to relevant physical properties of the land surface, such as NDVI and thermal inertia.

However, a sufficient number of diurnal acquisitions are only available in coarse spatial resolution. An alternative approach involves monitoring of the annual temperature cycle (ATC) which shows comparable deterministic variation in the irradiation (not due to the Earth's rotation but the inclination of the Earth's axis to the ecliptic and the resulting seasons). The idea to likewise fit a model to the ATC and derive annual cycle parameters (ACP) was first proposed in [34] and tested for its robustness in [35]. Like the DTC models, the ACP are related to important thermal surface characteristics and allow a continuous description of the thermal surface behavior independent of clouds which can be considered as "thermal landscapes". Further case studies have shown their relevance in different parts of the world [36,37]. However, the applicability of ACPs remains limited to small areas on the order of 100 s to 1000 s of km^2 and so far has not been tested on continental or global scale.

In this study a new global climatology dataset of annual land surface temperature based on MODIS LST on 1 km resolution is presented and tested. ACPs were calculated based on the MOD11A1 product twice a day based on all acquisitions from 2000 to 2013. The paper is organized as follows. In Section 2, the datasets and the annual cycle fitting methods are presented. In Section 3, the derived parameters are presented and analyzed and selected potential applications are highlighted and discussed. Conclusions are given in Section 4.

2. DATA AND METHODS

2.1. Land Surface Temperatures MOD11A1

The MODIS Land Surface Temperature and Emissivity products provide temperature and emissivity values in swath-based and grid-based global products of different resolutions from the satellites Terra and Aqua. For this study the daily level 3 global 1 km grid product from Terra (MOD11A1) [38] was used, which has 1 km spatial resolution and is produced with a generalized split window approach using the MODIS bands 31 and 32 [17], which separates ranges of atmospheric column water vapor and lower boundary air surface temperatures into tractable sub-ranges. The data are freely available and are delivered by the U.S. Geological Survey (USGS) Land Processes Distributed Active Archive Center (LP DAAC), which is a component of NASA's Earth Observing System Data and Information System. Here, version 5 of the product was used which contains a number of major refinements compared to the versions 4 and 4.1 [39]. Unfortunately, version 6 was not yet available at the time the study was conducted, but the ACP will be reprocessed as soon as the refined MOD11A1 dataset is available.

These data are delivered in gridded HDF files for each tile which contain a number of scientific data sets; LST_Day_1km, QC_Day, Day_view_time, Day_view_angl, LST_Night_1km, QC_Night, Night_view_time, Night_view_angl, Emis_31, Emis_32, Clear_day_cov, Clear_night_cov which are: the LST data itself, a Quality control flag, the viewing time and angle, and the number of cloud free acquisitions used to calculate the value (which might be more than one due to overlapping orbits) for day and night respectively. Further, the emissivity values for the two bands are given, which are based on the classification based approach of [19] for 18 land cover types with some small modifications [39].

The accuracy of the LST product has been tested and was found to mostly fall within 1 K [39,40,41]. In [42] the version 5 product was validated again using the superior radiance-based method as well as additional ground sites and it was found that LST was underestimated by more than 2 K at certain bare soil and desert sites. Besides the errors were also within 2 K and within 1 K in most cases. For this study, both day- and nighttime LST were used to derive ACP and the quality control flags were considered as a filter (see Section 3.1). Due to the large number of hundreds to thousands of observations, the effects of slightly varying overpass times and angles are assumed to be negligible. Therefore, these parameters were not evaluated. Terra is on a sun-synchronous, near-polar, circular orbit with equatorial crossing times around 1030 h local solar time in the descending (daytime) orbit and about 2230 h local solar time in its ascending (nighttime) orbit.

The MOD11A1 products are delivered in a sinusoidal grid by mapping the level-2 LST product (MOD11_L2) on 1 km (precisely 0.928-km) grids (Figure

1). The tiles are numbered horizontally and vertically starting at h00v00 (horizontal tile number, vertical tile number) in the upper left corner and proceeding right (horizontal) and downward (vertical). The tile size is 10° by 10° at the equator. The projection is an equal area projection and has large distortions at higher latitudes and longitudes.

The V5 MOD11A1 product's acquisition range started 5 March 2000. For this study all acquisitions from then until the 31 December 2013 were used.

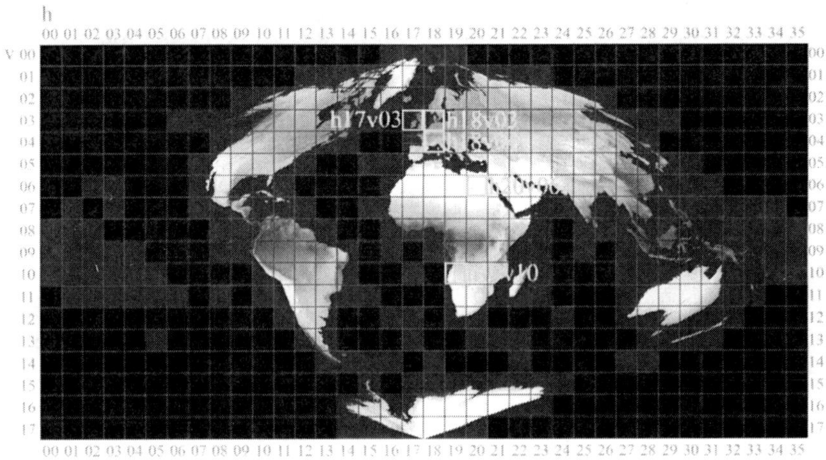

Figure 1. Tiles and projection of MODIS land products. Tiles selected as examples for this study are highlighted.

2.2. MODIS Land Cover and Urban Areas

The MODIS Land Cover product MCD12Q1 version 5.1 provides a classification of the global land cover in different classification systems. The land cover properties are contained within 500 m sinusoidal grids and are derived from one year of observations from the satellites Terra- and Aqua. Version 5.1 are produced with revised training data and several algorithm refinements [43]. The product contains five classification schemes with the primary land cover scheme identifying 17 land cover classes defined by the International Geosphere Biosphere Programme (IGBP). The scheme consists of 11 natural vegetation classes, three developed and mosaicked land classes, and three non-vegetated land classes. Here the IGBP classification scheme was utilized unless otherwise stated.

In this study, the land cover dataset was used to extract the surface urban heat island. Therefore, urban land cover was resampled to the 1000 m grid using nearest neighbor interpolation. Major urban centers were extracted by a morphological opening with a disk shaped structuring element of two pixel radius which means smaller urban areas (for instance, towns and villages) were excluded from the analysis. The corresponding rural areas were defined as all pixels within 15 pixels (~15 km) distance from the urban core areas that were not classified as urban in the original classification. Since only mean values for the urban and rural areas were analyzed and substantial change was not expected in the selected region (namely Great Britain), the land cover data from approximately the middle of the period (2006) were used.

2.3. Annual Cycle Parameters

The fundamental idea of ACP is the use of multi-temporal thermal data from long time periods to generate a pixel-wise simplified climatology of LST. The derived LST is applicable for acquisitions during largely cloud free conditions. Hence, every measurement is divided into a long-term average annual cycle of LST (which reflects the irradiation and climatology) as well as a day-specific anomaly (determined by meteorological, vegetation and soil conditions, as well as the exact overpass time), hereafter referred to as residuals. This climatology is periodic and hence can be approximated by a series of harmonic functions [35].

For the ACP model, the annual cycle is approximated by a constant term plus a sine function

$$f(d) = MAST + YAST \sin(d 2\pi / 365 + \theta) \quad (1)$$

where *d* is the day of the cycle and the three ACPs are the mean annual surface temperature (MAST), the yearly amplitude of surface temperature (YAST) and the phase shift θ. The phase shift accounts for the periodic boundary conditions of the surface energy balance which are shifted relatively to the radiative forcing (*i.e.*, air temperature) and hence can be seen as a measure of the latency of heat uptake. It has to be acknowledged that between the tropic of Cancer and the tropic of Capricorn the model is less meaningful since the irradiation is characterized by two annual maxima instead of one especially at very low latitudes.

Since the ACPs were first developed for the Northern hemisphere and the larger percentage of land areas are on this hemisphere, the phase shift was defined relative to the spring equinox on 21st of March. Since a phase shift of π (respectively 182.5 days) is equivalent to a multiplication of the amplitude by −1, the YAST was set ≥ 0 per definition and the phase shift as $-\pi \leq \theta < \pi$ respectively -182.5 days $\leq \theta < 182.5$ days. Since the full cycle was set to 365 days, intercalary days were ignored.

As additional parameters the root mean squared error (RMSE) of the fit and the number of clear sky acquisitions (NCSA) were calculated. Since the RMSE is the quadratic mean of the residuals, it is not only an accuracy measure but also a climatic characteristic providing an integrated measure of both the inter-diurnal and inter-annual variability of LST. NCSA on the other hand, is a measure for frequency of cloud occurrence, although it is strongly influenced by the orbital characteristics (areas in overlapping orbits have a larger number of cloud free observations). The principle is illustrated in Figure 2.

For a given series of LST measurements, the ACP for every pixel were estimated with an unconstrained nonlinear optimization algorithm [44] minimizing the square sum of the residuals. This iterative direct search method without gradients is based on a simplex (essentially an *N*-dimensional triangle) of parameter vectors. At each step a new point in or near the current simplex is created, its function value is compared with the function's values at the vertices, and a new simplex of the best remaining values is created until the diameter of the simplex falls below a predefined threshold.

The minimization algorithm was already implemented in MATLAB and the ACPs were computed on a supermicro-server with 4 sockets of AMD Opteron (Abu-Dhabi) 6386 SE (4 × 16 cores, 1.4 to 2.8 GHz), 256 GB RAM, and a 100 TB external general parallel file system. Since the entire set of acquisitions (about 5000 scenes) for one tile was held in memory for processing, each tile demanded for approximately 14 GB of memory. Thus, the day and night-time ACP each took about one week of processing using 15 cores or 30 CPU-weeks in total respectively.

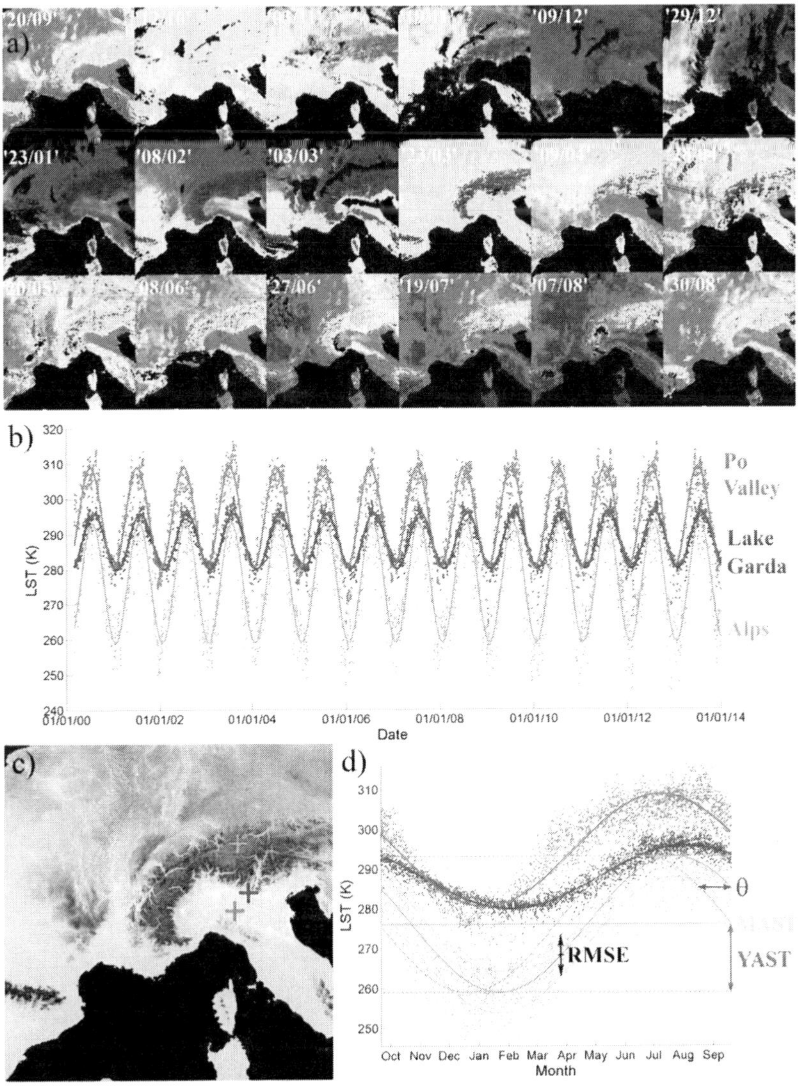

Figure 2. Principle of the annual cycle parameters. (**a**) Sequence of acquisitions from tile h18v04 (covering Northern Italy, Switzerland, as well as large parts of France and Southern Germany) from different seasons (and also years); (**b**) Full time series for three selected pixels with different annual LST characteristics (Po Valley, Lake Garda, and Alpine mountains). (**c**) RGB color composite from the fitted parameters YAST, MAST and phase shift θ. Selected pixels are marked with +; (**d**) Annual cycle of the selected pixels and explanation of the parameters YAST, MAST, θ, and RMSE for the Alpine pixel.

3. RESULTS AND DISCUSSIONS

3.1. Cloud Quality Control

An accurate cloud masking plays an important role in the determination of multi-temporal LST parameters. Particularly, there is a tradeoff between avoidance of cloud contaminated pixels and a sufficiently large number of acquisitions for a precise fit of the model. The MOD11A1 product is already cloud filtered and delivers additional information about the quality of each produced data. The mandatory quality assessment (QA) flag within the product consists of two bits. A value of 00 indicates "LST produced, good quality, not necessary to examine detailed QA" while a value of 01 means "LST produced, unreliable or unquantifiable quality" [38]. Values of 10 and 11 mean "LST not produced due to cloud effects" and "other reasons than clouds" respectively. Hence, for this study the relevant decision was whether the unreliable or unquantifiable quality observations should be included in the ACP modelling. Figure 3 shows the number of available observations for the years 2011–2012 for MODIS tile h18v04 in total (a) and (b) with QA flag 00 only. While the number of available scenes generally decreased from an average of 231 to 131 by employing this criteria, the decrease is highly dependent on the land cover type. This especially is the case for snow and urban areas, which are frequently miss-classified as clouds or routinely not assessed as good quality. Large urban agglomerations like Paris or Milan can even be seen in the total number of produced acquisitions.

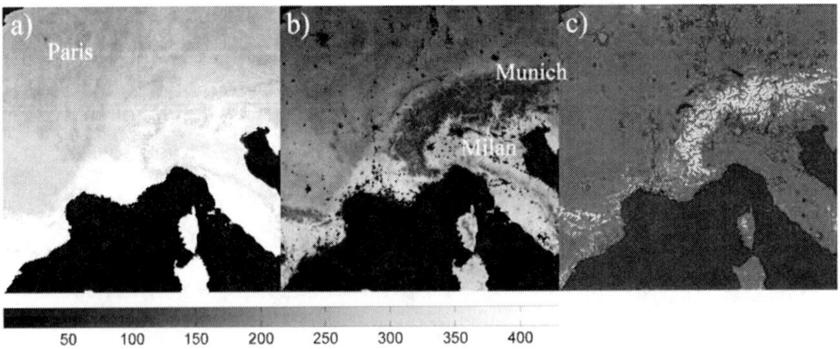

Figure 3. Number of cloud free scenes for the period 2011–2012 for tile h18v04 for (**a**) all produced acquisitions; (**b**) acquisitions with good ("00") quality flag; (**c**) Urban areas (red) according to MODIS land cover dataset.

This selection also has implications on the quality of the accuracy of the produced ACP. Table 1 shows the mean values of the ACP produced with *all* acquisitions and the *good* acquisitions (QA = "00") only for the same tile and period. Both, MAST and YAST are higher for the good acquisitions, which indicates a certain cloud contamination in the full dataset. Unfortunately, this variation is systematic (not displayed) with smaller differences in the Mediterranean areas (South) than in the region closer to the North Sea, which is likely related to the frequency of cloud cover and unfiltered contamination. Otherwise, the Alps (East) also show comparably high differences which are assumed to be a result of a systematic filtering of days with snow cover. Hence, it cannot inevitably be concluded that the ACP from the QA filtered product are more accurate. While for MAST both datasets are very well correlated for YAST, the R value amounts to only 0.87, which indicates that the parameter is more sensitive to outliers. The phase shift θ seems least sensitive to the selection of cloud free scenes. Since urban areas are one of the major applications of the product the entire LST data set was used without consideration of the QA flag.

Table 1. Mean ACP for tile $h_{18}v_{04}$ for 2011–2012 for all produced values, good QA only, and correlation coefficient between both.

Parameter	All	Good	R
YAST (K)	12.0	12.6	0.87
MAST (K)	289.4	291.0	0.99
θ	−0.35	−0.34	0.95

3.2. Accuracy of ACP Retrieval

To assess the dependence of the accuracy on the number of available acquisitions, a bootstrapping type random experiment was conducted. Therefore, from the years 2001–2013 repeatedly scenes were randomly drawn that correspond to the expected numbers from 1, 2, and 4 years respectively. For each sample size ten different ACP sets were calculated. Figure 4 shows the spatial distribution of the standard deviation of the ten MAST values calculated from the different sets. The accuracy of the calculated parameters was seen to clearly increase with larger sample size. The median standard deviation of all pixels was 0.47 K for 1 year, 0.29 K for

2 years and 0.20 K for 4 years of data. Similarly, the median standard deviation of YAST was 0.66 K for 1 year, 0.46 K for 2 years and 0.30 K for 4 years of randomly chosen data which again shows that YAST is more sensitive towards the selection of scenes. It should be noted that the chosen sets are not entirely independent since some scenes are drawn in several sets. However, statistically the overlap between two sets was less than 1% for 1 year and still less than 10% for 4 years. On the other hand, the variability within the 13 years of data can be assumed to be higher than within one real year, hence the real accuracy may be higher. Although the 4 years of data already result in a reasonable accuracy the error due to the random selection of scenes is still considerable. Hence, it is recommended to calculate ACP from time series of several years if available. For this study, the entire period from 2000 to 2013 was used to achieve the highest possible accuracy even if this increases the computational costs.

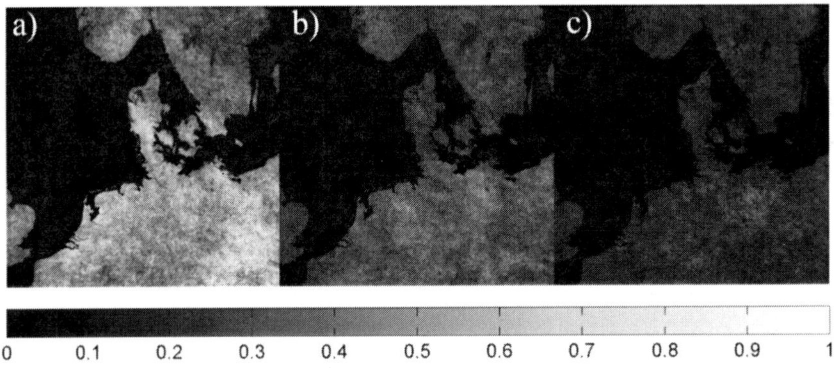

Figure 4. Standard deviation for MAST of 10 randomly sets of scenes equivalent to (**a**) one, (**b**) two, and (**c**) four years of data in Kelvin for tile h18v03.

3.3. Global Climatology of LST

Figure 5 shows some results of the global ACP modelling experiment. MAST is dominated by large scale temperature gradients. Roughly, the tropics and subtropics appear in red, temperate climates in orange, continental and especially boreal climate in green and artic climates in blue. The topography is also evident, in particular highland areas such as the Tibetan Plateau are seen as considerably colder than surrounding areas in the annual mean LST, as well as the Alps and the Andes, both of which are clearly visible.

Vegetation decreases MAST, as for instance can be seen in the areas of tropical rain forests. The pattern shows effects of the general (atmospheric) circulation such as the high LST values evident in the subtropical desserts at the down-welling edges of the Hadley cells. To a certain extent the effects of the Thermohaline (ocean) circulation can also be seen as for instance the comparably warmer temperatures in Europe than in America at comparable latitudes caused by the Gulf Stream.

The annual amplitude YAST increases from the equator to the poles which demonstrates increasing seasonality associated with higher latitudes. Moreover YAST increases with continentality (e.g., in Siberia). The eastern parts of America and Eurasia show a more pronounced increase in YAST below the Ferrel cell, which might result from the frequent passage of Westerly disturbances and according oceanic influence in the west. Large water bodies especially at higher latitudes (e.g., Lake Baikal, Great Lakes) and river deltas (Nile, Amazon) show low YAST and partly seem to cause decreasing YAST values in the surrounding areas. Besides extreme variations in the irradiation forcing, Artic areas show smaller annual amplitude than boreal areas due to their perennial snow coverage.

The phase shift θ especially reflects the shifted seasons on the northern and southern hemispheres. In the northern hemisphere the lag (compared to the irradiation maximum) amounts to a couple of weeks and slightly increases from about 15 days at 25° N to about 29 days at 70° N. Large water bodies and high snow covered mountainous regions exhibit higher θ. Climates in low latitudes that are dominated by dry and wet seasons rather than the irradiation show a positive θ (e.g., in India and Southeast Asia, likely related to Monsoon effects). These values should be treated with caution since they result from the unequal distribution of cloud-free observations and the sine function presumably does not fit the real annual course of LST well. Nevertheless, e.g., the θ of about 50 days which was found in the Gulf of Mexico is consistent with the characteristic warmest Months April and May in Mexico-City. In the Southern hemisphere, θ is shifted by half a year and hence also a few weeks after solar irradiation. Around the equator a mixed “transition zone” with discontinuous θ can be found, however this occurs at low amplitudes. As stated above the ACP should be treated carefully between the tropics since the annual irradiation is not well approximated by the sine function at low latitudes. A general influence of the continentality cannot be detected in these regions. The noticeably increased phase shifts at the African and South American West coasts might be associated with the cold ocean currents Benguela Current and Peru Current.

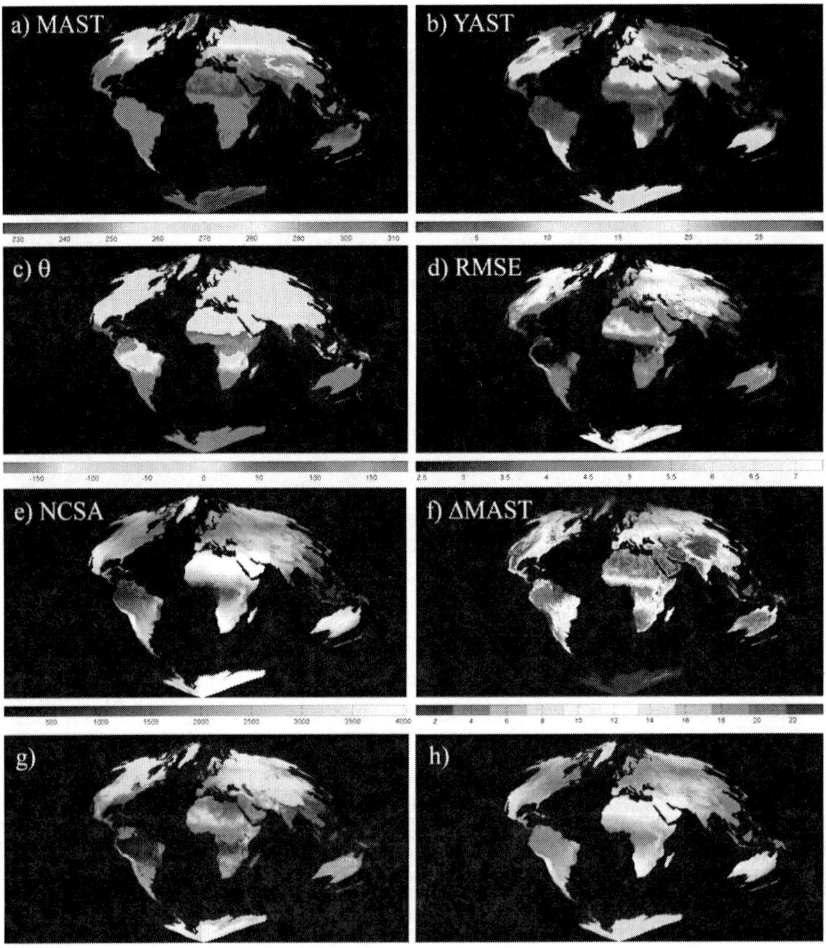

Figure 5. Results of the global ACP modelling experiment including all observations from 2000–2013. (**a**) Daytime mean annual surface temperature (MAST) in K; (**b**) Daytime yearly amplitude of surface temperature (YAST) in K; (**c**) Daytime phase shift θ in days; (**d**) Daytime RMSE of fit in K; (**e**) Number of daytime clear sky acquisitions (NCSA); (**f**) ΔMAST between day- and nighttime as measure for the amplitude of the diurnal temperature cycle in K; (**g**) RGB color composite of θ-RMSE-YAST; (**h**) Composite of MAST-NCSA-$\Delta YAST_{day\text{-}night}$.

The variability parameter RMSE is also increasing with latitude and likewise shows a strong impact of seasonality and continentality. Generally, the large scale RMSE pattern is mostly consistent with YAST. This makes sense, since large annual variation is likely to promote larger inter-diurnal and inter-annual variation. Further, substantial impact of land cover can be seen. Especially, tropical forests show very low variability. Otherwise, the Sahel zone shows a comparably high RMSE which is likely resulting from the high inter-annual variability.

The number of cloud free acquisitions essentially reflects the cloud climatology with the lowest values in the tropics and the highest values in sub-tropical deserts. Large rivers like Amazon and Danube and Lakes can be seen as reduced NCSA, which might either result from increased water availability for convective cloud formation or be an artefact of the cloud filtering algorithm which is likely sensitive to the water absorption spectrum.

The parameter ΔMAST denotes the difference between daytime and nighttime MAST and hence is a measure for the amplitude of the DTC. It shows low values in the tropical, boreal, and artic zones and high values in arid climates. Since the latitudinal gradient is comparably smaller for this "relative" parameter than for the absolute parameters MAST and YAST, the impact of land cover is much more pronounced. This results in a strong signal of rivers and water bodies, forest, and geology. Regarding the day-night differences of the other parameters (not displayed) YAST and RMSE generally are also lower during night which coincides with the reduced heat in the system. The phase shift is in general very similar at nighttime although the transition around the equator is a bit smoother for the nighttime acquisitions. The NCSA is generally higher during day in the tropics and higher during night in higher latitudes which might be related either to the different diurnal cloud characteristics or the acquisition schedule. It should be noted that the pattern is not useful without correction, since it is largely disturbed by the day and nighttime orbits.

If several parameters are combined as a RGB color composite the thermal landscapes can be visualized with even higher information density. Two examples are given in subfigures (g) and (h) which are RGB color composite of θ-RMSE-YAST and MAST-NCSA-$\Delta YAST_{day\text{-}night}$ respectively. Here the interpretation of the single parameters is more difficult but the composite gives a good intuitive impression of the thermal characteristics of the landscapes and the scope of the parameter set.

Besides real surface characteristics, the ACPs also contain some artefacts. As already discussed, the orbit patterns can be seen in NCSA especially at low latitudes. This inconsistency is partly inherited by RMSE since a higher number of (averaged) daily acquisitions smoothens the time series and reduces their variability. Another inconsistency can be found in NCSA and likewise in RMSE between the vertical tiles v02 and v03 in Siberia and North America especially at night-time.

Table 2 shows the correlation coefficients between the different parameter patterns (calculated from a smaller dataset which was resampled to 20% of the size). High correlations were found between YAST and RMSE due to the concurrence of high annual amplitudes and higher inter-diurnal amplitudes. The high correlations between YAST and MAST at daytime and the respective differences to the night-time values are trivial. A strong negative correlation was found between YAST and MAST due to the opposed latitudinal gradients and between YAST and RMSE which is a combined effect of the previous two. Generally, the low correlations indicate a low redundancy and high information content of the parameters. However, the small scale correlation might differ. For instance YAST and MAST show strong positive correlation at local scale where the differences are controlled by the thermal characteristics of the surface rather than differential irradiation. Hence, larger heat conductivity results in both, higher temperatures and larger diurnal and annual cycles. Consequently, the ACP must also be studied at a higher spatial scale in order to fully capture the local variation and utility of the parameters.

3.4. Thermal Landscape of Annual Cycle Characteristics

Due to the large number of more than 300 tiles and 10 parameters per tile (without combinations) only a very limited selection can be shown here. To increase the information density, color composites of 3 parameters are displayed. Figure 6shows four such composite of single tiles. The first tile (a) shows a θ-YAST-RMSE composite of tile h18v03 which comprises mostly the Netherlands, Northern Germany as well as parts of neighbor countries. The phase shift (red) is strongest closer to the North Sea and in snow covered parts of Norway. YAST (green) is higher in Scandinavia and the East which indicates the increasing seasonality with latitude and continentality. The RMSE (blue) also shows a gradient from west to east (only partly visible in the composite). Urban areas are clearly distinguishable in yellow which somewhat resembles nightlight imagery.

Table 2. Correlation between YAST, MAST, θ, NCSA respective day-night differences.

R	$YAST_{day}$	$MAST_{day}$	$theta_{day}$	$RMSE_{day}$	$NCSA_{day}$	$\Delta YAST_{day\text{-}night}$	$\Delta MAST_{day\text{-}night}$
$YAST_{day}$	1.00	−0.46	0.00	0.71	0.13	0.66	0.01
$MAST_{day}$	−0.46	1.00	−0.33	−0.54	−0.13	−0.10	0.66
$theta_{day}$	0.00	−0.33	1.00	0.16	0.26	0.06	−0.14
$RMSE_{day}$	0.71	−0.54	0.16	1.00	0.36	0.39	0.08
$NCSA_{day}$	0.13	−0.13	0.26	0.36	1.00	0.07	0.38
$\Delta YAST_{day\text{-}night}$	0.66	−0.10	0.06	0.39	0.07	1.00	0.21
$\Delta MAST_{day\text{-}night}$	0.01	0.66	−0.14	0.08	0.38	0.21	1.00

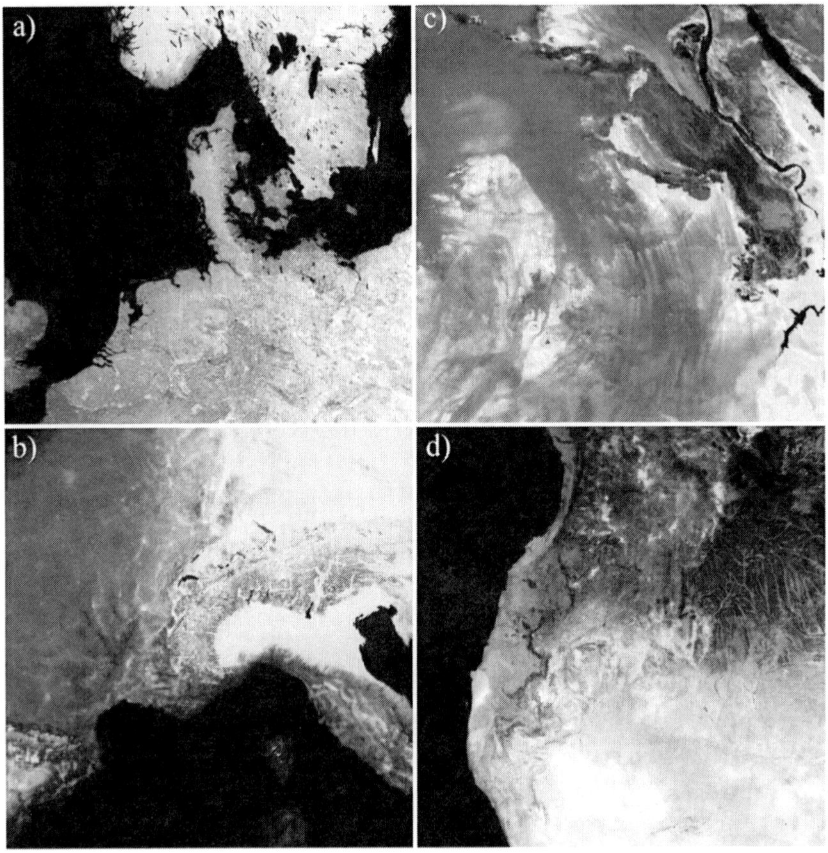

Figure 6. Selected ACP color composite of single tiles ($h_{xx}v_{yy}$) (**a**) θ-YAST-RMSE of ($h_{18}v_{03}$); (**b**) Nighttime RMSE-YAST-θ ($h_{18}v_{04}$); (**c**) RMSE-MAST-YAST ($h_{20}v_{06}$); (**d**) MAST-YAST-RMSE ($h_{19}v_{10}$).

The second composite (b) is the adjacent tile to the south encompassing Northern Italy, Switzerland, as well as large parts of France and Southern Germany. This is a composite of three nighttime parameters, namely RMSE, YAST and θ. The (inter-diurnal and inter-annual) variation is especially high in the Alps and the Pyrenees which likely is an effect of different snow covers. The annual amplitude (green) shows a gradient from west to east with substantially larger values in the Po Valley south of it. MAST (blue) is lowest in the high mountainous regions (which together with the high RMSE results in the yellow color) and shows a strong gradient from the North towards the Mediterranean coast. Again, MAST shows a substantial impact of urban areas, most prominently the metropolitan region of Paris in the upper left of the image.

The third tile (c) shows a composite of RMSE, MAST and YAST of the Nile delta (upper left) and parts of the Sahara which mostly belong to Egypt as well as partly to Libya and Sudan. The delta is clearly visible in green while the rest of the composite is dominated by the geology which also leaves a large fingerprint in the diurnal differences (not shown). RMSE contains slight artefacts of NCSA in Northern Africa which can be seen as stripes in the upper left.

Tile (d) shows the West coast of Southern Africa with Angola in the north and Namibia in the south. The composite of daytime MAST, YAST and RMSE shows strong mean values (red) in the desert in the West, lower values in Kalahari Basin east of it, and much lower values in the Savannas in the north. The amplitude shows a strong gradient from north to south (green). While at the coastline geological influences can be seen, in the north-east the source areas of the Okavango (although not the delta) are very pronounced. The Etosha pan inside the Kalahari Basin can be recognized as homogenous very light blue area in the south.

3.5. Urban Climatology: A New Way of Mapping the Surface Urban Heat Island

The urban heat island (UHI) is of large interest and has been investigated over decades [45,46]. Thermal imagery plays a key role in UHI assessment. It offers the chance to directly measure the surface temperature which is crucial for the urban energy balance and modulates the air temperature of the lowest atmospheric layer. Further, the spatial structure of the surface temperature patterns is directly related to surface characteristics and also used to study the energy balance and the relation between atmospheric and surface UHI (SUHI). However, only the SUHI can be directly observed and there are various problems [7] but also progress [47,48,49] involved in estimating air temperatures from TIR data. Another common shortcoming of remote sensing SUHI studies is that they are frequently based on a single or a few acquisitions only and hence are strongly influenced by actual meteorological and soil conditions. If for instance clouds have shaded the ground before the observation this might result in substantial gradients which cannot easily be interpreted or eliminated. Since the ACP provides LST characteristics from a very large number of acquisitions they might be seen as a new paradigm for SUHI studies, as they allow researchers to analyze the effect and its annual course independent of clouds and instantaneous surface conditions. Further, the analysis can easily be conducted not only for a single city but for larger areas at once.

Figure 7 shows an example for Great Britain. The first image displays nighttime MAST and the second a composite of nighttime MAST, daytime MAST, and $\Delta MAST_{day-night}$. Urban areas are clearly visible due to their high values in all three parameters. Besides, coastlines also show high values of nighttime MAST due to the large heat capacity of the adjacent water bodies. Figure 7c shows the daytime SUHI average values for the urban centers (according to the MODIS land cover data) and the average values for the surrounding areas (defined by a 15 km buffer without any urban pixels). All cities show clear mean annual daily SUHIs with the largest effect in London. Likewise, cities show greater YAST (not displayed) but also smaller (absolute) phase shift, which indicates that no heat storage in the urban fabric for periods longer than the diurnal cycle can be seen in the parameter.

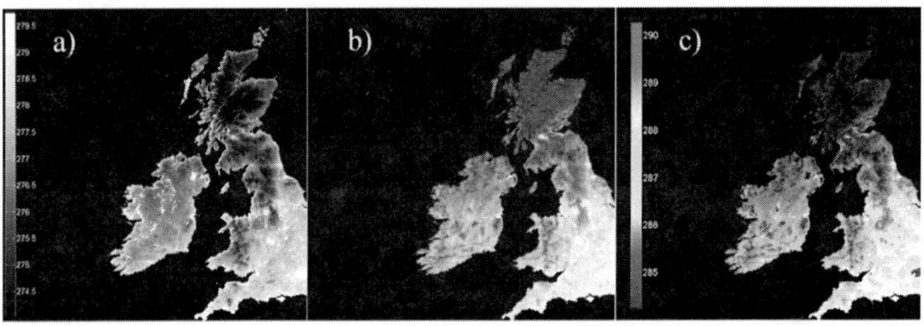

Figure 7. Results of the surface urban heat island analysis for Great Britain. (**a**) Nighttime MAST; (**b**) Composite of nighttime MAST, daytime MAST, and $\Delta MAST_{day-night}$; (**c**) Daytime surface urban heat island of metropolitan areas. Mean MAST of urban and surrounding non-urban area.

Since the urban extents from the MODIS mask are smaller than they would be estimated from the thermal landscape, ACP could possibly also improve urban mapping. Similarly, is has been demonstrated before that ACP are useful for the classification of intra-urban structural types like local climate zones [50,51]. However, urban areas are only one example and the global ACP dataset can be used for mapping purposes for different land covers.

3.6. Land Cover Mapping and Effective Climate Classification

Thermal data is rarely included in global land cover mapping methods due to cloud gaps and its high temporal variability between subsequent acquisitions. Since they provide a robust long term estimate this is not the case for the ACPs. To get an initial indication of their utility for land cover mapping, a k-means cluster analysis ($k = 30$) of selected normalized ACPs (day: YAST, MAST, RMSE and NCSA; night: YAST, MAST as well as ΔMAST) was conducted. The ESA land cover map (Figure 8a) corresponds well with the result (Figure 8b in mostly random colors) although the classification algorithm is unsupervised and hence the class borders are not necessarily consistent with real classes. However, this does not necessarily mean that the land cover determines the LST characteristics since the large scale land cover also strongly reflects the climatic conditions. Hence, the classification likewise shows similarities (but also distinct differences) with the effective climate classification in Figure 8c, which is based on *in situ* observations. Since such temperature observations are sparse in some regions, they are frequently interpolated using LST [52,53]. Therefore, the ACPs could also be of added value since they provide a gap-free virtual LST pattern for every day of the year. Figure 8d shows the ratio between ΔMAST and YAST as measure for diurnal (red) and seasonal (blue) climates.

Since the ACP are clearly affected by both climate and land cover they might have potential to contribute to respective classifications, but this will require further investigations of the most useful parameters and algorithms.

4. CONCLUSIONS

In this study a new global dataset of annual land surface temperature based on MODIS LST on 1 km resolution was presented and potential applications were highlighted. The ACPs: YAST, MAST, θ, RMSE, and NCSA for daytime and nighttime were calculated from all acquisitions from MODIS/Terra from 2000 to 2013. The parameters provide a new climatology of the important

Earth system parameter LST, and characterize the thermal landscape extensively and free of cloud gaps. More specifically, they allow estimating the LST patterns at different scales for every place on Earth and for every day in the year and additionally deliver an estimate of the accuracy of both inter-diurnal and inter-annual variability. The parameters are robust, but the LST estimates are limited to acquisition times and largely cloud free conditions. If cloud free conditions are unlikely to occur, for example in wet seasons, the estimated LST must be treated with caution. Furthermore, between the tropics the validity of the ACP is limited since the annual cycle of irradiation is not well approximated by the sine function at low latitudes. The accuracy due to the random selection of scenes was tested and it was found that several years of data should ideally be used. The MODIS time series is certainly long enough to provide sufficient data to achieve a high accuracy. However, this might result in errors for areas that underwent land cover change. Despite a few artefacts from the orbit characteristics and level 3 product processing, the parameters generally reflect real surface conditions. The analyses correlations between most parameters indicated their high information content and low redundancy.

The parameters revealed different important influencing factors including climate, land cover, vegetation phenology, anthropogenic effects, and geology. Accordingly, the number and scope of potential applications are various and only a few could briefly be highlighted in this study. Besides the classification of local climate zones that was already tested, ACP could also be used for classification of land cover, geology, or effective climate zones. Further, they have been found to be good predictors for the LST downscaling of LST [26,55] and they could be used for change detection if the thermal characteristics show discontinuous or systematic deviations. Finally, a comprehensive evaluation of global SUHIs based on the ACP is currently conducted.

Since the parameters now have reached a stable level and are assumed to be of use for several applications, testers and applicants are welcome. After final processing the ACP datasets will published at the CliSAP Integrated Climate Data Center [56]—until then please contact the author. Besides, future work will include improvements of the parameters. This might include a more detailed analysis of the residuals and their temporal distribution to distinguish varying inter-daily variability in different seasons, inter-annual variability throughout the years, the effect of the seasonal cloud distribution, as well as systematic changes. Moreover, better cloud filtering by using neighborhood relations or excluding outliers in an iterative algorithm could improve the cloud-free fit and might even allow estimations of LST at conditions with a certain amount of convective clouds. A more detailed analysis of the QA flags and their dependency on land cover and LST also deserves further attention. However in this regard, it might make sense to wait for future anticipated products. For instance, version 6 of MOD11A1 will be available in the near future. This also applies to the new MOD21 product, which is based on a temperature emissivity separation algorithm with the MODIS bands 29, 31, and 32 [57,58]. Besides, the respective Aqua product MYD11A1 is currently processed which will increase the climatology to four times a day and will already allow for some DTC analysis.

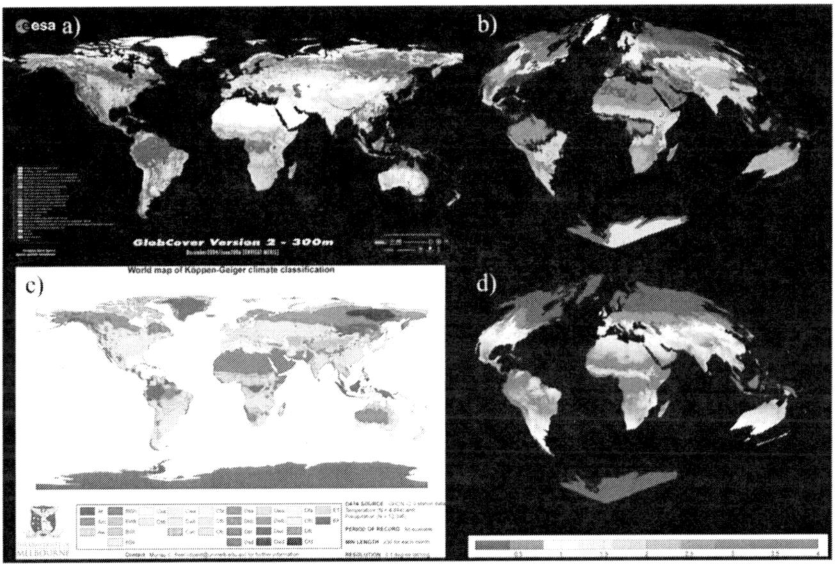

Figure 8. (**a**) ESA GlobCover. Map may be used freely, © ESA/ESA GlobCover Project, MEDIAS France/Postel; (**b**) k-means clustering of ACP with 30 clusters in random color-map; (**c**) Effective climate classification after Köppen & Geiger by Murrat Peal and colleagues [54] © CC BY-SA 3.0 (**d**) diurnal (red) and seasonal (blue) climates.

ACKNOWLEDGMENTS

I thank all data providers, especially the NASA and the USGS. Further, I thank Eleonore Schenk and Jürgen Böhner for acquisition and access to the computing resources as well as Thomas Orgis and Hinnerk Stüben for technical support. Eventually, I want to thank Manfred Bechtel, Lars Gerlitz, and Paul Alexander for proof reading as well as Otto Bechtel for his dedicated overall help with my work. This work was financed by the Cluster of Excellence CliSAP (EXC 177), University of Hamburg funded through the German Science Foundation (DFG).

REFERENCES

1. Dash, P.; Göttsche, F.-M.; Olesen, F.-S.; Fischer, H. Land surface temperature and emissivity estimation from passive sensor data: Theory and practice-current trends. *Int. J. Remote Sens.* **2002**, *23*, 2563–2594.

2. Li, Z.-L.; Tang, B.-H.; Wu, H.; Ren, H.; Yan, G.; Wan, Z.; Trigo, I.F.; Sobrino, J.A. Satellite-derived land surface temperature: Current status and perspectives. *Remote Sens. Environ.* **2013**, *131*, 14–37.

3. Bastiaanssen, W.G.M.; Menenti, M.; Feddes, R.A.; Holtslag, A.A.M. A remote sensing surface energy balance algorithm for land (SEBAL). 1. Formulation. *J. Hydrol.* **1998**, *212*, 198–212.

4. Frey, C.M.; Parlow, E. Flux measurements in Cairo. Part 2: On the determination of the spatial radiation and energy balance using ASTER satellite data. *Remote Sens.* **2012**, *4*, 2635–2660.

5. Kalma, J.D.; McVicar, T.R.; McCabe, M.F. Estimating land surface evaporation: A review of methods using remotely sensed surface temperature data. *Surv. Geophys.* **2008**, *29*, 421–469.

6. Li, Z.; Jia, L.; Lu, J. On uncertainties of the Priestley-Taylor/LST-Fc feature space method to estimate evapotranspiration: Case study in an arid/semiarid region in Northwest China. *Remote Sens.* **2014**, *7*, 447–466.

7. Voogt, J.A.; Oke, T.R. Thermal remote sensing of urban climates. *Remote Sens. Environ.* **2003**, *86*, 370–384.

8. Weng, Q. Thermal infrared remote sensing for urban climate and environmental studies: Methods, applications, and trends. *ISPRS J. Photogramm. Remote Sens.* **2009**, *64*, 335–344.

9. Fabrizi, R.; Bonafoni, S.; Biondi, R. Satellite and ground-based sensors for the urban heat island analysis in the city of Rome. *Remote Sens.* **2010**, *2*, 1400–1415.

10. Liu, L.; Zhang, Y. Urban heat island analysis using the Landsat TM data and ASTER data: A case study in Hong Kong.*Remote Sens.* **2011**, *3*, 1535–1552.

11. Prata, A.J.; Caselles, V.; Coll, C.; Sobrino, J.A.; Ottle, C. Thermal remote sensing of land surface temperature from satellites: Current status and future prospects. *Remote Sens. Rev.* **1995**, *12*, 175–224.

12. Sobrino, J.A.; Jiménez-Muñoz, J.C.; Paolini, L. Land surface temperature retrieval from LANDSAT TM 5. *Remote Sens. Environ.* **2004**, *90*, 434–440.

13. Qin, Z.; Karnieli, A.; Berliner, P. A mono-window algorithm for retrieving land surface temperature from Landsat TM data and its application to the Israel-Egypt border region. *Int. J. Remote Sens.* **2001**, *22*, 3719–3746.

14. Jimenez-Munoz, J.C.; Sobrino, J.A. A single-channel algorithm for land-surface temperature retrieval from ASTER data. *IEEE Geosci. Remote Sens. Lett.* **2010**, *7*, 176–179.

15. Sobrino, J.A.; Li, Z.-L.; Stoll, M.P.; Becker, F. Multi-channel and multi-angle algorithms for estimating sea and land surface temperature with ATSR data. *Int. J. Remote Sens.* **1996**, *17*, 2089–2114.

16. Price, J.C. Land surface temperature measurements from the split window channels of the NOAA 7 advanced very high resolution radiometer. *J. Geophys. Res. Atmos.* **1984**, *89*, 7231–7237.

17. Wan, Z.; Dozier, J. A generalized split-window algorithm for retrieving land-surface temperature from space. *IEEE Trans. Geosci. Remote Sens.* **1996**, *34*, 892–905.

18. Berk, A.; Anderson, G.P.; Bernstein, L.S.; Acharya, P.K.; Dothe, H.; Matthew, M.W.; Adler-Golden, S.M.; Chetwynd, J.H.J.; Richtmeier, S.C.;

Pukall, B.; *et al.* MODTRAN4 radiative transfer modeling for atmospheric correction. *Proc. SPIE* **1999**, *3756*, 348–353.

19. Snyder, W.C.; Wan, Z.; Zhang, Y.; Feng, Y.Z. Classification-based emissivity for land surface temperature measurement from space. *Int. J. Remote Sens.* **1998**, *19*, 2753–2774.

20. Valor, E.; Caselles, V. Mapping land surface emissivity from NDVI: Application to European, African, and South American areas. *Remote Sens. Environ.* **1996**, *57*, 167–184.

21. Sobrino, J.A.; Jiménez-Muñoz, J.C.; Sòria, G.; Romaguera, M.; Guanter, L.; Moreno, J.; Plaza, A.; Martínez, P. Land surface emissivity retrieval from different VNIR and TIR sensors. *IEEE Trans. Geosci. Remote Sens.* **2008**, *46*, 316–327.

22. Gillespie, A.; Rokugawa, S.; Matsunaga, T.; Cothern, J.S.; Hook, S.; Kahle, A.B. A temperature and emissivity separation algorithm for Advanced Spaceborne Thermal Emission and Reflection radiometer (ASTER) images. *IEEE Trans. Geosci. Remote Sens.* **1998**, *36*, 1113–1126.

23. Tomlinson, C.J.; Chapman, L.; Thornes, J.E.; Baker, C. Remote sensing land surface temperature for meteorology and climatology: A review. *Meteorol. Appl.* **2011**, *18*, 296–306.

24. Inamdar, A.K.; French, A.; Hook, S.; Vaughan, G.; Luckett, W. Land surface temperature retrieval at high spatial and temporal resolutions over the southwestern United States. *J. Geophys. Res.* **2008**, *113*.

25. Zakšek, K.; Oštir, K. Downscaling land surface temperature for urban heat island diurnal cycle analysis. *Remote Sens. Environ.* **2012**, *117*, 114–124.

26. Bechtel, B.; Zakšek, K.; Hoshyaripour, G. Downscaling land surface temperature in an urban area: A case study for Hamburg, Germany. *Remote Sens.* **2012**, *4*, 3184–3200.

27. Chen, X.; Li, W.; Chen, J.; Rao, Y.; Yamaguchi, Y. A combination of TsHARP and thin plate spline interpolation for spatial sharpening of thermal imagery. *Remote Sens.* **2014**, *6*, 2845–2863.

28. Keramitsoglou, I. Advanced earth observation methodologies for the study of the thermal environment of cities. In Proceedings of the 2012

Second International Workshop on Earth Observation and Remote Sensing Applications (EORSA), Shanghai, China, 8–11 June 2012; pp. 11–15.

29. Lu, L.; Venus, V.; Skidmore, A.; Wang, T.; Luo, G. Estimating land-surface temperature under clouds using MSG/SEVIRI observations. *Int. J. Appl. Earth Obs. Geoinf.* **2011**, *13*, 265–276.

30. Göttsche, F.-M.; Olesen, F.S. Modelling of diurnal cycles of brightness temperature extracted from METEOSAT data.*Remote Sens. Environ.* **2001**, *76*, 337–348.

31. Göttsche, F.M.; Olesen, F.S. Modelling the effect of optical thickness on diurnal cycles of land surface temperature.*Remote Sens. Environ.* **2009**, *113*, 2306–2316.

32. Duan, S.-B.; Li, Z.-L.; Wang, N.; Wu, H.; Tang, B.-H. Evaluation of six land-surface diurnal temperature cycle models using clear-sky *in situ* and satellite data. *Remote Sens. Environ.* **2012**, *124*, 15–25.

33. Zhou, J.; Chen, Y.; Zhang, X.; Zhan, W. Modelling the diurnal variations of urban heat islands with multi-source satellite data. *Int. J. Remote Sens.* **2013**, *34*, 7568

34. Bechtel, B. Multitemporal Landsat data for urban heat island assessment and classification of local climate zones. In Proceedings of the 2011 Joint Urban Remote Sensing Event (JURSE), Munich, Germany, 11–13 April 2011; pp. 129–132.

35. Bechtel, B. Robustness of annual cycle parameters to characterize the urban thermal landscapes. *IEEE Geosci. Remote Sens. Lett.* **2012**, *9*, 876–880.

36. Weng, Q.; Fu, P. Modeling annual parameters of clear-sky land surface temperature variations and evaluating the impact of cloud cover using time series of Landsat TIR data. *Remote Sens. Environ.* **2014**, *140*, 267–278.

37. Bechtel, B. Die Hitze in der Stadt Verstehen—Wie sich Die Jahreszeitliche Temperaturdynamik von Städten aus Dem All Beobachten Lässt. In *Globale Urbanisierung—Perspektive aus dem All*; Taubenböck, H., Wurm, M., Esch, T., Dech, S., Eds.; Springer Spektrum: Berlin, Germany, 2015; in press.

38. Wan, Z. *Collection-5 MODIS Land Surface Temperature Products Users' Guide*; University of California: Santa Barbara, CA, USA, 2007.

39. Wan, Z. New refinements and validation of the MODIS Land-Surface Temperature/Emissivity products. *Remote Sens. Environ.* **2008**, *112*, 59–74.

40. Wan, Z.; Zhang, Y.; Zhang, Q.; Li, Z.-L. Quality assessment and validation of the MODIS global land surface temperature. *Int. J. Remote Sens.* **2004**, *25*, 261–274.

41. Wang, K.; Liang, S. Evaluation of ASTER and MODIS land surface temperature and emissivity products using long-term surface longwave radiation observations at SURFRAD sites. *Remote Sens. Environ.* **2009**, *113*, 1556–1565.

42. Wan, Z. New refinements and validation of the collection-6 MODIS land-surface temperature/emissivity product.*Remote Sens. Environ.* **2014**, *140*, 36–45.

43. Friedl, M.A.; Sulla-Menashe, D.; Tan, B.; Schneider, A.; Ramankutty, N.; Sibley, A.; Huang, X. MODIS Collection 5 global land cover: Algorithm refinements and characterization of new datasets. *Remote Sens. Environ.* **2010**, *114*, 168–182.

44. Lagarias, J.; Reeds, J.; Wright, M.; Wright, P. Convergence properties of the Nelder—Mead simplex method in low dimensions. *SIAM J. Optim.* **1998**, *9*, 112–147.

45. Yow, D.M. Urban heat islands: Observations, impacts, and adaptation. *Geogr. Compass* **2007**, *1*, 1227–1251.

46. Arnfield, A.J. Two decades of urban climate research: A review of turbulence, exchanges of energy and water, and the urban heat island. *Int. J. Climatol.* **2003**, *23*, 1–26.

47. Bechtel, B.; Wiesner, S.; Zaksek, K. Estimation of dense time series of urban air temperatures from multitemporal geostationary satellite data. *IEEE J. Sel. Top. Appl. Earth Obs. Remote Sens.* **2014**, *7*, 4129–4137.

48. Nichol, J.E.; Fung, W.Y.; Lam, K.; Wong, M.S. Urban heat island diagnosis using ASTER satellite images and "*in situ*"air temperature. *Atmos. Res.* **2009**, *94*, 276–284.

49. Pichierri, M.; Bonafoni, S.; Biondi, R. Satellite air temperature estimation for monitoring the canopy layer heat island of Milan. *Remote Sens. Environ.* **2012**, *127*, 130–138.

50. Bechtel, B.; Daneke, C. Classification of local climate zones based on multiple earth observation data. *IEEE J. Sel. Top. Appl. Earth Obs. Remote Sens.* **2012**, *5*, 1191–1202.

51. Bechtel, B.; Alexander, P.J.; Böhner, J.; Ching, J.; Conrad, O.; Feddema, J.; Mills, G.; See, L.; Stewart, I. Mapping local climate zones for a worldwide database of the form and function of cities. *ISPRS Int. J. Geo-Inf.* **2015**, *4*, 199–219.

52. Prihodko, L.; Goward, S.N. Estimation of air temperature from remotely sensed surface observations. *Remote Sens. Environ.* **1997**, *60*, 335–346.

53. Cresswell, M.P.; Morse, A.P.; Thomson, M.C.; Connor, S.J. Estimating surface air temperatures, from Meteosat land surface temperatures, using an empirical solar zenith angle model. *Int. J. Remote Sens.* **1999**, *20*, 1125–1132.

54. Peel, M.C.; Finlayson, B.L.; McMahon, T.A. Updated world map of the Köppen-Geiger climate classification. *Hydrol. Earth Syst. Sci.* **2007**, *11*, 1633–1644.

55. Bechtel, B.; Böhner, J.; Zakšek, K.; Wiesner, S. Downscaling of diurnal land surface temperature cycles for urban heat island monitoring. In Proceedings of the Urban Remote Sensing Event (JURSE), Sao Paulo, Brazil, 21–23 April 2013.

56. CliSAP Integrated Climate Data Center. Available online: http://icdc.zmaw.de/modis_lst_annualcycle.html (accessed on 31 December 2014).

57. Hulley, G.C.; Hook, S.J. Generating consistent land surface temperature and emissivity products between ASTER and MODIS data for earth science research. *IEEE Trans. Geosci. Remote Sens.* **2011**, *49*, 1304–1315.

58. Hulley, G.; Hook, S.; Hughes, T. Moderate Resolution Imaging Spectroradiometer (MODIS) MOD21 Land Surface Temperature and Emissivity Algorithm Theoretical Basis Document. Available online: http://lst.jpl.nasa.gov/emissivity.jpl.nasa.gov/downloads/examples/documents/MOD21_LSTE_ATBD_Hulley_v2.1.pdf (accessed on 31 December 2014).

Chapter 4

Evaluation of Precipitation Climatology Derived from TRMM Multi-Satellite Precipitation Analysis (TMPA) Monthly Product over Land with Two Gauge-Based Products

Zhong Liu

Center for Spatial Information Science and Systems (CSISS), George Mason University, 4400 University Drive, Fairfax, VA 22030, USA

ABSTRACT

The NASA/JAXA Tropical Rainfall Measuring Mission (TRMM) has been in operation for over 17 years since 1997. The length of TRMM is far shorter than those from ground observations, raising a question as to whether TRMM derived climatology products are good enough for research and applications. In this study, three climatologies derived from a blended product (the TRMM Multi-Satellite Precipitation Analysis (TMPA) monthly product or 3B43) and gauge-based ground observations (Global Precipitation Climatology Center (GPCC) and Willmott and Matsuura (WM)) are compared over land on a global scale (50° N–50° S) to assess the performance and weaknesses of the TMPA-derived climatology. Results show that the 3B43 climatology matches well with the two gauge-based climatologies in all seasons in terms of spatial distribution, zonal means as well as seasonal variations. However, large variations are found in light rain (<0.5 mm/day) regions such as the Sahara Desert. At high rain rates, large negative biases (3B43 < WM < GPCC) are found in orographically influenced regions such as windward sides of monsoon mountains in JJA, which is associated with underestimation of shallow orographic rain rates in microwave radiometer algorithms and sparse gauge data for bias correction. Nonetheless, biases in 3B43 exist and vary with geographic locations due to a number of factors such as gauge availability and algorithm issues.

KEYWORDS

climatology; precipitation; TRMM; gauges; satellites; comparisons; droughts

1. INTRODUCTION

Precipitation is a very important physical variable in our daily lives [1]. However, as a double-edged sword, droughts and floods can occur each year in different regions around the world [2,3,4], often causing considerable damage to crops, properties, human fatalities, *etc.* In addition to precipitation, derived anomaly information can be useful in drought detection, monitoring and management activities by providing additional information compared to the normal condition which is also known as climatology. Precipitation is notoriously difficult to measure and forecast both over land and ocean. In remote regions such as mountains, deserts, *etc.* it is difficult to obtain precipitation information from ground measurements. Satellite-based precipitation products take an increasingly important role in filling data gaps in those regions and are widely used in research and decision support or monitoring applications.

In recent years, precipitation products that utilize multi-satellite and multi-sensor datasets have been gaining more and more popularity than products from a single sensor or satellite. Adjusted for bias with gauge and ground radar products, satellite-based precipitation products have been significantly improved over the years. An example is the TRMM Multi-Satellite Precipitation Analysis (TMPA) monthly product (3B43), a popular near-global (50° N–50° S) precipitation product that is widely used around the world. However the history of satellite precipitation products is relatively short compared to the length of 30 years in the World Meteorological Organization (WMO) definition for climatology. For example, the NASA/JAXA (Japan Aerospace Exploration Agency) Tropical Rainfall Measuring Mission (TRMM) [5,6] has been in operation for over 17 years since 1997. The length of TRMM is far shorter than those from ground observations, raising a question as to whether TRMM-derived climatology products are good enough for research and applications.

Previously, there have been a few efforts to compare and evaluate TRMM-derived climatologies [7,8]. Based on surface types, Wang *et al.* [8] compared the TRMM Composite Climatology (TCC) with several other datasets. Over ocean, TCC consists of the data products from TRMM Microwave Imager (TMI, 2A12), TRMM Precipitation Radar (PR, 2A25), and the combined TMI and PR (2B31). Data from atolls were used for comparison. Over land, the TMPA monthly product (3B43) is picked for TCC and surface observations from the Global Precipitation Climatology Center (GPCC) were used for comparison. However, the study areas over land from Wang *et al.* [8] are only limited to the southeastern United States and eastern China. No information on seasonal variations is provided as well. Furthermore, the study domain is limited to the TRMM coverage (38° N–38° S), but in fact, the full spatial coverage of 3B43 is between 50° N–50° S. Some of the regions outside the TRMM coverage are heavily populated such as southern Europe. Including such regions can be helpful for research and applications as well. Therefore, to better understand how TMPA 3B43 performs in the rest of the world and in different seasons, additional studies are needed. In this study, two climatology products from gauge-based observations are compared with the TMPA 3B43 climatology to assess the performance and weaknesses of the satellite-based climatology product. Furthermore, this study looks into ancillary data to explain biases found in the comparison. It needs to be emphasized that validation or verification of satellite-based precipitation products still heavily relies on individuals who often can obtain ground measurements with higher spatial and temporal resolutions than those from GPCC. For such reason, this study is not intended for ground validation for 3B43 and the task should be left to those individual investigators. This article is organized as, Section 2 describes the data; Section 3 the results and Section 4 summary and discussion.

2. DATA

Three global monthly precipitation products (to derive their climatologies) were used in this study. Table 1 summarizes the basic information.

The purpose of the Version 7 TMPA 3B43 algorithm [9,10,11,12] is to produce the best-estimate precipitation rate and root-mean-square (RMS) precipitation-error estimates from TRMM and other data sources. The 3B43 algorithm [9,10,11,12] consists of multiple independent precipitation estimates from the TMI (TRMM Microwave Imager), Advanced Microwave Scanning Radiometer for Earth Observing Systems (AMSR-E), Special Sensor Microwave Imager (SSMI), Special Sensor Microwave Imager/Sounder (SSMIS), Advanced Microwave Sounding Unit (AMSU), Microwave Humidity Sounder (MHS), microwave-adjusted merged geo-infrared (IR), and monthly accumulated rain gauge analysis from GPCC for bias correction.

The preprocessing [12] of 3B43 is as follows: (a) all input microwave products mentioned above are intercalibrated to TRMM Combined Instrument (TCI) precipitation estimates (TRMM product 3B31); (b) the IR estimates are computed using monthly matched microwave-IR histogram matching; (c) then missing data in individual 3-hourly merged-microwave fields are filled with the IR estimates. When the preprocessing is complete, the 3-hourly multi-satellite fields are summed for the month and combined with the monthly gauge analysis using inverse-error-variance weighting to form the best-estimate precipitation rate and RMS precipitation-error estimates [12].

According to Huffman [9], the accuracy of 3B43 can be broken into two parts, systematic and random departures from the true answer. The former is also called bias and the latter random error [9]. Random errors can be averaged out in climatological averages; however for bias, it is not the case and it cannot be averaged out in climatological averages, posing the biggest problem for climate studies and applications. In this study, the bias issue is the main concern since random errors have been averaged out during climatological averages. Furthermore, the 3B43 ancillary data (*i.e.*, gauge weight) are analyzed to gain insight on biases.

There were only a few issues [13] in Version 6 3B43, such as a V-shape trough in the global averaged time series, which has been largely corrected in Version 7. Version 7 3B43 data in this study were downloaded from Mirador [14] at the NASA Goddard Earth Sciences Data and Information

Services Center (GES DISC). There have been a few processing issues [12] before, but all the 3B43 data used in this study are current.

Table 1. Monthly precipitation products used in this study. As of this writing, the 3B43 monthly product for December 2014 was not available.

Product	Data Sources	Spatial Resolution	Spatial Coverage
TMPA 3B43 Monthly Precipitation Version 7	Satellites and gauges (blended) (1998–2013)	0.25 deg. × 0.25 deg.	50°N–50°S
GPCC 2011 Precipitation Climatology	WMO GTS (Global Telecommunication System) gauges and historical data (1951–2000)	0.25 deg. × 0.25 deg.	90°N–90°S
Willmott and Matsuura (WM) Monthly Precipitation Version 1.02	GHCN (Global Historical Climatology Network) gauges (1950–1999)	0.5 deg. × 0.5 deg. (regridded to 0.25 deg.)	90°N–90°S

The Version 1.02 Willmott and Matsuura (WM hereafter) global land-only monthly precipitation dataset [15] was developed based on the Version 2 Global Historical Climatology Network (GHCN) and Legates and Willmott's station records [16,17] of monthly and annual mean total precipitation. The Version 3 GHCN is currently available, but the WM monthly product has not been updated yet and in this study, Version 1.02 is used instead. There were 20,599 GHCN stations, varying from about 1870 to 16,360, used for the precipitation dataset. The DEM-assisted Legates and Willmott archive [18] (Version 2.02) was used as the background field for the Climatologically Aided Interpolation (CAI) and the traditional interpolation algorithm is based on the spherical version of Shepard's distance-weighting method [19,20]. For this study, the 0.5 deg. WM dataset was regridded to 0.25 deg., using a box average method, to match the spatial resolution of the other two datasets.

The Version 2011 monthly land-surface precipitation climatology [21] from GPCC was derived from rain-gauges built on WMO GTS (Global Telecommunication System)-based and historic data. The climatology [21] is based on the 67,200 stations world-wide that feature record durations of 10 years or longer, for the target reference period between January 1951 and December 2000. Meyer-Christoffer *et al.* [21] used earlier or later data records of 10 years or longer if no station data were available for the target period, thereby yielding a high station density and temporal stability for the climatology.

To facilitate intercomparison of these climatologies, the NASA GES DISC has developed an online intercomparison tool [22] in the TRMM Online Visualization and Analysis System (TOVAS) [13,23,24], as a member of the GES DISC Geospatial Interactive Online Visualization ANd aNalysis Infrastructure (Giovanni) [23,25,26,27]. Giovanni is an online or web-based application that provides a simple and intuitive way to visualize, analyze, and access vast amounts of Earth science remote sensing data mainly archived at the GES DISC without having to download data and software. Basic and customized comparison tasks (*i.e.*, scatter plot, correlation, time series, histogram, *etc.*) can be done in TOVAS [13] or through offline analysis by downloading customized data from TOVAS. Because it is difficult to cover every place on Earth in this study and 3B43 biases vary with geographic locations, it is encouraged to use the online tool for comparison in one's own area of interest. For this study, all comparisons were done offline because not all the analytical functions are available in TOVAS.

3. RESULTS

Evaluation of the 3B43 climatology on a global scale was done by comparing two gauge-based climatologies. Ensemble statistics [28], zonal means, scatterplots and Pearson's correlation coefficients were computed in four seasons. Biases against gauge measurements were estimated in different seasons as well as at different rain rates. Gauge weighting in 3B43 as one of the possible influencing factors is analyzed to explain differences in comparisons.

3.1. Ensemble Statistics, Pearson's Correlation Coefficients, Gauge Weighting and Zonal Means

To understand the differences among the three climatologies, means (by averaging rain rates in three products) and relative (to the ensemble mean) standard deviations are computed for each season as a three-member ensemble [28]. Seasonal ensemble mean maps are shown in Figure 1. It is seen that the major monsoon regions in the Tropics and sub-tropics such as the Asian monsoons, the African monsoons, and their seasonal changes are adequately shown in the rainfall ensemble maps (Figure 1). Heavy rain regions are concentrated in the Tropics. Regions of heaviest rainfall are found in orographically influenced regions such as along the coasts of southwest India, Burma (Myanmar), *etc.* where the orographic uplifting effect takes an important role in producing heavy precipitation on the windward sides. Seasonal variations are also shown in Figure 1 where heaviest rainfall regions are found in the summer (JJA) monsoon regions in the Northern Hemisphere and in contrast, little rain is received there during boreal winter (DJF).

Figure 2 contains the maps of seasonal relative (with respect to or divided by the ensemble mean in Figure 1) standard deviations in percentage. In general, small variations (<~15%) are found in major monsoon regions or regions with moderate amount of rainfall (see Figure 1); however, in some heavy monsoon rain regions such as the coasts of southwest India, northwest Colombia, *etc.* variations are over 30% in JJA (Figure 2), suggesting the variations in rainfall estimates there are quite large and uncertain. More will be discussed in later sessions.

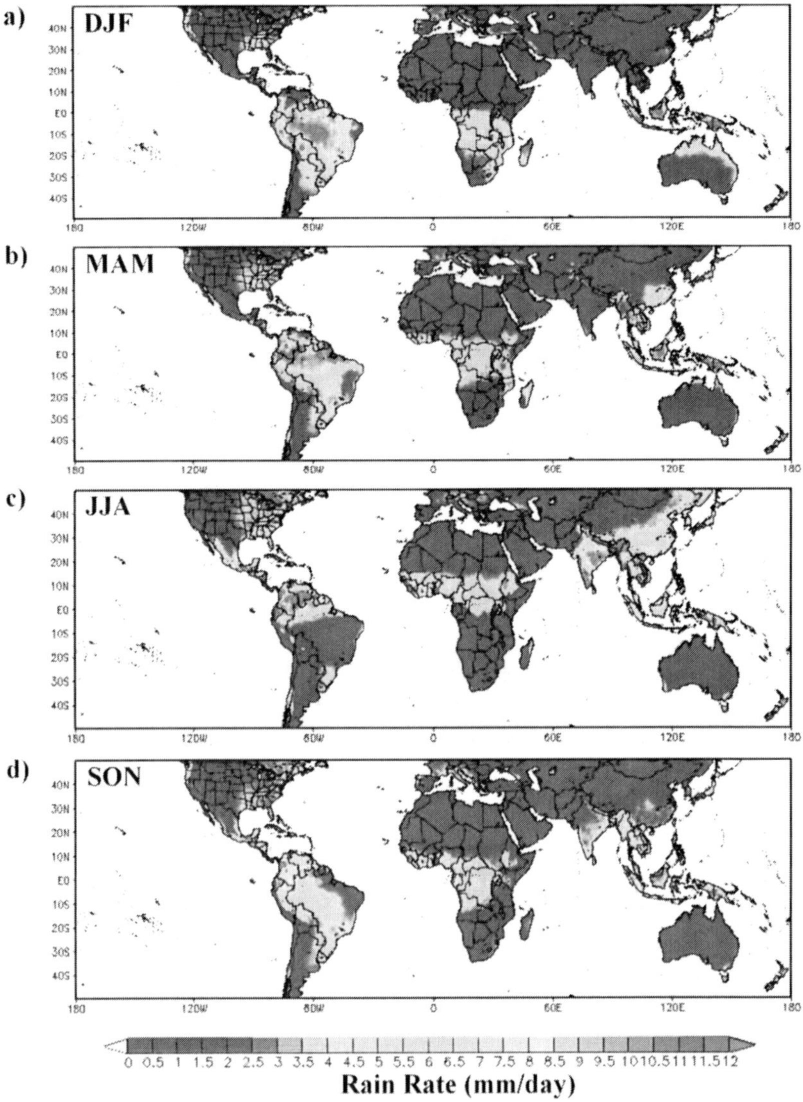

Figure 1. Mean seasonal precipitation (mm/day) of the 3-member ensemble for (**a**) DJF; (**b**) MAM; (**c**) JJA; and (**d**) SON.

Large variations (>15%) in standard deviation are also found in light rain (rain rate <0.5 mm/day, Figure 1) regions that mainly consist of remote, less populated and mountainous regions such as the Sahara Desert, the Arabian Peninsula, the Tibetan Plateau, the Andes, *etc.* It is worth noting that light rain rate here does not mean light rain events in those regions. More on this will be discussed later. Variations in those regions can exceed over 90% in all seasons (Figure 2). In boreal winter (DJF), the regions of large variations extend further north in the Northern Hemisphere compared to JJA, in regions such as the northern Eurasia, northwest United States (Figure 2). In boreal summer (JJA), the most prominent region (Figure 2c) is southern Africa, which does not show up in the other three seasons. Large variations are found in Western Australia in all seasons, especially in SON and DJF. Further investigation reveals that large variations are found between the two ground-based climatologies as well (not shown). More on this will be discussed next. In short, 3B43 behaves well compared to the other two climatologies in regions with rain rates that are above 0.5 mm/day.

Figure 3a is generated to understand the relationship between relative standard deviation and ensemble rain rate on a global scale in four seasons. Due to larger variations associated with light rain, as mentioned earlier, only rain rates that are greater than 0.5 mm/day are plotted. In Figure 3a, it is seen that in SON, DJF, and MAM, the maximum available rain rate is 25 mm/day which is much lower than that (40 mm/day) in JJA. Meanwhile, the relative standard deviations in the three seasons are fluctuating and are confined between ~7% and ~13% of the ensemble mean. In contrast, in JJA, the relative standard deviation first experiences a brief decrease from 0.5 mm/day to 5 mm/day and then begins to climb up from there. The relative standard deviation in JJA can reach nearly 18% at the rain rate of 35 mm/day. From Figure 1, it is seen that regions of high rain rates in JJA are confined mainly in orographically influenced regions such as the windward side of southwest India. In short, all three products do not quite agree with each other (between ~7% and ~13% of the ensemble mean) and even more so in JJA than in the other three seasons. Larger variations are found in JJA and they increase with rain rates. Relatively smaller variations are found in SON, DJF and MAM.

Rain gauge data are used in the bias correction in the 3B43 algorithm [9,10,11,12]. Figure 3b shows a map of the gauge weighting percentages averaged between 1998 and 2013. It is seen gauge weighting percentages are not evenly distributed around the world, with higher percentages in developed countries and populated regions where more gauge data are available, such as North America, Europe, *etc.* than those in developing countries and less populated/remote regions (the Sahara Desert, the Amazon basin). The lack of gauge data in deserts is likely to be one of the contributing factors in the large standard deviations mentioned earlier in addition to the difficulty of observing light rain events from satellite-based sensors. More on this will be found in the discussion section.

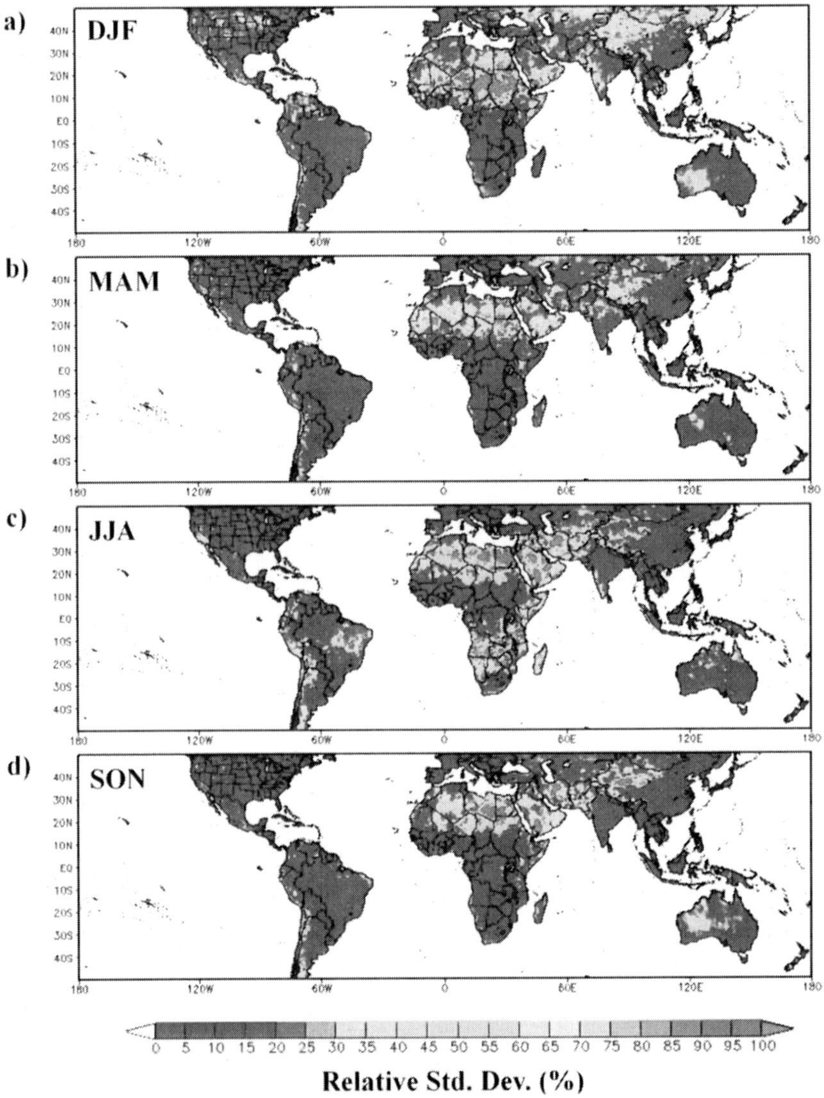

Figure 2. Similar to Figure 1, except for relative (to the ensemble mean in Figure 1) standard deviation in percentages (%). (**a**) DJF; (**b**) MAM; (**c**) JJA; and (**d**) SON.

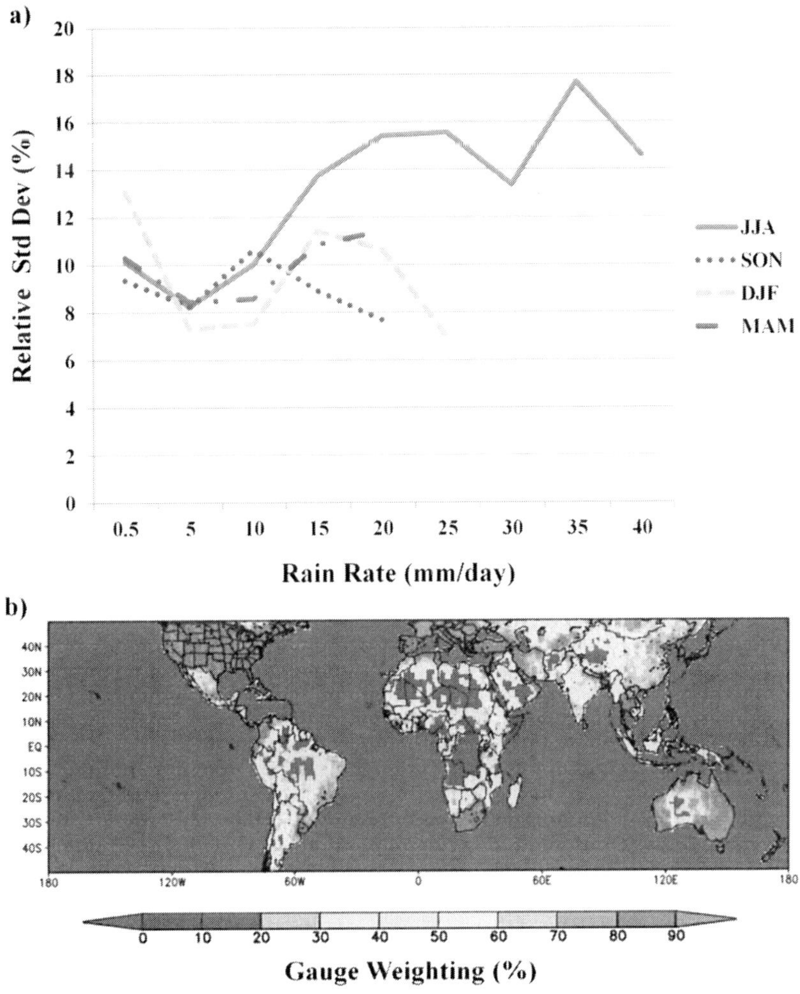

Figure 3. (**a**) Relative (with respect to the ensemble mean) standard deviation in percentages (%) as a function of rain rate. (**b**) Gauge weighting in percentage averaged between 1998 and 2013.

To examine whether these three climatologies share the similar seasonal variations, Pearson's correlation coefficients were computed from January to December (12 months). Figure 4 shows the maps of Pearson's correlation coefficients for each pair of climatology. In general, high correlation coefficients (>0.9) are found in most regions in all three pairs (Figure 4). However, significantly low positive (<0.7) or even low negative correlation coefficients are found in regions of light rain (<0.5 mm/day) such as the Sahara Desert, the Arabian Peninsula, Southern Australia, the Andes, the Southwestern United Sates,*etc.* As indicated earlier, large variations and uncertainty in light rain regions as well as other factors to be discussed later can be the main reasons for the observed low correlation coefficients in those regions. The two gauge-based climatologies are more closely related than 3B43 and the two gauge climatologies (Figure 4), respectively. In summary, the 3B43 climatology can adequately catch the seasonal variations in many regions around the world except for the light rain regions where even the two gauge climatologies do not quite agree.

Zonal means can be another indicator for assessing whether 3B43 can successfully catch the meridional distribution of precipitation in all seasons. Figure 5 contains the zonal means from the three climatologies for each season. It is seen that the 3B43 climatology can successfully reproduce the meridional distribution of precipitation and catch the seasonal shifts of the heavy rain band in the Tropics (Figure 5). Three zonal means in Figure 5 are quite similar with very little differences in rain rates in all seasons between 40° N and 40° S; however, it is seen that rain rates from WM are in general slightly lower than GPCC in some latitudes in particular in JJA and SON (Figure 5). Likewise, 3B43 rain rates are higher than the other two, especially in DJF. The bias issue associated with rain rates will be further discussed in the next section. Outside this belt, the differences in the three climatologies are more visible and 3B43 rain rates are higher than the rest especially in DJF and MAM; however, the rain rates there in general are much smaller than those in the tropical and sub-tropical regions. These differences appear to be related to precipitation types such as mixed or frozen precipitation. To conclude, 3B43 performs considerably well between 40°N and 40°S.

3.2. Bias Assessment

Figure 6 contains scatterplots between the three climatology pairs in four seasons. Higher rain rates (Figure 6c,g,k) are found in JJA, as shown earlier. In JJA, points spread wider than in other seasons (Figure 6). In addition, more points in high rain rates are found above the 1:1 line in both 3B43 *vs.* WM (Figure 6c) and 3B43 *vs.* GPCC (Figure 6g), suggesting large negative biases (3B43 < WM and 3B43 < GPCC) at high rain rates. Such large biases are only found in JJA and biases in other seasons are relatively small. Compared to the other two pairs, the biases are smaller between GPCC and

WM (Figure 6k); however the points (Figure 6k) spread similarly to the other two pairs (Figure 6c,g) along the 1:1 line, suggesting that significant differences exist in the two gauge-based climatologies as well.

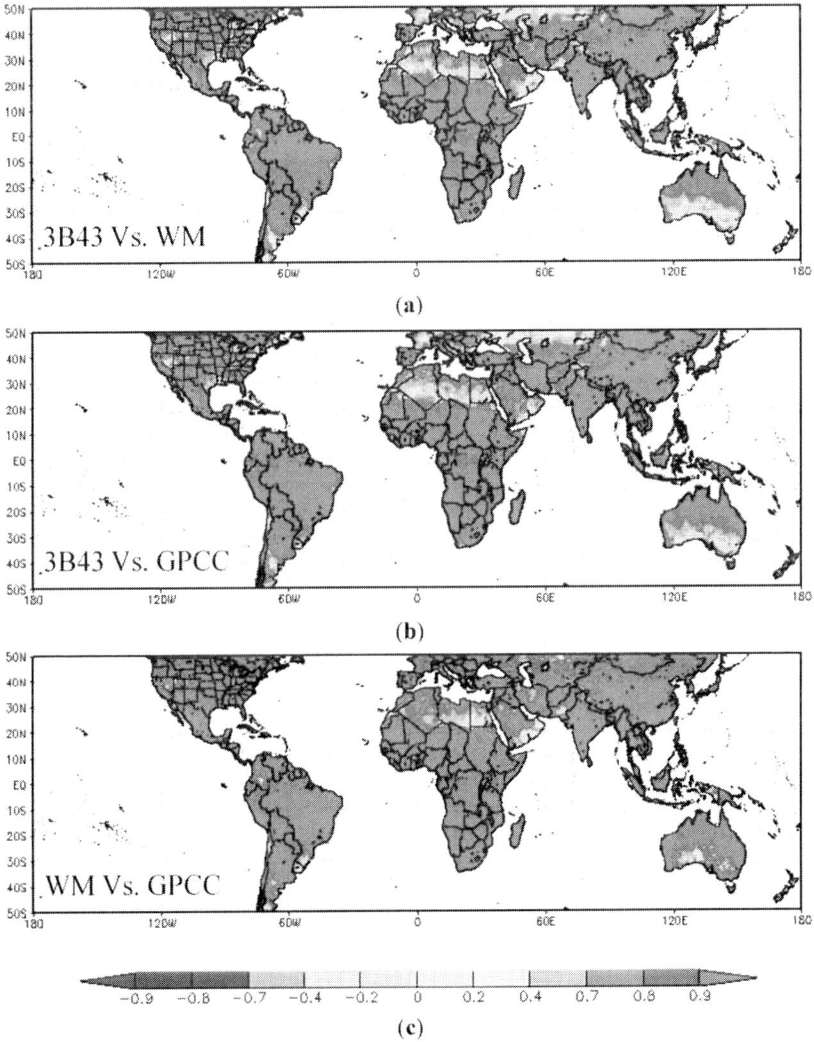

Figure 4. Pearson's correlation coefficients between (a) 3B43 and WM; (**b**) 3B43 and GPCC; and (**c**) WM and GPCC.

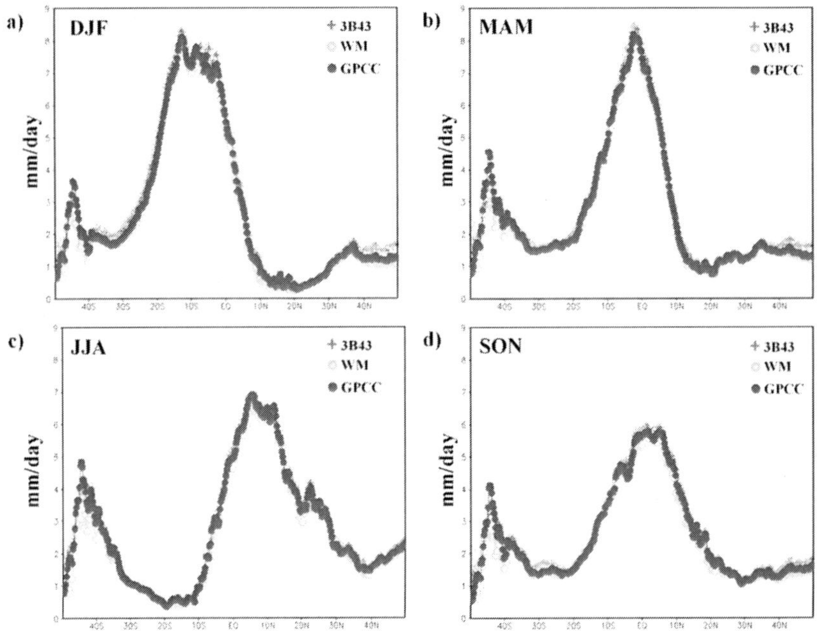

Figure 5. Zone means of 3B43, WM and GPCC for (**a**) DJF; (**b**) MAM; (**c**) JJA; and (**d**) SON.

Figure 6 only shows that biases exist among the three climatology pairs. Figure 7 shows the quantified biases among 3B43, WM and GPCC at different rain rates in four seasons. For all seasons, both 3B43 *vs.* WM (Figure 7a) and 3B43 *vs.* GPCC (Figure 7b) pairs all show positive biases (*i.e.*, 3B43 > WM) at low rain rates and negative biases at high rain rates. The latter confirms the findings in Figure 6. In JJA when the rain rates are the highest in all seasons, it is seen that the negative bias between 3B43 and WM can reach near −20% of the mean at the highest rain rate (Figure 7a). A much larger (~−36%) bias can be found in the 3B43 *vs.* GPCC pair (Figure 7b), which is much larger than that between 3B43 and WM. The differences between WM and GPCC (Figure 7c) are smaller than the other two pairs.

In MAM, large biases at high rain rates are found between 3B43 and WM (Figure 7a) as well as between WM and GPCC at 20 mm/day (Figure 7c), which is not found in those between 3B43 and GPCC (Figure 7b). In contrast, large biases at high rain rates between 3B43 and GPCC at 15 mm/day and above (Figure 7b) and between WM and GPCC at 15 mm/day and 20 mm/day are found in DJF (Figure 7c), but not found between 3B43 and WM (Figure 7a). The biases between the WM *vs.*GPCC pair (Figure 7c) are, in general, smaller than the other two pairs, as shown in the scatterplot in Figure 6, but for some reason, the differences at 20 mm/day are larger than the other two pairs. Negative systematic differences (Figure 7c) prevail in all seasons, except in MAM and SON at 20 mm/day. Nonetheless, 3B43 biases vary with rain rates with positive biases at low rain rates and much larger negative biases at high rain rates, especially in JJA, MAM and DJF.

To better understand the high biases at high rain rates in JJA shown in Figure 6c,g,k, it is necessary to examine regions of high rain rates and the spatial distribution of bias. Figure 8a shows the regional 3B43 rain rate climatology for JJA in Southeast Asia and India where high rain rates are located. In Figure 8a, it is seen that the regions of high rain rates are found in orographically influenced regions such as the Western Ghats of India, the coasts of Thailand and Burma (Myanmar). As mentioned earlier, these regions are heavily influenced by the orographic uplifting effect [29], characterized with high/low rain rates along the coastal windward/leeward sides of the mountains. The relative (with respect to the mean in Figure 8a) difference map (Figure 8b) between 3B43 and GPCC shows that the biases are not evenly distributed: large negative biases (3B43 < GPCC) are found in high rain rate regions along the coastal windward sides (Figure 8b). In contrast, large positive biases are found in the light rain rate regions such as the leeward side of the Western Ghats (Figure 8b).

Gauge data are an important data source for bias correction in satellite-based precipitation products such as 3B43 [9,10,11,12]. Therefore, the availability of gauge data can have a significant impact on data quality. Numerous validation studies show that data quality can be linked closely to the availability of gauge data in a study area. Figure 8c is a map of gauge weighting in percentage averaged between 1998 and 2013 for JJA. High percentages suggest that more gauge data are available for bias correction in the algorithm. It is seen that the weighting percentages along the coastal windward sides are in general not very high, especially in Burma where the percentages range only between 10%–30%. In short, the availability of gauge data can vary greatly (Figure 8c), which needs to be considered when assessing data quality of 3B43. Meanwhile, density of a gauge network can also influence the quality of a gauge-based precipitation product.

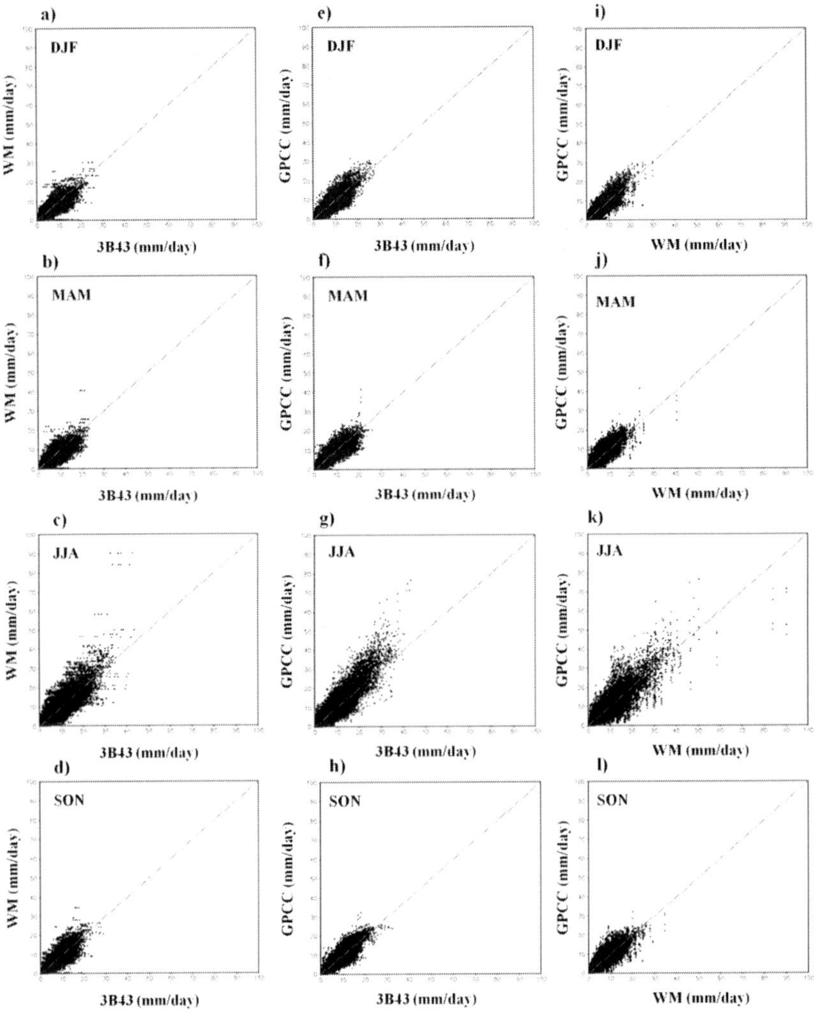

Figure 6. Scatterplots between: WM and 3B43 for (**a**) DJF; (**b**) MAM; (**c**) JJA; and (**d**) SON; GPCC and 3B43 for (**e**) DJF; (**f**) MAM; (**g**) JJA; and (**h**) SON; and GPCC and WM for (**i**) DJF; (**j**) MAM; (**k**) JJA; and (**l**) SON. Each black dot represents the rain rate of a grid point in each dataset. The red dashed line is the 1:1 line.

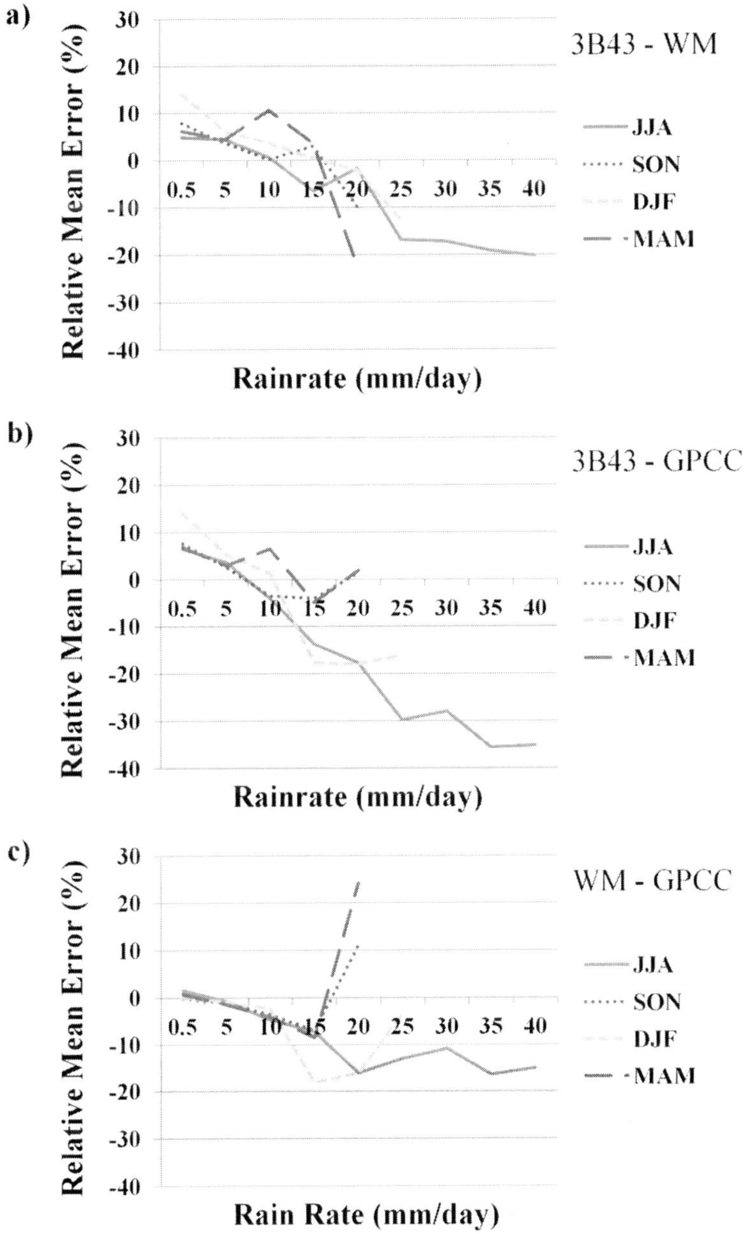

Figure 7. Relative (with respect to the mean) mean error between: (a) 3B43 and WM; (**b**) 3B43 and GPCC; and (**c**) WM and GPCC.

4. SUMMARY AND DISCUSSION

A satellite-based precipitation climatology, derived from the popular TMPA monthly product (3B43), was evaluated over land on a near-global scale with two gauge-based products (GPCC and WM). Statistical results show that the 3B43 climatology, in general, performs well against the two gauge-based climatologies, namely, all three products are quite similar and their variations are less than ~15% of the means. However, the variations are not evenly distributed around the world. When rain rates are lighter than 0.5 mm/day in remote and less populated regions such as the Sahara Desert, the Arabian Peninsula, the Tibetan Plateau, the Andes, *etc.* large variations in rain rates are found, suggesting that large uncertainties exist, including the two gauge-based climatologies as well. It is worth noting that light rain rate here does not mean light rain events in those regions. For example, studies (*i.e.*, [30]) show that Saharan rainfall can be highly irregular and can be heavy during a MCS (mesoscale convective system) event and as a result, light rain appears in the climatology after a long time period (for 3B43, 16 years) of averaging. Seasonal analysis shows that the largest variation (~18%) is found at 35 mm/day in JJA and the variations for other seasons range between ~7% and ~13% of the ensemble means at different rain rates.

Seasonal variations were compared using Pearson's correlation coefficients. Results show that high correlation coefficients (>0.9) are found in most regions except in the light rain (<0.5 mm/day) regions where large variations in rain rates are found as well. Nonetheless, the 3B43 climatology can correctly catch the seasonal variations in many regions, except the light rain regions.

The meridional distribution of precipitation was examined by computing the zonal means of the three climatologies in four seasons. Overall, 3B43 performs very well by reproducing the meridional distribution of precipitation with small differences between 40°N–40°S. The differences are more noticeable predominantly in cold seasons outside the 40°N–40°S belt where light and mixed precipitation prevails.

Finally, biases between 3B43 and the two gauge-based products were investigated. Basically, the biases between the two gauge-based climatologies are in general smaller than 3B43 *vs.* WM and 3B43 *vs.* GPCC, respectively. However, differences do exist in these two gauge-based datasets. In JJA, precipitation estimates are more spread out along the 1:1 line than in the other seasons. At high rain rates, 3B43 has low biases against GPCC and WM. The biases are not consistent, with positive biases (3B43 > WM and 3B43 > GPCC) found at low rain rates and much larger negative biases (3B43 < WM and 3B43 < GPCC) at high rain rates, especially in JJA. The negative biases can be as high as 20% of the ensemble means for WM and 36% for GPCC; however, the regions with such high rain rates are confined in orographically influenced regions such as along the windward coasts. This orographically associated issue has been discussed by several investigators [12,31,32,33] and their results show that shallow orographic rain rates are underestimated in microwave radiometer algorithms due to weak ice scattering signatures [31] as well as the lack of gauge data for bias correction. As a result, new methods have been developed to correct the said issue [12,31,32,33].

As mentioned earlier, many verification and validation activities have been done at local and regional scales with dense rain gauge networks. For example, Prakash *et al.* [34] and Kim *et al.* [35] have conducted seasonal intercomparison analysis with gridded rain gauge data from the India Meteorological Department (IMD) due to its high gauge density. Several precipitation products were compared including the GPCC product. Their results show that most precipitation products had difficulties in representing rainfall over orographic regions including the Western Ghats mountains, Northeast India and the Himalayan foothills, implying that even gauge-based products like the GPCC may still have serious issues. Joss and Lee [36] discussed gauge network density issues in precipitation product validation in complex terrain and they found a large number of gauges are needed in order to correctly depict a rain event.

In addition to the orographic factor, other factors need to be considered as well, such as deficiencies in space as in rain gauge measurement which cause issues in measuring convective rain which is highly inhomogeneous as well as measuring rainfall during high wind conditions.

Wang *et al.*, (2014) compared TCC (mainly 3B43, mentioned earlier) with GPCC over land in southeastern China and southeastern United States, respectively. They found that, in general, TCC estimates are smaller than GPCC, especially at high rain rates, which is consistent with the results in this study. However, TCC behaves a little bit differently in the two regions. In southeastern China, negative biases (TCC < GPCC) prevail; in contrast, the negative biases are not that obvious with positive biases at low rain rates in southeastern United States. As seen earlier, spatial variations in bias do exist; therefore, a separate investigation is needed for a different area of interest to ensure that more accurate biases can be analyzed.

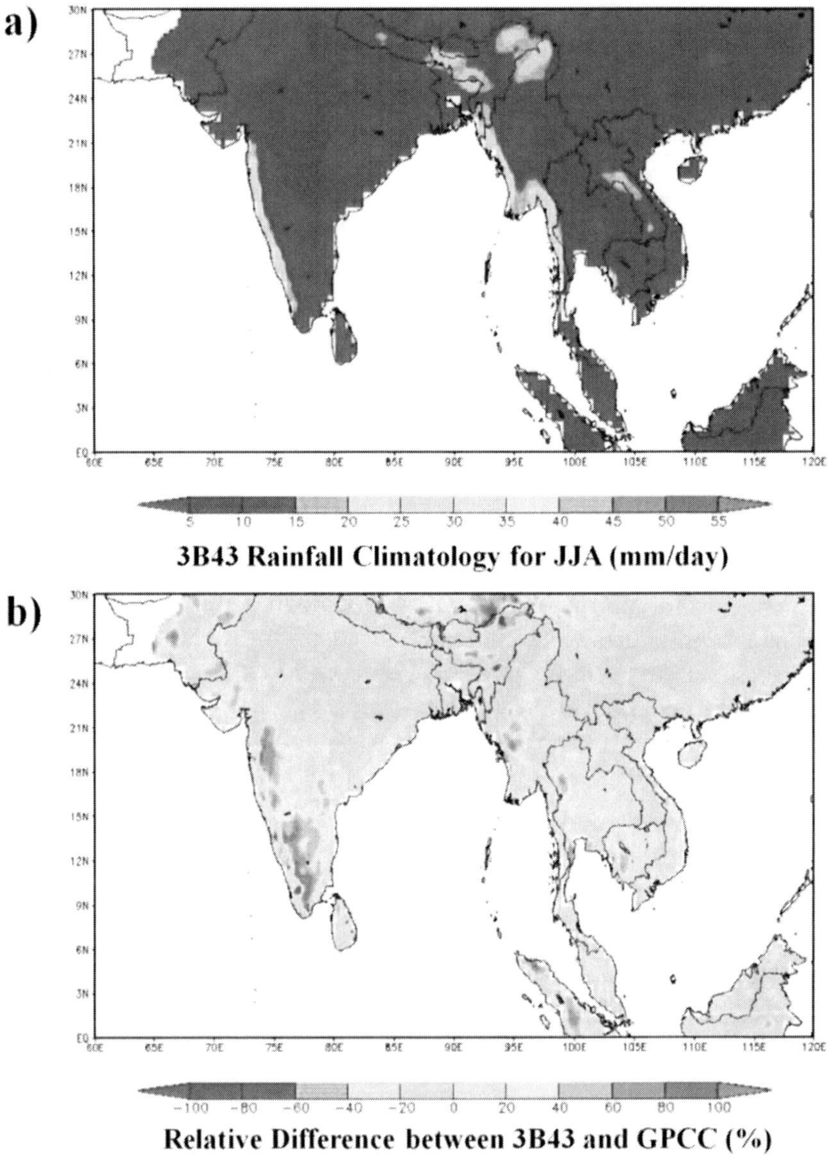
a)
30N
27N
24N
21N
18N
15N
12N
9N
6N
3N
EQ
60E 65E 70E 75E 80E 85E 90E 95E 100E 105E 110E 115E 120E
5 10 15 20 25 30 35 40 45 50 55
3B43 Rainfall Climatology for JJA (mm/day)
b)
30N
27N
24N
21N
18N
15N
12N
9N
6N
3N
EQ
60E 65E 70E 75E 80E 85E 90E 95E 100E 105E 110E 115E 120E
−100 −80 −60 −40 −20 0 20 40 60 80 100
Relative Difference between 3B43 and GPCC (%)

c)

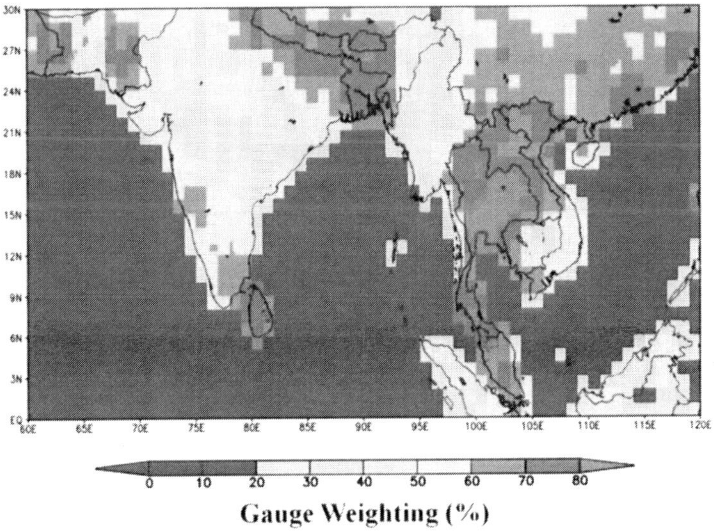

Figure 8. (**a**) 3B43 rainfall climatology in mm/day for JJA; (**b**) Relative (with respect to the mean in (a)) rain rate difference between 3B43 and GPCC in percentages; (**c**) Gauge weighting (in percentages) averaged between 1998 and 2013 for JJA.

As mentioned in Section 2, the 3B43 algorithm includes the monthly gauge analysis using inverse-error-variance weighting. From the GPCC product information page [37], the GPCC monitoring product that is used in the 3B43 algorithm for bias correction contains quality-controlled gauge data from 7000–8000 stations. By contract, the GPCC product used in this study contains much more quality-controlled gauge data from 67,200 stations. The difference in ground observation stations may explain the previously mentioned difference between 3B43 and GPCC, because more gauge data can increase gauge weighting percentages and biases can be corrected more adequately. Nonetheless, it is a challenge to conduct bias correction under the condition of limited gauges. As mentioned earlier, another way to combat this issue is to continue improving the algorithm, making it less relying on gauge-based bias correction.

Estimates for light rain still remain a challenge not only for satellite-based algorithms [10,28], but also for gauge-based measurements, as seen in this study. Although the problematic regions are remote and less populated, in order to better understand the global climate, it is necessary to be able to obtain correct information no matter what rain rate it may be for any missing or incorrect information can create difficulties in understanding the global hydrological cycle. This similar situation may apply to oceans as well where there has been no gauge-based bias correction so far.

ACKNOWLEDGMENTS

This project is supported by NASA Research Opportunities in Space and Earth Science-2010 (ROSES-2010), NNH10ZDA001N-ESDRERR, Appendix A.32: "Earth System Data Records Uncertainty Analysis". Special thanks to the GES DISC and the Giovanni development team. The TMPA data were provided by the NASA Goddard Space Flight Center's Mesoscale Atmospheric Processes Laboratory and Precipitation Processing System (PPS), which develop and compute the TMPA as a contribution to TRMM. The author would like to thank two anonymous reviewers for their thought-provoking and constructive comments.

REFERENCES

1. Liu, Z. Comparison of global precipitation climatology products derived from ground and satellite-based measurements. *Proc. SPIE* **2014**.

2. Houze, R.A.; Rasmussen, K.L.; Medina, S.; Brodzik, S.R.; Romatschke, U. Anomalous atmospheric events leading to the summer 2010 floods in Pakistan. *Bull. Am. Meteorol. Soc.* **2011**, *92*, 291–298.

3. Tripoli, G.J.; Medaglia, C.M.; Dietrich, S.; Mugnai, A.; Panegrossi, G.; Pinori, S.; Smith, E.A. The 9–10 November 2001 Algerian Flood: A numerical study. *Bull. Am. Meteorol. Soc.* **2005**, *86*, 1229–1235.

4. Hoerling, M.; Kumar, A.; Dole, R.; Nielsen-Gammon, J.; Eischeid, J.; Perlwitz, J.; Quan, X.; Zhang, T.; Pegion, P.; Chen, M. Anatomy of an extreme event. *J. Clim.* **2013**, *26*, 2811–2832.

5. Garstang, M.; Kummerow, C.D. The Joanne Simpson special issue on the tropical rainfall measuring mission (TRMM).*J. Appl. Meteorol.* **2000**, *39*, 1961.

6. Liu, Z.; Ostrenga, D.; Teng, W.; Kempler, S. Tropical rainfall measuring mission (TRMM) precipitation data and services for research and applications. *Bull. Am. Meteorol. Soc.* **2012**, *93*, 1317–1325.

7. Adler, R.F.; Wang, J.-J.; Gu, G.; Huffman, G.J. A ten-year tropical rainfall climatology based on a composite of TRMM products. *J. Meteorol. Soc. Jpn.* **2009**, *87*, 281–293.

8. Wang, J.-J.; Adler, R.F.; Huffman, G.J.; Bolvin, D. An updated TRMM composite climatology of tropical rainfall and its validation. *J. Clim.* **2014**, *27*, 273–284.

9. Huffman, G.J. Estimates of root-mean-square random error for finite samples of estimated precipitation. *J. Appl. Meteorol.* **1997**, 1191–1201.

10. Huffman, G.J.; Adler, R.F.; Bolvin, D.T.; Gu, G.; Nelkin, E.J.; Bowman, K.P.; Hong, Y.; Stocker, E.F.; Wolff, D.B. The TRMM multi-satellite precipitation analysis: Quasi—Global, multi-year, combined-sensor precipitation estimates at fine scale. *J. Hydrometeorol.* **2007**, *8*, 38–55.

11. Huffman, G.J.; Adler, R.F.; Bolvin, D.T.; Nelkin, E.J. The TRMM multi-satellite precipitation analysis (TMPA). In*Satellite Rainfall Applications for Surface Hydrology*; Hossain, F., Gebremichael, M., Eds.; Springer Verlag: Berlin, Germany, 2010; pp. 3–22.

12. Huffman, G.J.; Bolvin, D.T. TRMM and Other Data Precipitation Data Set Documentation. Available online: ftp://meso-a.gsfc.nasa.gov/pub /trmmdocs/3B42_3B43_doc.pdf (accessed on 5 August 2015).

13. Liu, Z.; Ostrenga, D.; Teng, W.; Kempler, S.; Milich, L. Developing GIOVANNI-based online prototypes to intercompare TRMM-related global gridded-precipitation products. *Comput. Geosci.* **2014**, *66*, 168–181.

14. Mirador. Available online: http://mirador.gsfc.nasa.gov (accessed on 5 August 2015).

15. Willmott, C.J.; Matsuura, K. Terrestrial Air Temperature and Precipitation: Monthly and Annual Time Series (1950–1999). Available online: http://climate.geog.udel.edu/~climate/html_pages/ README. ghcn_ts2.html (accessed on 5 August 2015).

16. Legates, D.R.; Willmott, C.J. Mean seasonal and spatial variability global surface air temperature. *Theor. Appl. Climatol.* **1990**, *41*, 11–21.

17. Legates, D.R.; Willmott, C.J. Mean seasonal and spatial variability in gauge-corrected, global precipitation. *Int. J. Climatol.* **1990**, *10*, 111–127.

18. Willmott, C.J.; Matsuura, K.; Legates, D.R. Global Air Temperature and Precipitation: Regridded Monthly and Annual Climatologies. Available online: http://climate.geog.udel.edu/~climate (accessed on 5 August 2015).

19. Shepard, D. A two-dimensional Interpolation function for irregularly-spaced data. In Proceedings of the 1968 ACM National Conference, Las Vegas, NV, USA, 27–29 August 1968; pp. 517–524. Available online: http://dl.acm.org/citation.cfm?id=810616 (accessed on 5 August 2015).

20. Willmott, C.J.; Rowe, C.M.; Philpot, W.D. Small-scale climate maps: A sensitivity analysis of some common assumptions associated with grid-point interpolation and contouring. *Am. Cartogr.* **1985**, *12*, 5–16.

21. Meyer-Christoffer, A.; Becker, A.; Finger, P.; Rudolf, B.; Schneider, U.; Ziese, M. GPCC Climatology Version 2011 at 0.25°: Monthly Land-Surface Precipitation Climatology for Every Month and the Total Year from Rain-Gauges Built on GTS-Based and Historic Data. Available online: ftp://ftp.dwd.de/pub/data/gpcc/html/gpcc_normals_v2011_doi_download.html (accessed on 5 August 2015).

22. Giovanni TOVAS. Available online: http:// gdata1.sci.gsfc.nasa.gov/daacin/G3/gui.cgi?instance_id=TRMM_Clim(accessed on 5 August 2015).

23. Liu, Z.; Rui, H.; Teng, W.; Chiu, L.; Leptoukh, G.; Vicente, G. Online visualization and analysis: A new avenue to use satellite data for weather, climate and interdisciplinary research and applications. Measuring precipitation from space—EURAINSAT and the future. *Adv. Glob. Change Res.* **2007**, *28*, 549–558.

24. Liu, Z.; Rui, H.; Teng, W.; Chiu, L.S.; Leptoukh, G.G.; Kempler, S. Developing an online information system prototype for global satellite precipitation algorithm validation and intercomparison. *J. Appl. Meteorol. Climatol.* **2009**, *48*, 2581–2589.

25. Giovanni. Available online: http://giovanni.gsfc.nasa.gov (accessed on 5 August 2015).

26. Acker, J.G.; Leptoukh, G. Online analysis enhances use of NASA earth science data. *Eos Trans. Am. Geophys. Union***2007**, *88*, 14–17.

27. Berrick, S.W.; Leptoukh, G.; Farley, J.D.; Rui, H. Giovanni: A Web service workflow-based data visualization and analysis system. *IEEE Trans. Geosci. Remote Sens.* **2009**, *47*, 106–113.

28. Tian, Y.; Peters-Lidard, C.D. A global map of uncertainties in satellite-based precipitation measurements. *Geophys. Res. Lett.* **2010**, *37*, L24407.

29. Whiteman, D. *Mountain Meteorology: Fundamentals and Applications*; Oxford University Press: Oxford, UK, 2000; p. 355.

30. Kelley, O.A. Where the least rainfall occurs in the Sahara Desert, the TRMM radar reveals a different pattern of rainfall each season. *J. Climate* **2014**, *27*, 6919–6939.

31. Shige, S.; Kida, S.; Ashiwake, H.; Kubota, T.; Aonashi, K. Improvement of TMI rain retrievals in mountainous areas. *J. Appl. Meteorol. Climatol.* **2013**, *52*, 242–254.

32. Shige, S.; Yamamoto, M.K.; Taniguchi, A. Improvement of TMI rain retrieval over the Indian Subcontinent. In *Remote Sensing of the Terrestrial Water Cycle*; Lakshmi, V., Alsdorf, D., Anderson, M., Biancamaria, S., Cosh, M., Entin, J., Huffman, G., Kustas, W., van Oevelen, P., Painter, T., *et al*, Eds.; John Wiley & Sons, Inc.: Hoboken, NJ, USA, 2014; pp. 27–42.

33. Yamamoto, M.K.; Shige, S. Implementation of an orographic/nonorographic rainfall classification scheme in the GSMaP algorithm for microwave radiometers. *Atmos. Res.* **2014**.

34. Prakash, S.; Mitra, A.K.; Momin, I.M.; Rajagopal, E.N.; Basu, S.; Collins, M.; Turner, A.G.; Achuta Rao, K.; Ashok, K. Seasonal intercomparison of observational rainfall datasets over India during the southwest monsoon season. *Int. J. Climatol.* **2015**, *35*, 2326–2338.

35. Kim, J.; Sanjay, J.; Mattmann, C.; Boustani, M.; Ramarao, M.V.S.; Krishnan, R.; Waliser, D. Uncertainties in estimating spatial and

interannual variations in precipitation climatology in the India-Tibet region from multiple gridded precipitation datasets. *Int. J. Climatol.* **2015**.

36. Joss, J.; Lee, R. The application of radar-gauge comparisons to operational precipitation profile corrections. *J. Appl. Meteorol.* **1995**, *34*, 2612–2630.

37. GPCC Product Information Page. Available online: http:// gpcc.dwd.de (accessed on 5 August 2015).

Chapter 5

Urban-Rural Temperature Differences in Lagos

Vincent N. Ojeh [1,2,*], A. A. Balogun [2,] and A. A. Okhimamhe [3,]

[1]The West African Science Service Centre on Climate Change and Adapted Land Use (WASCAL GRP-WACS), Federal University of Technology, Akure 340252, Nigeria

[2]Department of Meteorology and Climate Science, Federal University of Technology, Akure 340252, Nigeria

[3]West African Science Service Centre on Climate Change and Adapted Land Use Master Research Program in Climate Change and Adapted Land Use (WASCAL CC&ALU), Federal University of Technology, Minna 920252, Nigeria

ABSTRACT

In this study, the hourly air temperature differences between City hall (urban) and Okoafo (rural) in Lagos, Nigeria, were calculated using one year of meteorological observations, from June 2014 to May 2015. The two sites considered for this work were carefully selected to represent their climate zones. The city core, City hall, is within the Local Climate Zone (LCZ 2) (Compact midrise) while the rural location, Okoafo, falls within LCZ B (Scattered Trees) in the south-western part on the outskirt of the city. This study is one of very few to investigate urban temperature conditions in Lagos, the largest city in Africa and one of the most rapidly urbanizing megacities in the world; findings show that maximum nocturnal UHI magnitudes in Lagos can exceed 7 °C during the dry season, and during the rainy season, wet soils in the rural environment supersede regional wind speed as the dominant control over UHI magnitude.

KEYWORDS

temperature difference; urban climatology; heat island; Lagos

1. INTRODUCTION

Urbanization is a shift in population from rural to urban, *i.e.*, the gradual increase in the proportion of people living in urban areas, and the ways in which each society adapts to the change. In most respects, urbanization is beneficial as it enables increased standards of living: Urban areas create more than 90% of global gross value added [1]. Unfortunately, urbanization also brings with it some negative environmental, social, and economic consequences. The rapid pace of urbanization has been shown to be a global problem present in most developing countries.

The population growth rate will continue to be particularly rapid in the urban areas of less developed regions, averaging 2.4 percent per year during 2002–2030, consistent with a doubling time of 29 years [2]. The concentration of population in cities is expected to continue such that by 2030, 84 percent of the inhabitants of more developed countries will be urban dwellers. The United Nations (UN) predicts that between 2011 and 2050, the world population is expected to increase by 2.3 billion, increasing from 7.0 billion to 9.3 billion [3]. At the same time, the population living in urban areas is projected to increase by 2.6 billion, from 3.6 billion in 2011 to 6.3 billion by 2050 [4]. Thus, the urban areas of the world are expected to absorb all the population growth expected over the next four decades while at the same time drawing in some of the rural population such that nearly all the population growth (64%) will be in the cities of developing countries.

This rapid population growth will largely drive the extent and rate of global environmental changes and many of these changes are related to the climate and atmospheric composition of cities, including the canopy layer

Urban Heat Island (UHI) [5], the observed warmth of the urban core compared to its rural surroundings, heat stress, and various forms of air pollution [6]. Climate change will impact living and working environments negatively and create health threats for millions of people in the next decade and beyond [7,8,9,10,11]. The average global temperature is increasing and it is estimated that it will increase further from 1.8 to 4.0 °C (estimated average of 3.0 °C) by the year 2100 [7], depending on actions to reduce greenhouse gas emissions. The extent of local climate change will vary depending on geographic and local meteorological conditions. Increasing local ambient temperature implies higher human exposure to heat such that during hot seasons in hot regions of the world, people who are not able to afford either the cost of air conditioning and other cooling methods or the cost of energy required to run them will be subjected to very severe heat stress and health risks. Both general living environments and working environments will be affected.

The latter may impact workers' health, productivity, and socio-economic development [9,11]. Workers in low and middle income tropical countries are likely to be at the highest risk of excessive heat exposure. Due to anthropogenic activities, the earth's surface is being significantly altered in some manner, and human presence on the earth and use of the land has had a profound effect on almost all meteorological/climate parameters. In order to make the environment more conducive, humans have built structures in urban areas to cater for their needs. These urban structures alter the surface energy budget [12], modify the vertical profile of various atmospheric properties and interact with both local and regional circulation [13].

Urbanization is a major land change driver that leads to significant, often rapid, and near irreversible changes in land systems and land properties. Rapid urbanization and industrialization and the development of more buildings and transportation systems to cater for Lagosians has led to the recognition of human impacts on microclimate, otherwise referred to as the Urban Heat Island phenomenon. The intensity of the Urban Heat Island (Tu-r) can be quantified by the difference in temperature between the central urban site and a rural site. The magnitude of this phenomenon depends on meteorological factors, land use pattern, the population density, and anthropogenic activities of the city [14]. The socio-economic and meteorological impacts of Urban Heat Island in low and high latitude countries are numerous. Some are beneficial while others are detrimental. With respect to human health or thermal comfort, there is a negative impact over low-latitudes, which includes aggravation of heat stress during prolonged heat waves. Heat stress in these regions and seasons are invariably unfavorable for human health. It adds to discomfort and even contributes to death, most especially among the aged already predisposed to certain illness as well as children up to five years old, which are very vulnerable to the vagaries of weather [15]. The Urban Heat Island effect over the cities, which results in heat waves, can also promote the chemical break-down of some building materials by weakening them [16]. The study of Urban Heat Island (UHI) is important as it affects health, comfort and energy consumption for heating and cooling urban buildings in cities.

1.1 Literature Review

Research on the modification of local climate through urbanization is not new. According to Landsberg [17], Luke Howard is credited to have carried out the first scientific study of inadvertent urban climate modifications in 1818, by comparing the temperature records of a London city weather station with that of Kew Gardens (then a rural station) and found that the city station was warmer.

The importance of the Urban Heat Island [18] has become increasingly recognized by planning authorities during recent years (e.g., Mayor of London, [19]. The temperature increment generated by European cities like London was found to lead to an increased mortality rate during hot summer periods and with climate change, will lead to greater demands for cooling the internal environment. However, the winter urban temperature increment has benefits: leading to lower mortality rates and lower energy consumption due to warmer nighttime temperatures and reduced heating. A review of urban climate research by Jauregui [20] shows that out of all of the UHI studies carried out during the period 1968–1980, those in tropical regions accounted for only 2%, this proportion rose to 4% in the mid-1980s. According to Wienert [21] out of the UHI studies carried out in 223 cities, only 17% were in the tropics. By the mid-2000s, Roth [22] opined that only basics of the physical processes operating in the atmosphere of tropical climates are available because of the limited range of urban morphologies of cities studied. The situation is not different in Nigeria, Africa's most populous country, as some studies have been conducted in this regard.

Back in the 1980s, Oguntoyinbo [23] was of the opinion that a slow rate of urbanization and industrialization and the general lack of expertise are what led to a lack of interest in urban climatology in the tropics.

Oguntoyinbo [24] examined the impact of urbanization on the climate of Ibadan. Air temperature and relative humidity data were collected for certain synoptic hours across the city during the daytime using simple Thermohygrographs and whirling hygrometers. These data were compared with data from the observatory at the Ibadan airport. The result shows that 3.0 °C higher air temperatures and 7% lower relative humidity in the city were recorded during the day time. In his later work, he used an Assman Aspirated psychrometer and a whirling psychrometer in assessing temperature and relative humidity observed by using mobile traverses in different seasons across the city of Ibadan. The study revealed that there is a significant relationship between UHI development and cloud cover and building density. A UHI of 2.2 °C was recorded in the afternoon in March (representing the dry season) in the heavily built-up areas in Oja'ba, Oje and Agbokojo; this dropped to 0.8 °C in the early morning. In August (representing the short dry season), a UHI of 3.3 °C was recorded in the afternoon and it decreased to 1.7 °C in the early morning. In November (representing the Harmattan period), 3.6 °C was recorded in the afternoon and it dropped to 1.6 °C in the early morning. The study concluded that elevation, vegetation and open space have negative effects on UHI development.

Ojo [25] studied the spatial and temporal variation of temperature across Lagos city using temperature data taken from transect across the city. His findings showed a UHI effect of 2 °C to 4 °C in the zone of dense traffic and main traffic corridors of Mushin/Oshodi areas of the city at noon or in the late afternoon. Land-use energy balance relationships in the city were also examined. Later, [20] in Lagos discovered that the UHI effect is higher in the morning on the islands adjacent to the Atlantic Ocean, due to marine influence. But for the rest of the city, it is conditioned by the characteristics of the land use type. Oguntoyinbo [23] discovered that UHI varies with seasonal variation and time of day in Ibadan. UHI was most marked at the height of the dry season (March), especially in the afternoon, ranging between 5 °C to 7.5 °C in the city center and dropping to 0.8 °C in the early morning. During the wet season, the recorded mean values of the UHI ranged from 1 °C to 3 °C, while in the Harmattan season it was about 1 °C to 5 °C.

In Nigeria, the first comprehensive urban climate study was that of Adebayo [26], who analyzed the spatial, diurnal and seasonal characteristics of global radiation, surface albedo, net longwave radiation and latent sensible fluxes of energy for Ibadan city. The study was based on data collected on a daily basis (06:00 h–18:00 h GMT) for one year from 20 stations located all over the city. It showed that there was a decrease of radiation (about 14%) in the urban center, which was attributed to the effects of pollution veil and the reduced sky view factor within the canopy. There was also a decrease in surface albedo and an increase in net radiation both at the city center when compared to those of the rural areas. The increase in net radiation is attributed to an increase in the amount of energy absorbed from surface and long wave radiation by atmospheric contamination. The study also indicated a decrease in relative humidity (about 5%) in the city center compared with that of the rural areas during the wet season, whereas during the Harmattan season, the decrease was about 8%. A UHI effect of 1 °C–1.5 °C was observed at the city center during the wet season; it ranged from 2.5 °C to 3 °C during the Harmattan while in the middle of dry season, a UHI effect of up to 8 °C was observed.

Adebayo [27,28,29] identified the following as causes for the city-country temperature differences (UHI effect): (i) Reduction in albedo (thermal reflection), which causes more heat to flow to and from the earth's surface; (ii) The presence of the air pollution veil; and (iii) Urban canyon, which traps the heat radiation. In their study in Akure, Balogun *et al.* [30] showed that the nocturnal heat island was more frequent than the daytime heat island as it exists from less intense to higher intensity categories throughout the study period. They also reported that the nocturnal heat Island intensity was stronger during the dry season and lower intensity daytime heat Island exists throughout the day except for a few hours in the months of November and December that exhibited a reverse thermal contrast. The daytime heat island was observed to be more intense in the wet months than the dry months, which may be caused by the evaporative cooling of wet surfaces and on average, the urban/rural thermal differences were positive, varying from 4 °C for nocturnal hours during dry months to an approximate value of 2 °C around noon during wet months [30,31]. More recently, Ayanlade and Jegede [32] explored the role of remote sensing on the intensity of the daytime surface Urban Heat Island of the Lagos metropolis using a time series of Landsat data, from 1984 to 2012, and reported that source landscape contributes positively to the intensity of SUHI in the Lagos metropolis. Their results further showed a general increase in mean LST during the periods of study from 1984 to 2012.

It is evident from the literature that apart from the studies of [25] and [23], there are virtually no studies dealing with the issue of air temperature differences/UHI in Lagos using observation data. This is not disconnected from the fact that Lagos, being a coastal city, also faces the challenge of urban flooding and sea level rise and therefore, research attention has been directed towards this challenge. For instance, in of all the papers presented during the Lagos State Government annual climate change summits for the past seven years (2009–2014) by both local and international scholars, the topical issue of the Urban Heat Island has never been addressed despite the huge climate, energy use and health implications of UHI. More so, the Lagos Climate Change Policy document (2012–2014) addressed the effects of climate change on flooding, water shortages, food insecurity, and increased disease incidences together with associated social disruption without consideration of UHI in the policy document.

In this study, we present urban-rural temperature differences and the effects of synoptic observations on the UHI in Africa's most urbanized coastal megacity of Lagos. Hitherto, there are only very few studies that have addressed this issue in the study area in particular and Nigeria in general using remotely sensed data or observations of meteorological parameters. This study aims at investigating the existence, seasonal variation and intensity of the Urban Heat Island; the excess of the urban over rural air temperatures in Lagos. The result of this study will provide useful input for considerations in the future urban development/renewal and planning policy of a climate-smart city of Lagos.

2. METHODS

2.1. Study Area

Lagos is the smallest state in Nigeria by landmass. The geographical area of Nigeria that is defined as metropolitan Lagos is unique in many ways. It is the 6th largest city in the World and the largest city in sub-Saharan Africa. It is composed of 20 local government areas (LGA), of which 16 form the high-density metropolitan region [33]. The city of Lagos lies in south-western Nigeria, on the Atlantic coast. It is located at longitude 2°42′E and 3°42′E and between latitudes 6°22′N and 6°42′N (see Figure 1). On this stretch, lies the Lagos Lagoon with long coastal sand spits or sand bars. Badagry Creek flows parallel to the coast for some distance before finding an exit through the sand bars to the sea. The southern boundary of the state is formed by the 180-km long Atlantic Coastline while its northern and eastern boundaries are shared with Ogun state.

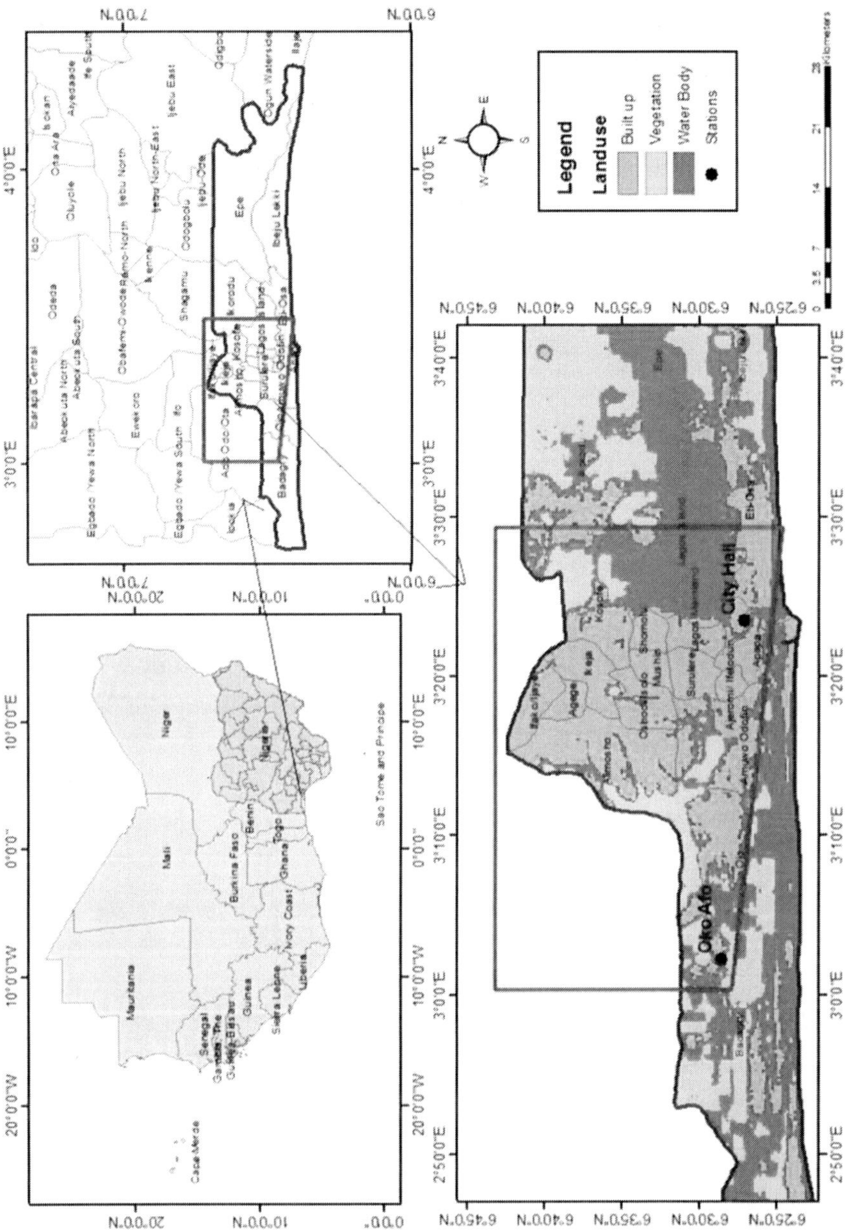

Figure 1. Map of the study location.

The region falls within the tropical rain forest belt and the eco-zones are predominantly wetlands and rain forest. The dominant vegetation of the State is the tropical swamp forest consisting the freshwater and mangrove swamp forests, both of which are influenced by the double rainfall pattern of the State, which makes the environment a wetland region, hence, the reference to Lagos as an environment of aquatic splendor. Its wetland environment is characterized by rich alluvial and terrallitic red-yellow soil, on which would be found dense luxuriant undergrowth, climbers, epiphytes and tropical hardwoods. The vegetation types found outside the urbanized area are mostly secondary forest, mangrove swamps, freshwater swamps and cultivated crops [34].

The interaction between the warm, humid maritime tropical air mass and the hot and dry continental air mass from the interior gives the state two seasons; a wet season from April to October and a dry season from November to March [35]. The mean annual rainfall of the state is roughly 1, 657 mm. The rain falls every month of the year with the highest amount in the month of June as a result of strong rain-bearing southerly wind prevailing in the area from the Atlantic Ocean. The air is very humid throughout the year, with monthly average maximum temperatures ranging from 28.6 °C in July/August to 33.7 °C in February/March (see Figure 2).

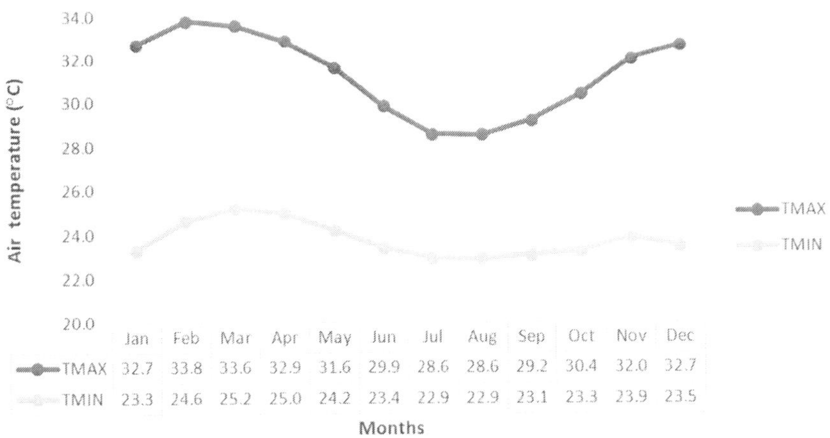

Figure 2. Mean monthly Air Temperature in Lagos (1981–2013).

Lagos has a wet equatorial type of climate influenced by its nearness to the equator and the Gulf of Guinea and is characterized by a deep and poorly drained soil. The drainage system of the State is characterized by a maze of lagoons and Waterways, which constitute about 22% or 787 sq km of the state's total landmass. The major water bodies are the Lagos and Lekki 8 lagoons, Yewa and Ogun Rivers. Others are Ologe Lagoon, Kuramo Waters and Badagry, Five Cowries and Omu Creeks. Water bodies and wetlands cover over 40 percent of the overall land area of the state and additional 12 percent is subjected to seasonal flooding [35]. The topography of Lagos consists of low-lying coastal beaches, extensive inland lagoons and depressions and marsh and mangrove wetlands at elevations of 0.2 m and upland areas with moderately drained soils and an elevation of 2 to 50 m above sea level [36].

Currently, Lagos has an estimated population of over 21 million people and this is expected to grow to over 23 million by 2015 [33,37] using an annual growth rate (AGR) of 3.2%. Over fifty per cent of Nigeria's electrical power generation is consumed by metropolitan Lagos. More than half the number of vehicles in Nigeria is concentrated on its network of roads. Eighty percent (80%) of the population of Lagos State resides in the 37% of the land and the State's economic activities are concentrated in the metropolis, which served as the seat of the former Federal capital of Nigeria until 1991 and is currently the seat of Lagos State Government.

According to the Master Plan for Metropolitan Lagos (MPML) the urban land area was approximately 172 km^2 in 1976 [38,39]. The major part (97.5%) of this land was in the contiguous built-up area, which generally fell within a broad triangle having a base of about 30 km along the Bight of Benin extending from Lekki, Maroko and Ikoyi westwards to Ojo and Ijanikin. The north-south part of the triangle covers a distance of about 26 km from Lagos Island to Alagbado at the southern and northern ends, respectively. A detached portion of the Lagos urban development area surrounds Ikorodu. This 2.5% of the total metropolitan area is separated from the contiguous area of the metropolis [39]. Lagos State has lots of housing estates comprising mainly midrise and high-rise buildings in the metropolitan areas with paved land cover. Construction materials include stone, brick, tile, and concrete materials. However, a major feature of Lagos' landscape is the proliferation of slums and informal settlements. These evolved in response to government's inability to provide adequate housing for the teeming population. These slums and informal settlements provide accommodation for the majority of the people and are usually the first point of call for new migrants [40]. As of 1991, the World Bank Urban

Renewal Project identified 42 slum communities in Lagos State [41]. The number of slums in Lagos is now estimated to have increased to more than 100 due to the inadequacy of private and public institutions to provide housing for the increasing population [40]. The slum areas are largely unpaved, low-rise buildings with thatch, wooden and zinc roofs and mud and brick walls. Most of the slums are located close to open water bodies such as the tributaries of rivers. Others are located very close to lakes, reservoirs, and lagoons, which abound in the state.

2.2. Weather Stations

A basic automatic weather station (BWS200) from Campbell Scientific was installed on the rooftop at Oshodi to collect synoptic data of the city. Measurement of daily air temperature with Thermocron ibutton data loggers logging at 15-min time intervals for one year (June 2014–May 2015) was carried out at City hall and Okoafo. The generated data were then computed into one-hour averages for ease of reporting. City hall is in the urban core and Okoafo is in the rural area close to Badagry, and the automatic weather station was located at Oshodi and was set up for synoptic observation (see Figure 3).

Table 1 shows the site characteristics of each of the stations classified based on the local climate zone classes [42]. The automatic weather station (AWS) (LCZ 2) is located on the rooftop of the official building of the Nigerian Meteorological Agency in Oshodi, central to the rural station (LCZ B) at Okoafo, a lightly wooded landscape of deciduous trees and tall grasses with mostly pervious (low plants) land cover. Urban station (LCZ 2) at City hall, is located in a dense mix of midrise buildings (3–9 stories) with few or no trees. Land cover is mostly paved and visible stone, brick, tile, and concrete construction materials.

The AWS was set up for synoptic observation of wind speed, wind direction, solar radiation, air temperature, Humidity, rainfall and barometric pressure were recorded at 15-min intervals over Lagos while the data loggers, Hygrochron ibuttons, and DS1923 model manufactured by maxim-ic were used to obtain the spatial temperature (Ta) and relative humidity (RH) data at the two selected sites. The data logger has a resolution of 0.05 with a measuring range of −20 °C to +85 °C and accuracy of ±0.5 °C. The stations were set up on the following dates, 20 May 2014 (Okoafo-rural), 21 May 2014 (city hall-urban) and 22 May 2015 (Oshodi-synoptic).

Figure 3. Datalogger sites in (**a**) Rural (Okoafo); (**b**) Urban (City Hall) and (**c**) Automatic Weather station (AWS) located on the rooftop of a one storey building at the Nimet premises in (Oshodi) used for the study (see Appendix for Google images of the locations).

Firstly, the monthly mean value of air temperature in the City hall and Okoafo area are presented. Secondly, the hourly temperature differences between the City hall and Okoafo are calculated, producing the hourly values of the Urban Heat Island, ΔTu-r. The Urban Heat Island was obtained for each day in each month to produce the diurnal pattern. Lastly, the synoptic data on wind speed and solar radiation were plotted against the UHI to ascertain their effects on the UHI intensity in the area.

Table 1. Summary of site characteristics.

Station	Site	Lat (N)	Long (E)	Elev Area (m)	Elev Sensors (m)	LCZ	Description
1	Oshodi (AWS)	06°32.40′	003°20.46′	25	31	5	Open Midrise
2	Okoafo (OK)	06°28.52′	003°02.15′	9	3	B	Scattered Trees
3	City Hall (CH)	06°27.02′	003°23.53′	11	4.5	2	Compact Midrise

* LCZ is a local climate zone, as defined by Stewart and Oke (2012).

3. RESULTS AND DISCUSSION

3.1. Distribution of Mean Temperature over Lagos (June 2014–May 2015)

Figure 4 shows the seasonal distribution of Tmean at City hall and Okoafo in (a) JJA, (b) SON, (c) DJF and (d) MAM. The air temperatures (Tmean) in the City hall area were 27.1 °C, 26.1 °C and 26.1 °C, respectively, during the JJA. It is seen that the mean air temperature was higher in the month of June than the remaining two months of the season, which recorded the lowest mean air temperature values for the study period. This could be attributed to the prevailing conditions of the sky and wind speed intensity. During the transition to the dry season period, SON, lower values of mean air temperature were recorded in City hall. The values recorded were 26.3 °C, 27.4 °C and 29.3 °C. The temperature pattern during the dry season period, DJF witnessed a slight increase in mean air temperature from the preceding season with values of 29.8 °C, 27.9 °C and 30.0 °C, respectively. The MAM period shows that mean air temperatures values were higher for this season during the study period, being 30.0 °C, 30.4 °C and 28.4 °C, respectively. Although the monsoon starts in this period, its effects are not fully felt on the air temperature until its peak between the months of July–August.

At Okoafo, the mean air temperature pattern shows a similar pattern to that recorded in City hall. Although the lowest values of mean air temperature were recorded during JJA, the month of June recorded a value higher than the months of July and August. The Tmean values for the period are 26.3 °C, 23.9 °C and 24.0 °C, respectively, for each month. During the transition period of SON, there was a slight increase in the Tmean values recorded. The Tmean values for SON were 25.2 °C, 26.1 °C and 27.0 °C. The mean air temperature showed another slight increase in the amount recorded during the dry season months (DJF) apart from the month of January, which recorded a lower value. The values of Tmean for DJF were 27.5 °C, 23.8 °C and 28.4 °C. The observation of low Tmean value for the month of January could be attributed to the effect of the Harmattan caused by the north-easterly wind prevailing over the season and the moisture content of the air because of the availability of vegetation. The MAM period recorded the highest mean air temperatures. The values recorded were 28.8 °C, 28.8 °C and 28.4 °C, respectively, for each month in the period. In both City hall and Okoafo, the air temperature pattern in the month of June being the first month in JJA shows that it still retains a characteristic pattern of the MAM period. The possible effect of the low moisture content from the sea to the land cannot be ruled out.

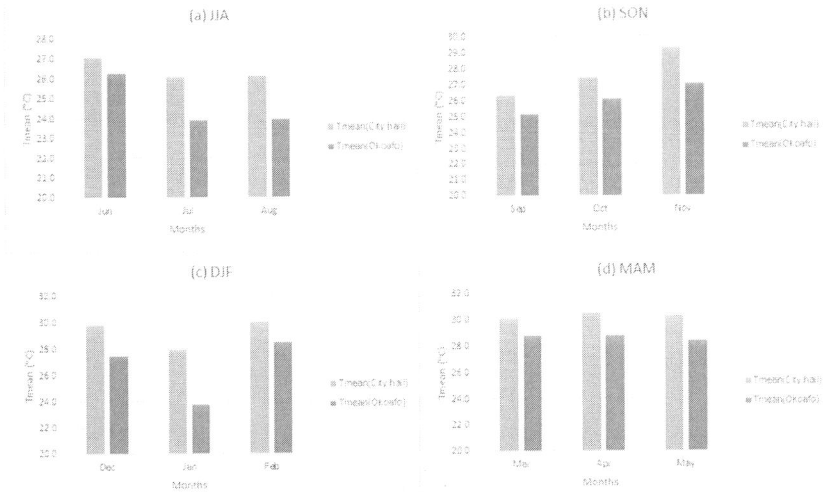

Figure 4. Seasonal distribution of Tmean at City hall and Okoafo in (**a**) JJA; (**b**) SON; (**c**) DJF and (**d**) MAM.

Figure 5 shows the diurnal pattern of Tmean at City hall and Okoafo in (a) JJA, (b) SON, (c) DJF and (d) MAM. In Figure 5, the diurnal pattern of the mean air temperature shows that generally temperatures begin to rise in the study area after 07:00 h until peaking at midday and gradually starts dropping until 18:30 h. However, the months of July and August during JJA did not quite follow this pattern during the period of study. This could also be attributed to the monsoon since it is at its maximum by August and the possibility of a cloudy sky is almost certain. This attenuates the effect of solar radiation on heating up of the ground surface, thereby reducing the values of the air temperature. Over Nigeria, the common consensus is that during the monsoon season, the country receives moisture from the tropical Atlantic via low-level southwesterly flow across her southern coast. At the surface, this moist southwesterly airstream can penetrate beyond the country as far as the southern fringes of the Sahara Desert near 20°N [43].

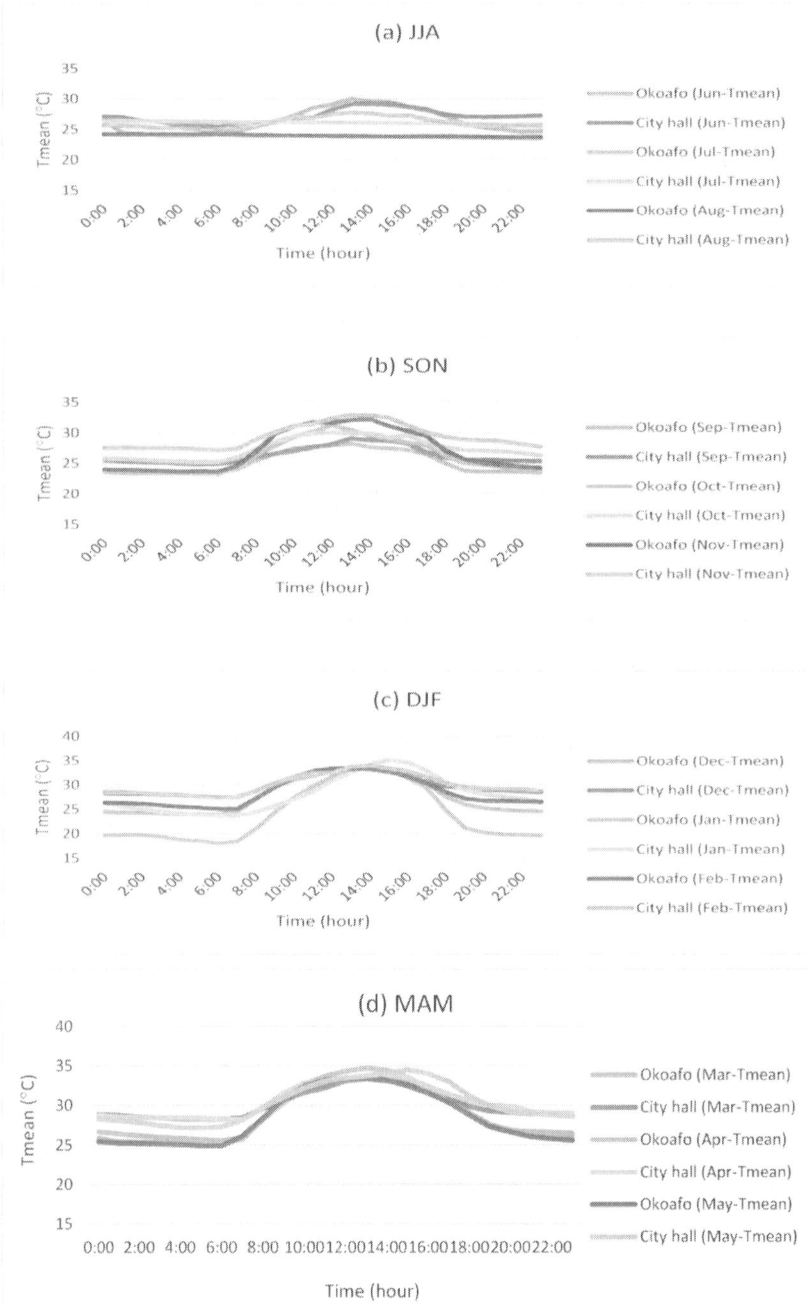

Figure 5. Diurnal pattern of Tmean at City hall and Okoafo in (**a**) JJA; (**b**) SON; (**c**) DJF and (**d**) MAM.

3.2. Seasonal Pattern of the UHI

Figure 6 shows the seasonal variation of mean UHI in (a) JJA, (b) SON, (c) DJF and (d) MAM. From Figure 6, it is evident that during JJA, the mean UHI in Lagos is higher during the day than the night period except for the month of June. This could be linked to excess soil moisture prevailing over the rural site in contrast to the urban site as the presence of wet soils and vegetation delays surface and near-surface heating, creating a large daytime heat island magnitude following sunrise. However, UCI (inverse heat island) is observed during the day in the month of June unlike the other rainy season months of July and August. The month of June also shows an unusual trend of delay between the rain and its effect on the UHI. This is due to the low moisture content moving from the Atlantic Ocean to the land because at this period, the highest mean UHI intensity values at during the day were 4.4 °C and 3.8 °C in July and August, respectively, while the cool island effect was −1.6 °C in June. During the night time, the highest mean UHI was 2.6 °C.

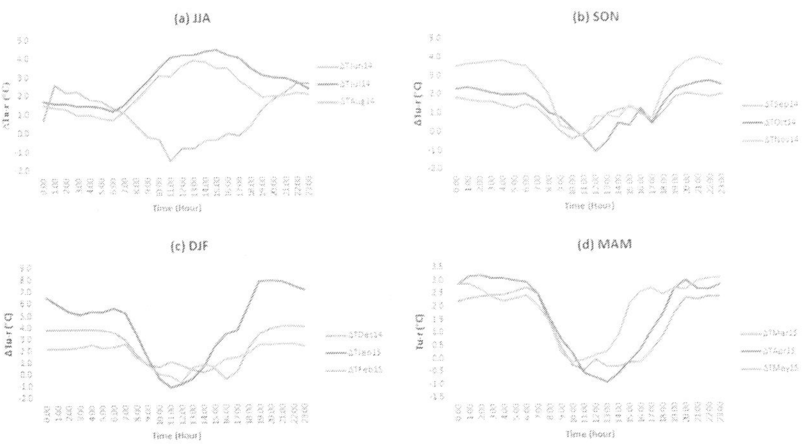

Figure 6. Seasonal variation of mean UHI in (**a**) JJA, (**b**) SON; (**c**) DJF and (**d**) MAM.

In SON, the day time was cooler than the night period. The daytime UCI was −1.1 °C at noon but there was a more prolonged heat island effect from

21:00 h to 04:00 h, which ranged between 3.8 °C to 3.9 °C. Being a transition to the dry season period, the UCL effect during the day could be linked to the presence of vegetal cover in the rural site, and modification of the moisture availability. The nighttime difference can be attributed to the geometry and urban form in the urban site.

DJF is the dry season period with the Harmattan. The Harmattan is a dry and dusty northeasterly trade wind that blows from the Sahara Desert over the West African subcontinent into the Gulf of Guinea between the end of November and the middle of March (winter). During this period, relative low mean UHI and UCI prevail during the day but high night time heat island is traceable to the urban built forms, pavement and construction materials and the dense mixture of tall buildings in the city centre; as such, the cool effect of the Harmattan is not felt at night. The stronger UHI intensity in the month of January especially from 19:00 h (7.8 °C) to midnight (7.0 °C) can be traced to effects of localised change in weather conditions. First, the Harmattan period is calmer than the wet period [44]; relative calmer conditions lead to the accumulation of heat within the atmospheric boundary layer [26]. Secondly, whereas the rainy season is cloudy, the Harmattan season is generally cloudless; cloudlessness allows for the direct uniform heating of the surface, therefore leading to higher heat island intensity [26]. Apart from the localised change in weather that introduced stronger UHI in January, the UHI during the nighttime in December and February was moderate with the strongest intensity being 3.9 °C and 2.3 °C in December and February, respectively. This result is congruent with Adebayo [26], who reported a higher amount of 'heat island' intensity during the Harmattan season than the rainy season in Ibadan metropolis, Nigeria in 1987.

MAM is another transition period leading to the rainy season in Nigeria with a less strong UHI pattern like the preceding season. Strongest UHI intensity was observed in the late hours of the day, *i.e.*, 18:00 h to midnight through early hours of the morning. Highest UHI intensity was between 2.9 °C to 3.0 °C for the period with a cool island effect of −1.0 °C at 13:00 h. The urban surface at this period is warmer than the rural site because of the attributes of the paved surface compared with the much-vegetated surface with relatively little or no paved surfaces This is congruent with Balogun [45], who observed maximum UHI in the dry season in Akure between 18:00 h to 22:00 h. However, the result also shows that maximum UHI occurs during the daytime in the rainy season at 12:00 h (noon) (6.9 °C) and 16:00 h (6.3 °C) local time with a gradual decrease between this period and from 4.8 °C at 17:00 h to 1.8 °C at 0:00 h (midnight). A similar result was obtained by Balogun *et al.* [43] in Akure; Oguntoyinbo [22] and Oke [46] in Mexico.

3.3. Noon and Midnight UHI Pattern

Figure 7 shows the noon and midnight variation of maximum UHI in Lagos. From Figure 7, the UHI effect is observed at midnight in all months of the year with the maximum of 6.5 °C in January (dry season) and a minimum of 0.7 °C in June (rainy season). However at noon, both the UHI and UCI effect is observed. UCI at noon was observed in six months with highest UCI of −1.1 °C in October and the UHI effect at noon was observed in the remaining six months of the year with the highest value of 3.6 °C in August. This is as a result of the excess heat retention of the urban form and pollution during the day, which is later released at night.

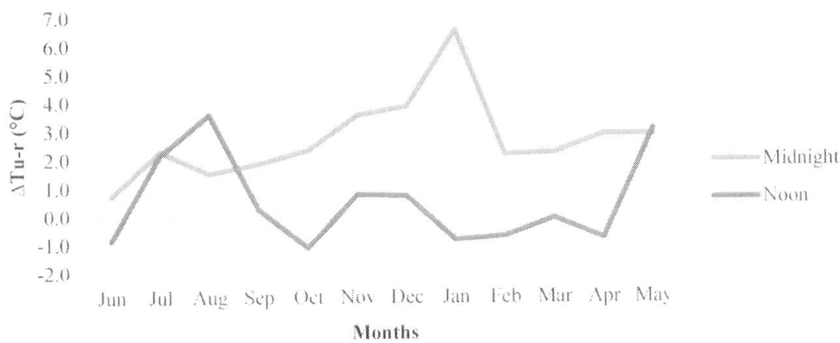

Figure 7. Noon and Midnight Variation of Maximum UHI in Lagos.

3.4. Effect of Synoptic Observation on UHI Pattern

Figure 8 shows the effect of solar radiation on the UHI pattern in (a) JJA, (b) SON, (c) DJF and (d) MAM. In Figure 8a–d, the solar radiation pattern is the same, sunrise at 7:00 h local time and sunset at 19:00 h local time. The time period between sunrise and sunset is about 12:00 h throughout the season. It should be noted that solar radiation is used here as a substitute variable for cloud cover (*i.e.*, low radiation values suggest greater cloud cover and vice versa). The cloud cover is observed universally closely correlated with UHI magnitude. It is also evident that apart from the monsoon months of July and August, solar radiation has an inverse relationship with the UHI pattern; daytime is when the solar radiation is highest (less cloud cover) but the daytime temperature difference between the urban and the rural site is lower, in most cases resulting in a cool island effect. However, the UHI intensity starts increasing about 4 h to sunset and this increases further after sunset, making the night time temperatures higher in the urban site. This is evidence that it is the heat that is trapped in the urban site due to the urban form, the building configuration patterns as well as contributions from dense, impervious surfaces that increase the amount of energy that is absorbed and eventually stored in the city; low-albedo surfaces, such as dark rooftops and asphalt roadways; anthropogenic heat sources such vehicular and building emissions could result from cooling systems [47,48,49].

Another visible pattern of solar radiation is that it has a bimodal peak reaching the first peak at 12:00 and the second peak at 16:00 h with a reduction at 14:00 h, which could be as a result of the presence of aerosol and pollutants that are capable of causing scattering of the incoming solar radiation and the dilution of heat due to intense vertical mixing in the midday boundary layer. In the four seasons presented, the intensity of the UHI is stronger at night with a cool island effect for a few hours in the day apart from July to August.

Figure 9 shows the effect of wind speed on the UHI pattern in (a) JJA, (b) SON, (c) DJF and (d) MAM. In Figure 9a–d, the seasonal pattern of the UHI and the effect of wind speed shows that a decrease in the amount of wind speed (light wind) favours the maximal mean heat island intensity and vice versa. Wind speed is strongest during the day but weaker at night in all seasons but lower UHI intensity and few hours of UCL is observed. The strong wind speed during the day at the urban site moderating the temperature at this time could be attributed to localized cool sea breeze from the Lagos Lagoon and the Atlantic Ocean unlike the rural site several kilometers inland. The peak of the mean wind speed is between 14:00 h to 16:00 h. In JJA season, a windy condition of 3.5 m/s resulted in a reduction of the cool island of −0.2 °C in June but with a light wind of 1.7 m/s, the UHI intensity was 2.7 °C and this was more evident at night. In SON season, September had the highest mean wind speed of 3.8 m/s with a corresponding low UHI intensity of 1.4 °C. A similar pattern is observed in DJF and MAM seasons where UHI is stronger at night with calm wind.

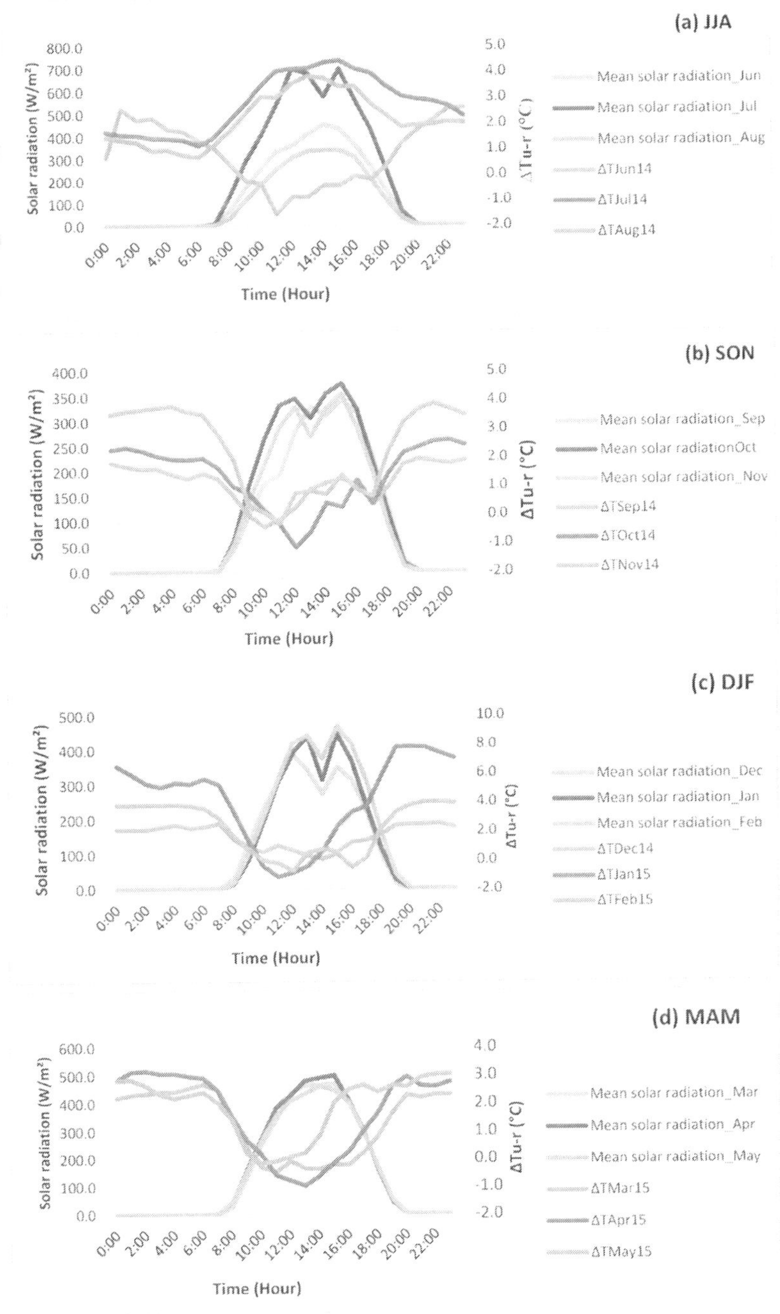

Figure 8. The effect of solar radiation on UHI pattern in (**a**) JJA; (**b**) SON; (**c**) DJF and (**d**) MAM.

4. CONCLUSIONS AND RECOMMENDATIONS

The urban-rural temperature differences, their variation in seasons as well as the effects of solar radiation and wind speed for Lagos have been investigated and discussed. Salient findings of the study are: (1) this study is one of very few to investigate urban temperature conditions in Lagos, the largest city in Africa and one of the most rapidly urbanizing megacities in the world; (2) maximum nocturnal UHI magnitudes in Lagos can exceed 7 °C during the dry season; and (3) during the rainy season, wet soils in the rural environment supersede regional wind speed as the dominant control over UHI magnitude. Projections of climate change suggest that Lagos State will experience a temperature rise of about 3 °C by 2100 [50]. We suggest that the effects of the heat island in Lagos on the thermal comfort, energy demand and health of its inhabitants should be considered as a study in the future but attention is drawn to the existence of the phenomenon in Lagos and possible reasons for its occurrence.

The implication of the study result is that policy towards making the city of Lagos less warm should be in place as the chances of mortality through heat stress especially among vulnerable populations are to be expected. Climate-smart urban planning is recommended with decentralization of economic activities from the city core and maintenance of urban gardens/parks and vegetation, which could ameliorate the heat effect. Moreso, cool and green roofs as well as the use of energy-efficient appliances and equipment could help in Lagos. However, there is a need to further this study by investigating the role of land/sea breeze on the UHI intensity in different seasons in Lagos.

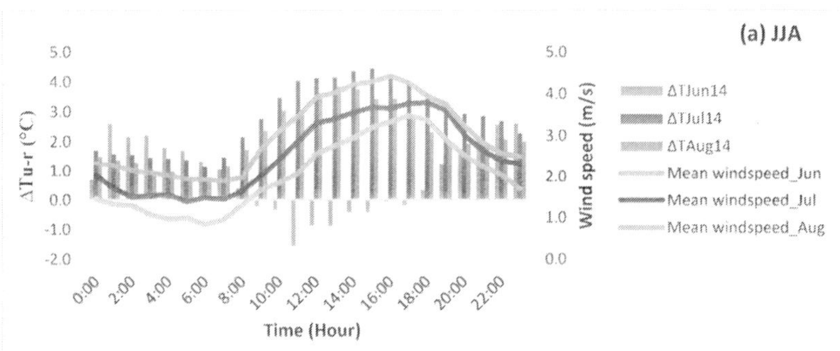

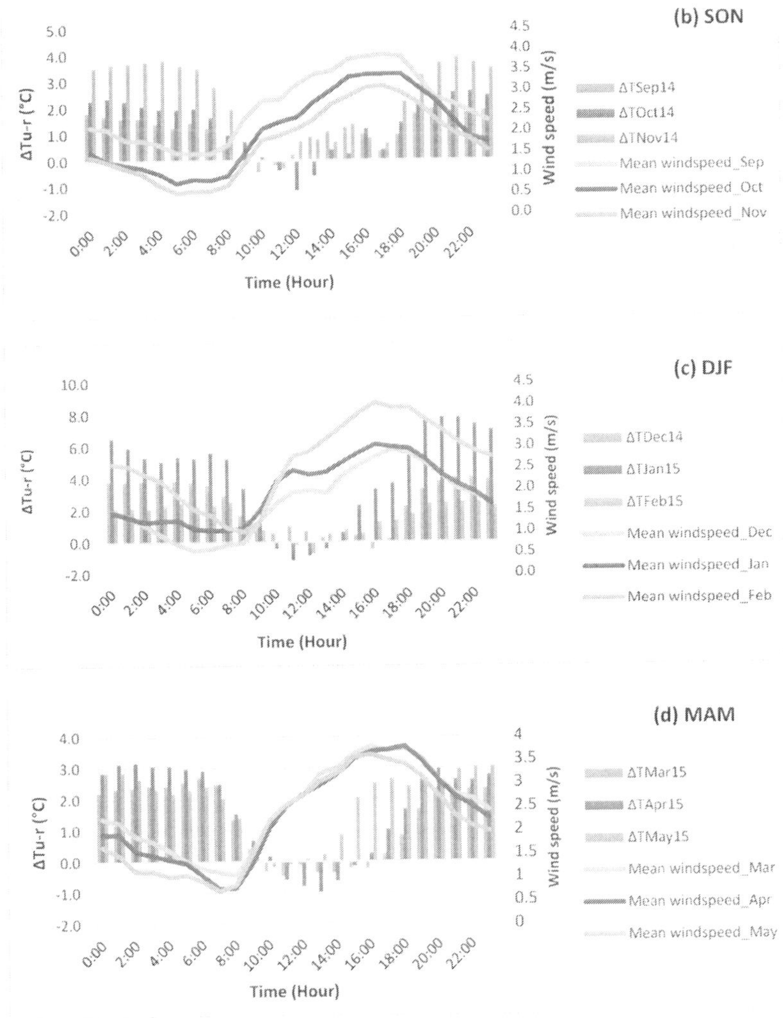

Figure 9. The effect of wind speed on UHI pattern in (**a**) JJA; (**b**) SON; (**c**) DJF and (**d**) MAM.

ACKNOWLEDGMENTS

This research is based on a study supported by the West African Science Service Centre for Climate Change and Adapted Land-use (WASCAL) of the Graduate Research Program in West African Climate Systems, (GRP-WACS), Federal University of Technology, Akure, Nigeria. The authors acknowledge the efforts of Iain Stewart for his useful contributions in the local climate zone landscape classification schemes that were used in the study. We also acknowledge the three independent reviewers of the original manuscript for their valuable comments.

AUTHOR CONTRIBUTIONS

VNO prepared the manuscript with contributions from all co-authors, AAB and AAO assisted in the data analyses. VNO and AAB designed the experiments and carried out the installations of the climate stations. All authors read and approved the final manuscript

REFERENCES

1. Fragkias, M.; Güneralp, B.; Seto, K.C.; Goodness, J. A synthesis of global urbanization projections. In *Urbanization, Biodiversity and Ecosystem Services: Challenges and Opportunities: A Global Assessment*; Elmqvist, T., Fragkias, M., Goodness, J., Güneralp, B., Marcotullio, P.J., McDonald, R.I., Parnell, S., Schewenius, M., Sendstad, M., Seto, K.C., *et al.*, Eds.; Springer: Berlin, Germany, 2013; pp. 409–435.

2. United Nations Population Division (UNPD). *World Urbanization Prospects: The 2001 Revision: Data Tables and Highlights*; United Nations: New York City, NY, USA, 2002.

3. United Nations. *World Urbanization Prospects: The 2011 Revision*; United Nations: New York City, NY, USA, 2012.

4. Oke, T.R. The distinction between canopy and boundary-layer urban heat islands. *Atmosphere* **1976**, *14*, 268–277.

5. Balogun, A.A.; Balogun, I.A.; Adeyewa, Z.D. Comparisons of urban and rural heat stress conditions in a hot humid tropical city. *Glob. Health Action* **2010**, *3*, 1–5.

6. Climate Change 2007: Synthesis Report. Contribution of Working Groups I, II and III to the Fourth Assessment Report of the Intergovernmental Panel on Climate Change. Available online:

https://www.ipcc.ch/pdf/assessmentreport/ar4/syr/ar4_syr_full_report.pdf (accessed on 28 February 2016).

7. Costello, A.; Abbas, M.; Allen, A.; Ball, S.; Bell, S.; Bellamy, R.; Friel, S.; Groce, N.; Johnson, A.; Kett, M.; *et al.*Managing the health effects of climate change. *Lancet* **2009**, *373*, 1693–1733.

8. Kjellstrom, T. Climate change, direct heat exposure, health and well-being in low and middle-income countries. *Glob. Health Action* **2009**, *2*.

9. Bowen, K.J.; Friel, S. Climate change adaptation: Where does global health fit in the agenda? *Glob. Health* **2012**, *8*, 1–7.

10. Kjellstrom, T.; McMichael, A.J. Climate change threats to population health and well-being: The imperative of protective solution that will last. *Glob. Health Action* **2013**, *6*, 1–9.

11. Grimmond, C.S.B.; Blackett, M.; Best, M.J.; Baik, J.-J.; Belcher, S.E.; Bohnenstengel, S.I.; Calmet, I.; Chen, F.; Coutts, A.;*et al.* Initial results from Phase 2 of the international urban energy balance model comparison. *Int. J. Climatol.* **2011**, *31*.

12. Argüeso, D.; Evans, J.P.; Pitman, A.J.; Di Luca, A. Effects of city expansion on heat stress under climate change conditions. *PLoS ONE* **2015**, *10*, e0117066.

13. Unger, J. Heat island intensity with different meteorological conditions in a medium-sized town: Szeged, Hungary.*Theor. Appl. Climatol.* **1996**, *54*, 147–151.

14. Rinner, C.; Patychuk, D.; Bassil, K.; Nasr, S.; Gower, S.; Campbell, M. The role of maps in neighborhood-level heat vulnerability assessment for the city of Toronto. *Cartogr. Geogr. Inf. Sci.* **2010**, *37*, 31–44.

15. Fasheun, T.A. Meteorological hazard in regional and urban settlement—On the destructive and deteriorating effect of tropical systems on building in Nigeria. In Proceedings of the International symposium on Meteorol, Hazards and development, Lagos, Nigeria, 22–24 May 1989.

16. Landsberg, H. *The Urban Climate*; Academic Press: New York, NY, USA, 1981; Volume 28.

17. Oke, T.R. The energetic basis of the Urban Heat Island. *Q. J. R. Meteorol. Soc.* **1982**, *108*, 1–24.

18. King's College London. *London's Urban Heat Island: A Summary for Decision Makers*; Greater London Authority: London, UK, 2006.

19. Jauregui, E. Tropical urban climates: Review and assessment. In *Urban Climatology and Its Applications with Special Regard to Tropical Areas*; Oke, T.R., Ed.; World Meteorological Organization: Geneva, Switzerland, 1986; pp. 26–45.

20. Wienert, U. Studies on Latitude and Climate Zone Depending on the Urban Heat Island—A Statistical Analysis. University of Duisburg-Essen: Essen, Germany, 2002; p. 211.

21. Roth, M. Review of urban climate research in (sub) tropical regions. *Int. J. Climatol.* **2007**, *27*, 1859–1873.

22. Oguntoyinbo, J.S. Some aspects of the urban climates of tropical Africa. In *Urban Climatology and Its Applications with Special Regard to Tropical Areas*; Oke, T.R., Ed.; World Meteorological Organization: Geneva, Switzerland, 1986; pp. 110–135.

23. Oguntoyinbo, J.S. Temperature and humidity traverses in Ibadan. In Proceedings of the 17th Annual Conference of Nigerian Geographical Association, Lagos, Nigeria, 19–20 April 1973.

24. Ojo, S.O. Land-Use Energy Balance Relationships and the Implications in Metropolitan Lagos. University of Lagos: Lagos, Nigeria, 1981.

25. Adebayo, Y.R. The Microclimatic Characteristics within the Urban Canopy of Ibadan. Ph.D. Thesis, University of Ibadan, Ibadan, Nigeria, 1985.

26. Adebayo, Y.R. Land-use approach to the spatial analysis of the Urban Heat Island in Ibadan, Nigeria. *Weather* **1987**, *42*, 273–280.

27. Adebayo, Y.R. Aspects of the variations in some characteristics of radiation budget within the urban canopy of Ibadan. *Atmos. Environ. B* **1990**, *24*, 9–17.

28. Adebayo, Y.R. Heat island in a humid tropical city and its relationship with potential evaporation. *Theor. Appl. Climatol.* **1991**, *43*, 137–147.

29. Balogun, I.A.; Balogun, A.A.; Adeyewa, Z.D. Observed Urban Heat Island characteristics in Akure, Nigeria. *Afr. J. Environ. Sci. Technol.* **2012**, *6*, 1–8.

30. Balogun, I.A.; Balogun, A.A. Urban Heat Island and bioclimatological conditions in a hot-humid tropical city: The example of Akure, Nigeria. *Ethiop. J. Environ. Stud. Manag.* **2014**, *145*, 3–15.

31. Ayanlade, A.; Jegede, O.O. Evaluation of the intensity of the daytime surface Urban Heat Island: How can remote sensing help? *Int. J. Image Data Fusion* **2015**, *6*, 348–365.

32. Lagos Bureau of Statistics. *Digest of Statistics 2010, Ministry of Economic Planning and Budget*; The Secretariat: Ikeja, Lagos, 2010; p. 242.

33. Nkeki, F.N.; Ojeh, V.N. Flood risks analysis in a littoral African city: Using geographic information system. In*Geographic Information Systems (GIS): Techniques, Applications and Technologies*; Nielson, D., Ed.; Nova Science Publishers: New York, NY, USA, 2014; pp. 279–316.

34. Building Nigeria's Response to Climate Change (BNRCC). Towards a Lagos State Climate Change Adaptation Strategy. Available online: http://www.nigeriaclimatechange.org/projectpublications.php (accessed on 28 April 2016).

35. Fasona, M.; Omojola, A.; Odunuga, S.; Tejuoso, O.; Amogu, N. An appraisal of sustainable water management solutions for large cities in developing countries through GIS: The case of Lagos, Nigeria. In Proceeding of the Symposium S2 Held during the 7th IAHS Scientific Assembly, Foz do Iguacu, Brazil, 3–9 April 2005; pp. 49–57.

36. United Nations Centre for Human Settlements (UNCHS). *The State of the World's Cities*; HABITAT: Nairobi, Kenya, 2001.

37. Lagos State Government. *Master Plan for Metropolitan Lagos 1980–2000*; Dioxadis Associates: Lagos, Nigeria, 1985.

38. Lagos State Government. *Review of the Lagos State Regional Plan*; John Asiyanbi and Associates: Lagos, Nigeria, 2002.

39. Stewart, I.D.; Oke, T.R. Local climate zones for urban temperature studies. *Bull. Am. Meteorol. Soc.* **2012**, *93*, 1879–1900.

40. Heinrich böll stiftung Nigeria; Fabulous Urban. Urban Planning Processes in LAGOS: Policies, Laws, Planning Instruments, Strategies and Actors of Urban Projects, Urban Development, and Urban Services in Africa's Largest City. Available online: https:// ng.boell. org/ 2016/02/12/ urban-planning-processes-lagos (accessed on 28 April 2016).

41. Adelekan, I.O. Vulnerability of poor urban coastal communities to climate change in Lagos, Nigeria. In Proceedings of the 5th Urban Research Symposium (URS'95), Marseille, France, 28–30 June 2009; pp. 1–18.

42. Olaniyan, E.; Afisimama, E.; Oni, F.; Lawal, K.A. Simulating the daily evolution of West African monsoon using high resolution regional Cosmo-model: A case study of the first half of 2015 over Nigeria. *J. Climatol. Weather Forecast.* **2015**,*3*.

43. Balogun, I.A.; Balogun, A.A.; Adeyewa, Z.D. A note on the effect of urbanization on air temperature and humidity of Akure, Nigeria. In Proceedings of the 7th International Conference of Urban Climate, Yokohama, Japan, 29 June–3 July 2009.

44. Ayoade, J.O. A note on spatial and seasonal variations in wind speed over Nigeria. *Singap. J. Trop. Geogr.* **1980**, *1*, 11–15.

45. Balogun, A.; Olaleye, J.O.; Adekunle, R.S.; Okpara, J.N.; Kayanga, J. The effect of urbanization on air temperature and humidity in Akure. *AARCHES J.* **1997**, *1*, 89–93.

46. Oke, T.R. Conclusions and recommendations of the conference. In Proceedings of the Technical Conference on Urban Climatology and Its Applications with Special Regards to Tropical Areas, Mexico D.F., Mexico, 26–30 November 1986.

47. Taha, H. Urban climates and heat islands: Albedo, evapotranspiration, and anthropogenic heat. *Energy Build.* **1997**, *25*, 99–103.

48. Hsieh, C.; Davis, A.P.; Needelman, B.A. Nitrogen removal from urban stormwater runoff through layered bioretention columns. *Water Environ. Res.* **2007**, *79*, 2404–2411.

49. Figuerola, P.I.; Mazzeo, N.A. Urban-rural temperature differences in Buenos Aires. *Int. J. Climatol.* **1998**, *18*, 1709–1723.

50. Lagos State Government; Ministry of the Environment. Climate Change Policy, 2012–2014. Available online: http://developmentdiaries.com/wp-content/uploads/2013/03/Lagos-State-Climate-Change-Policy-First-Draft_27.03.2012.pdf (accessed on 28 April 2016).

Chapter 6

Influence of Northwest Cloudbands on Southwest Australian Rainfall

Nicola Telcik and Charitha Pattiaratchi

School of Civil, Environmental and Mining Engineering, The UWA Oceans Institute, The University of Western Australia, 35 Stirling Highway, Crawley, WA 6009, Australia

ABSTRACT

Northwest cloudbands are tropical-extratropical feature that crosses the Australian continent originating from Australia's northwest coast and develops in a NW-SE orientation. In paper, atmospheric and oceanic reanalysis data (NCEP) and Reynolds reconstructed sea surface temperature data were used to examine northwest cloudband activity across the Australian mainland. An index that reflected the monthly, seasonal, and interannual activity of northwest cloudbands between 1950 and 1999 was then created. Outgoing longwave radiation, total cloud cover, and latent heat flux data were used to determine the number of days when a mature northwest cloudband covered part of the Australian continent between April and October. Regional indices were created for site-specific investigations, especially of cloudband-related rainfall. High and low cloudband activity can affect the distribution of cloudbands and their related rainfall. In low cloudband activity seasons, cloudbands were mostly limited to the south and west Australian coasts. In high cloudband activity seasons, cloudbands penetrated farther inland, which increased the inland rainfall. A case study of the southwest Australian region demonstrated that, in a below average rainfall year, cloudband-related rainfall was limited to the coast. In an above average rainfall year, cloudband-related rainfall occurred further inland.

1. INTRODUCTION

The northwest cloudband (NWCB) is one of three tropical-extratropical cloudbands that crosses the Australian continent [1]. It originates from Australia's northwest coast (Figure 1(a)) and develops a NW-SE orientation, with no immediate connection to southern frontal systems [2]. Extratropical cloudbands are not unique to Australia; they also occur over Southern Africa [3], over South America, and in the Northern Hemisphere [4]. The Northern Hemisphere equivalent, the southwest (SW) cloudband, extends from Central America to the North American east coast and from Indochina to Japan and is most pronounced in winter [1]. Kuhnel [4] defined 14 tropical-extratropical cloudbands around the globe (seven in each hemisphere). The NWCB is the third most frequently occurring cloudband globally and the fourth most frequently occurring cloudband in the Southern Hemisphere.

The NWCB is a large cloud structure, which can reach a length of 8000 km from northwest (NW) to southeast (SE) [5]. It can connect a cold front at its SE end to the intertropical convergence zone at its NW end [6]. If the SE end reaches the Pacific Ocean, the NWCB can also connect to the South Pacific convergence zone [6].

Cloudbands are prominent in Australia during autumn and early winter when the ocean waters around Australia are warm and the subtropical jet stream is strong [1]. The decline in cloudbands in late winter and early spring has been attributed to the cooling of the surrounding oceans [5]. An interannual signal is also present with cloudbands, which may be related to the oceanic conditions in the cloudband formation region [5]. NWCBs are formed through the broad-scale ascent of moist, poleward, and tropical air by a midlatitude baroclinic and could transport latent heat and moisture into higher latitudes [1].

It has been estimated [1] that cloudband-related rainfall provides up to 80% of the annual rainfall for NW Australia and up to 40% of the annual rainfall for SW Australia. Recent positive trends in winter rainfall over northwest Western Australia and Central Australia have been attributed to increased northwest cloudband activity [7]. Southern Australia receives most of its cloudband-related rainfall through convective activity, such as cloudbands interacting with cold fronts [1]. Frontal systems produce most of the rainfall that occurs over SW Australia; however, the interaction of these systems with NWCBs increases the rainfall resulting from the frontal system [1]. Thus knowledge of NWCBs and their interactions with other weather phenomena is important for seasonal rainfall forecasting in Australia.

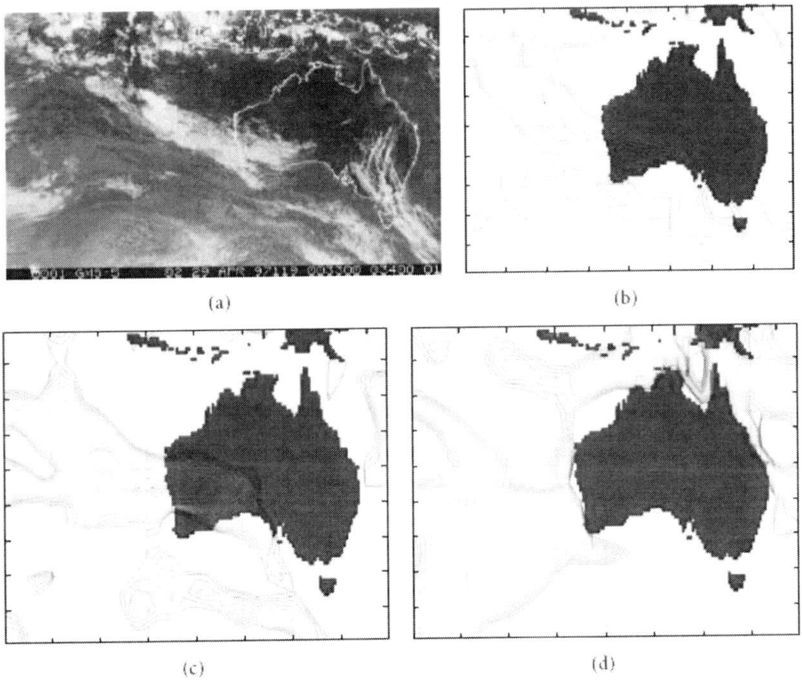

Figure 1. (a) Satellite image of a NWCB on 29 April 1996 (image source: Australian Bureau of Meteorology) and corresponding (b) OLR contours (140–250 W/m^2), (c) TCC contours (70–100%), and (d) latent heat flux contours (110–300 W/m^2).

Wright [1] defined "interaction" as the "capture of low-latitude moisture (north of 23°S), as represented by cloud masses in the circulation of a mid and low-latitude front or depression, and the subsequent amalgamation of mid and low-latitude cloud systems." An interaction is not defined until amalgamation of the mid- and low-latitude cloud systems has occurred or if there is evidence of cyclonic curvature at the SW end of the cloudband, where there are suspected cloudband-associated interactions [1].

1.1. Climate Influences on Northwest Cloudbands

The Indian Ocean is not an isolated water body. Teleconnections with other oceans through atmospheric and oceanic circulations affect the Indian Ocean (e.g., [8, 9]). Meehl [10] studied the precipitation, sea level pressure, and sea surface temperature (SST) in the Indo-Pacific to reveal a dynamically coupled ocean-atmosphere system. Variations in this system can affect NWCB formation; however, Webster et al. [11] found no connection between SST and the outgoing longwave radiation (OLR) to the NW of the NWCB formation region in the Indian Ocean. Webster et al.'s findings suggest that atmospheric dynamical processes or instabilities, not ocean-atmosphere interactions with warm SSTs, mainly produce the convection that occurs in this region.

Two direct influences on NWCB activity are the Madden-Julian oscillation (MJO) and the Indian Ocean dipole. Both phenomena have been observed in the NWCB formation region in the Eastern Indian Ocean. The El Niño southern oscillation also affects NWCBs through tropical forcing.

The MJO, or intraseasonal oscillation, occurs during fluctuations in tropical rainfall on the scale of 30–60 days [12, 13]. These fluctuations are visible in the wind, SST, precipitation, and cloudiness fields [3, 14]. The MJO moves east across the tropics towards the Australian continent every 30–50 days and can set up the synoptic conditions (winds, pressure, convection, and cloudiness) conducive to NWCB development. The MJO also has an influence on cloudbands across Southern Africa [3].

The Indian Ocean dipole (IOD), which occurs every four to five years, is an oscillation of opposing, anomalous SSTs off Sumatra and in the Western Indian Ocean, with accompanying wind and precipitation anomalies [15]. Strong coupling between the air-sea processes is unique to the Indian Ocean; however, this coupling is not completely independent of the El Niño southern oscillation [16]. Ummenhofer et al. [8] found that major droughts in southeast Australia were related to periods when the IOD remained persistently "positive" or "neutral" whilst during a negative phase of the

IOD, unusually wet conditions dominated across southern regions of Australia. The dipole mode index [15] is based on monthly SST anomalies between the Western Indian Ocean (50–70°E, 10–10°N) and the Eastern Indian Ocean (90–110°E, 10–0°S).

2. METHODOLOGY

2.1. Data

Several data sets were used in this study of NWCBs and their effects on Australian rainfall:(1)satellite images for 1997 and 1999 from the Australian Bureau of Meteorology,(2)National Centers for Environmental Prediction (NCEP) reanalysis data for 1950–1999 from the Climate Diagnostics Center, Colorado, including:(i)daily total cloud cover (TCC) and latent heat flux on a 2 × 2° grid,(ii)six-hour mean sea level pressure; wind fields at 250, 300, 400, 500, 600, and 700 hPa; geopotential heights at 500 and 700 hPa on a 2.5 × 2.5° grid,(3)daily interpolated OLR on a 2.5 × 2.5° grid for 1974–1998 from the Climate Diagnostics Center, Colorado,(4)Australian daily rainfall data for 1950–1999 from the Australian Bureau of Meteorology,(5)one-degree Australian daily rainfall data for 1950–1999 from the Australian Bureau of Meteorology Research Centre (Drosdowsky, pers. comm.); the gridded rainfall was based on the interpolation of the data obtained with the rainfall gauges,(6)Reynolds reconstructed mean monthly SST for 1950–1999 from the Climate Diagnostics Center, Colorado.

2.2. Northwest Cloudband Activity

Satellite data from April to October in 1997 and 1999 were used to analyse NWCBs over Australia. All the days between April and October that had some portion of a NWCB were collated. Cloudband activity days were isolated to determine the atmospheric conditions conducive to cloudband formation (see also [3]). Fifty-three days of cloudband activity were recorded and used for the composite of NWCB activity.

The wind fields, mean sea level pressure, and geopotential heights at 500 hPa over the Eastern Indian Ocean were used to identify the NWCBs; however, it was hard to identify NWCBs using these parameters because NWCBs have similar baroclinic features to a strong frontal system that crosses SW Australia, although the NWCBs did not appear to be connected to this southern frontal system [2]. This preexisting baroclinicity, which the NWCB moves through, is reinforced by the passage of a second cold front through SE Australia [2]. Bell [17] found that this baroclinic instability over Australia caused NWCBs. The OLR distributions were more useful for identifying NWCBs because NWCBs were visible in the OLR fields and distinguishable from other cloud features within the range of 140–250 W/m^2.

When the OLR distributions satisfied the prescribed NWCB criteria and some portion was located over the Australian continent the corresponding date was recorded as a "cloudband day" and the NWCB was defined as "mature." The NWCB index is a representation of cloudband days influencing the study region. Surges in tropical energy can reinforce NWCBs before the NWCBs dissipate [4], which makes it hard to discern whether a cloudband has dissipated and a new one has formed or whether the original cloudband persists. A cloudband's lifetime can vary considerably; thus it was more useful to include the number of days in which a NWCB was over part of Australia than how many cloudbands crossed over the continent.

The criteria for a cloudband day were defined from the studies of Tapp and Barrell [5], Kuhnel [18], and Wright [1] and included the dimensions, orientation, composition, and the formation region as follows:(1)a cloudband in a NW-SE orientation with a length greater than 2000 km and a width of at least 450 km; this orientation could range from NNW-SSE to WNW-ESE;(2)a generating or initiation region over the Eastern Indian Ocean (or Northern Australian waters east to 130° E);(3)a continuous cloud structure, the grid field was 2.5 × 2.5°, but the NWCB structure was much larger than the OLR or TCC grid fields and could be located as a continuity in these fields;(4)a section of the cloudband was located over the Australian continent.

Contour lines of equal OLR or TCC were used to detect NWCBs. These contour lines also had to meet the above criteria before the day was recorded as a cloudband day. The confirmed cloudband days were used to produce a NWCB index for 1974–1999, which was verified with satellite images from 1997 and 1999. To extend the index to 1950, the TCC was used as a proxy for the OLR.

A satellite image of a NWCB on 29 April 1996 and the corresponding OLR contours are shown in Figures 1(a)and 1(b), respectively. The TCC is presented as a percentage between 0 and 100 and treats the atmosphere as one layer; thus NWCBs can be difficult to locate because the cloudband's coherent structure can be obscured. More cloud activity was revealed over the Southern Indian Ocean with the TCC data than with the OLR data, which made it hard to distinguish a clear cloudband perimeter; however, when the latent heat flux data for the Southern Indian Ocean were used, most of this cloud "noise" was screened. The screen was based on a latent heat flux of >100 W/m^2 in the region of 80–105°E, 22.5–35°S; thus the TCC data for these grid points were removed.

Kuhnel [4] defined the cloudband formation region as 90–145°E; however, the analysis of satellite and OLR data showed that this region did not capture all the mature NWCBs present over Australia. The east end of this region (145°E) captured cloudbands that did not have the characteristic NW-SE orientation of a NWCB. These cloudbands were most likely the continental cloudbands that Wright [1] described. Therefore this study focused on NWCBs in the region of 80–130°E.

2.3. Northwest Cloudband Activity and Rainfall

Transects of the cloudband starting at 110°E and moving east were extracted to identify the NWCBs over Australia. The centre of a NWCB was defined as the local minimum in the OLR field and the local maximum in the TCC field and defining the OLR and TCC to the north and south of this point. When the values were within the OLR range (i.e., 140–250 W/m^2) and TCC range (60–100%) for a NWCB and generally decreased (OLR) or increased (TCC) from the local minimum, frontal disturbances that were not interacting with the cloudband showed a decrease in OLR and an increase in TCC. The OLR field (with NWCB coverage shaded) for 10 June 1998 is shown in Figure 2(a), and the corresponding land mask of cloudband activity is shown in Figure 2(b). In Figure 2(b), "1" denotes that a NWCB was recorded in the 2.5° grid, and "0" denotes that no NWCB was recorded in that grid.

Two data sets were obtained from the Australian Bureau of Meteorology to analyse the cloudband-related rainfall. The first data set was a daily gridded (1°) rainfall data set (Drosdowsky, pers. comm.), which was used for the large-scale analysis of the cloudband-related rainfall. The second data set consisted of daily rainfall station data. If part of a NWCB was detected in a particular grid, the rainfall that occurred in the corresponding grids was recorded as cloudband-related rainfall. The rainfall gauge data and cloudband-related rainfall data were collated for each month and each NWCB season (April–October) to study the spatial and temporal trends of the NWCBs. If NWCBs were detected in the 2.5° grid containing the gauge, any rainfall recorded that day was classified as cloudband-related. Cloudband-related rainfall data for the NWCB season were collated for each year and compared with the total April–October rainfall data for that gauge.

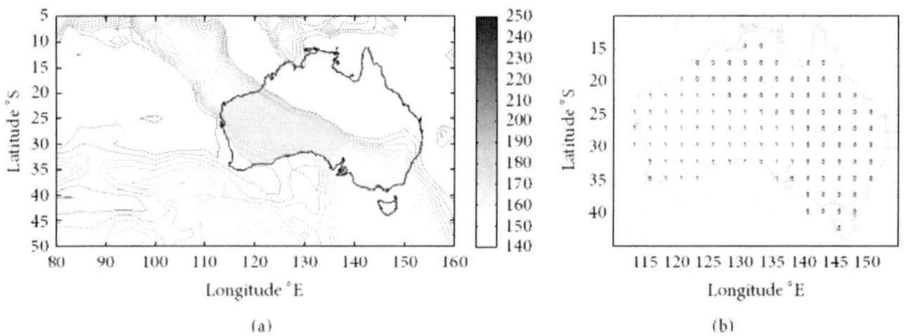

Figure 2. (a) The OLR field (with NWCB coverage shaded) for 10 June 1998 (OLR in W/m^2). (b) The land mask of NWCB activity across Australia used to create the regional NWCB indices. "1" denotes NWCB coverage for that grid.

3. RESULTS AND DISCUSSION

3.1. Northwest Cloudband Climatology

The synoptic conditions associated with a NWCB across Australia generally showed a high-pressure cell over Central and Southern Australia and another pressure cell over the Southern Indian Ocean at around 80°E. Downey et al. [2] also observed these high-pressure features. Between the two pressure cells, a deepening trough was located over the west Australian coast, which connected with a midlatitude depression off the SW coast (see Figure 3(a), which is based on a composite of 327 events of NWCBs between 1979 and 1999). High-pressure systems over the Indian Ocean and Central/SE Australia framed this tropical-extratropical connection of low pressures. A concentration of surface latent heat fluxes was also observed in the NWCB formation region (80–100°E, 10–20°S) (Figure 3(b)).

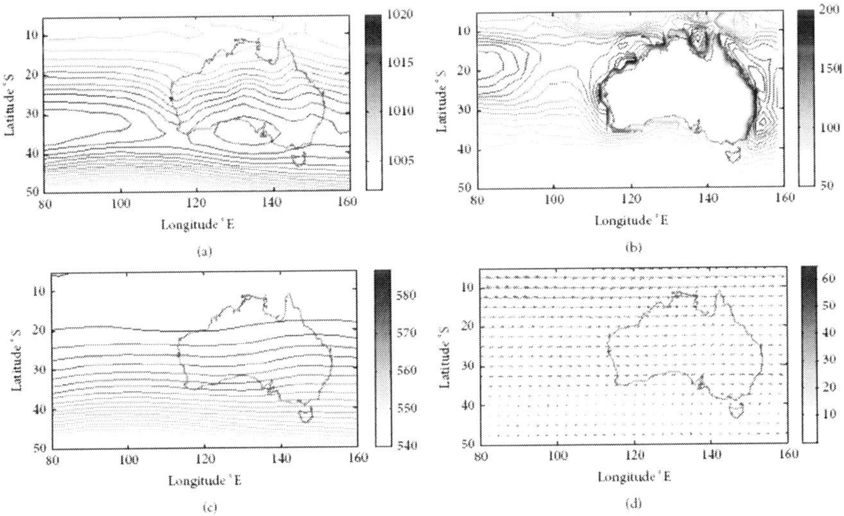

Figure 3. Composites of 327 NWCB events between 1979 and 1998 showing conditions associated with NWCB activity: (a) mean sea level pressure in hPa; (b) latent heat fluxes in W/m^2; (c) geopotential height at 500 hPa in dm; (d) upper-level wind fields at 500 hPa.

NWCBs are also associated with an upper-level trough in the Southern Indian Ocean at 80–100°E. This trough is not apparent in the composite of geopotential heights at 500 hPa in Figure 3(c) because of the longwave trough's differing amplitudes. The upper-level wind fields at 500 hPa showed a NW airstream over Australia from the NWCB formation region (Figure 3(d)), which was related to the upper-level trough. The upper-level winds formed a NW-SE "jet" near the NWCB formation region, along which the cloudband followed.

NWCB development is associated with the leading edge of a midlatitude upper-level trough. Downey et al. [2], Bell [17], and Tapp and Barrell [5] found that this upper-level trough, or cut-off low, was generally located off SW Australia. Wright [19] found that the trough's amplitude was correlated with the southern oscillation index (an indicator of the El Niño southern oscillation), an average winter Eastern Indian Ocean cloudiness, and average NW Victorian rainfall. The amplitude and position of this upper-level trough can vary; hence it is not well defined in Figure 3(d). Kuhnel [18] stated that not all upper-level troughs were well defined and that two out of ten upper-level troughs might have small amplitudes. We found that four out of ten upper-level troughs had a NW-SE orientation, which is characteristic of NWCBs.

3.2. A Northwest Cloudband Index

The OLR, TCC, and latent heat flux data were used to create a NWCB index for 1950–1999. The seasonal variability of the NWCB index is shown in Figure 4. The five-year, centred, weighted, running mean is included in the figure. Between April and October of 1950–1999, a mature cloudband was recorded over Australia on about 30 days. These NWCBs lasted from a day to more than a week (see Figure 5).

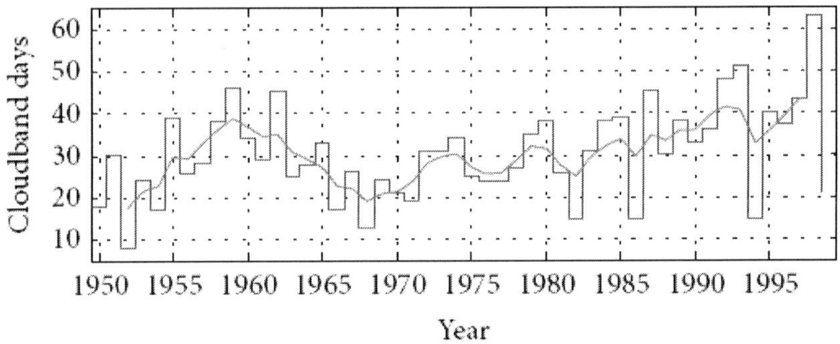

Figure 4. The NWCB index for Australia. The index represents the number of days in a season (April–October) in which a NWCB was located over part of Australia. A five-year, weighted, running mean was fitted to this index.

Several extreme NWCB events shown in Figure 4 were related to the El Niño southern oscillation. For example, NWCB activity was low in 1982 and 1986, which corresponded to El Niño years (1982-1983 and 1986-1987); however, NWCB activity was high in 1998, which corresponded to La Niña years (1998-1999). Indonesian throughflow-induced SST anomalies [20] in the NWCB formation region may have contributed to this relationship between the El Niño southern oscillation and NWCB activity.

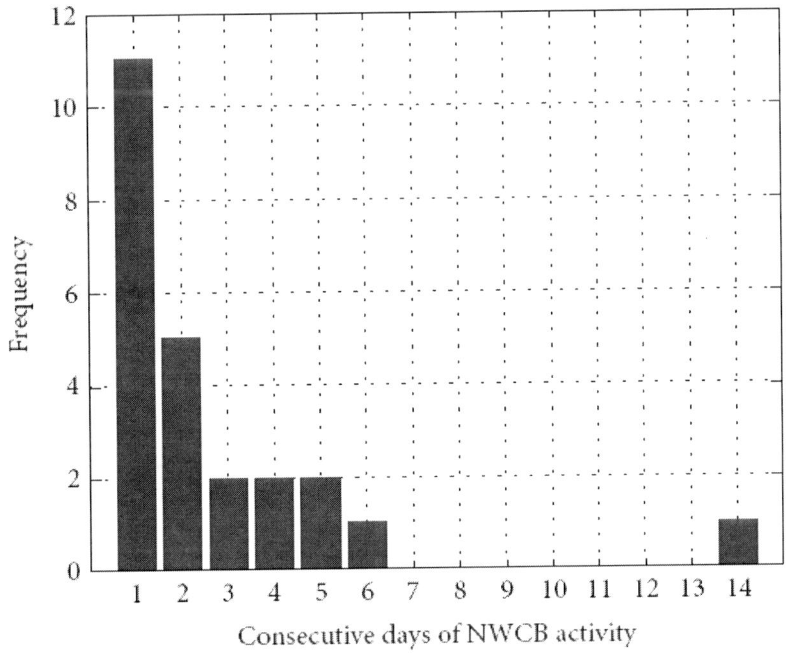

Figure 5. The frequency of individual NWCBs over Australia in 1998.

Kuhnel [18] showed that Western Australia experienced about 15 cloudband days a year and Central and South Australia experienced about nine cloudbands a year, which showed that cloudband activity over Australia decreased from west to east. The mean seasonal NWCB frequencies over Australia for 1979–1999 are shown in Figure 6. The figure shows that the NWCB activity had a NW-SE bias. The state average in this study was slightly higher than that in Kuhnel's [18] study. The high cloudband activities in 1992, 1993, and 1998 caused this increase in the average.

Hirst and Godfrey [21] showed that the SSTs in the Eastern Indian Ocean were linked to the Pacific Ocean via subsurface baroclinic waves. They found regions of convective mixing or strong upwelling that were susceptible to small changes in the throughflow. Variations in the throughflow affect the pattern and response time of the subsurface (and therefore surface) temperature. During El Niño, cool SSTs feed through the Indonesian throughflow into the Eastern Indian Ocean [22], whereas, during La Niña, warm SSTs feed into the Eastern Indian Ocean [22]. These SST anomalies were observed in the NWCB formation region.

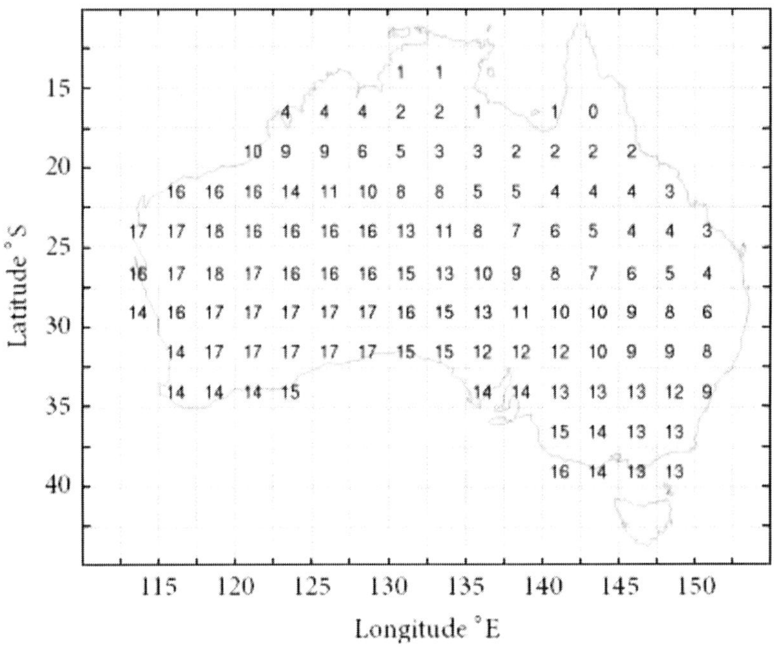

Figure 6. The mean seasonal (April–October) NWCB frequencies for 1979–1999.

The warmer SSTs during La Niña may increase cloudband activity through enhanced convection; conversely, the cooler SSTs during El Niño may decrease cloudband activity through reduced convection. Wright [1] showed that enhanced cloudband activity was associated with warm SST anomalies in the NWCB formation region and cool SST anomalies directly to the SW of this region. This reflects similar regions in an Indian Ocean dipole defined by Nicholls et al. [25].

3.3. Factors Influencing Northwest Cloudband Activity

Meehl [23] showed that SST anomalies on the scale of a season, associated with wind anomalies, affected air-sea interactions (especially evaporation) and therefore the latent heat release. Strong winds and warm SSTs increase convection and evaporation and decrease the sea level pressure. These actions, which are characteristic of monsoons, encourage NWCB formation. Weak winds then decrease evaporation and increase the sea level pressure the following year. This quasi-biennial feature was sometimes observed in the NWCB index (1985 to 1990 in Figure 4). Meehl [24] noted that a tropospheric biennial mechanism involving air-sea interactions was connected with the South Asian or Indian monsoon. This biennial mechanism is quasi-biennial, occurring intermittently or over several years (also apparent in Figure 4 as a noncontinuous biennial feature).

The SST anomalies in the NWCB formation region of 90–100°E, 10–15°S, in the Eastern Indian Ocean, are referred to as GEN. The correlation between NWCB activity and GEN was 0.61 in April—one of the highest NWCB activity months. The atmospheric dynamical processes might have caused this correlation, as the SST and OLR in a region next to the NWCB formation region were uncorrelated [11].

Kuhnel [4] found a correlation of −0.36 between NWCB activity and the southern oscillation index (SOI); however, this correlation was based on a slightly different study region and a single year of cloudband activity. The April–October NWCB activity accounted for about two-thirds of the cloudband activity that Kuhnel [4] described. The correlation between NWCB activity and the southern oscillation index in April was −0.52 (Table2). This inverse relationship suggested that a positive southern oscillation index in April was linked to less NWCB activity and a negative southern oscillation index was linked to more NWCB activity. The SST anomalies in the eastern equatorial Pacific Ocean (Nino12: 80–100°W, 5°S–5°N) in April were also related to NWCB activity (correlation of 0.66). These results showed that the El Niño southern oscillation in April was linked to cloudband activity in the eastern Indian Ocean.

Kuhnel [4] defined cloudband activity as a function of the southern oscillation index, which determines the strength and position of the main convective regions. Fluctuations in convective regions affect NWCB formation. The NWCB formation region is located where the 40–50 day oscillation (i.e., the MJO) has been observed.

Tapp and Barrell [5] and Kuhnel [4] stated that the NWCB season was between April and October. The results showed more cloudbands were usually present in the first half of the season (April–July). The mean monthly NWCB days for 1950–1999 are shown in Table 1. Cloudband activity generally peaked in May and then decreased over the following months. These results agreed with those of Kuhnel [4] and Wright [1]. Cloudband activity occurred in March but was more irregular and not usually associated with rainfall. Cloudbands present in the warmer months (November–March) are associated with monsoonal depressions or tropical cyclones NW of Australia [4] and are not considered NWCBs.

Table 1. Mean monthly (April–October) NWCB days over Australia for 1950–99.

	Mar.	Apr.	May	June	July	Aug.	Sep.	Oct.
NWCB days	5.5	7.9	7.9	5.9	3.7	2.3	1.0	1.8

Table 2. The maximum monthly cross-correlations between NWCB activity and the southern oscillation index (SOI), GEN, the dipole mode index (DMI), the Madden-Julian oscillation (MJO), and Nino12.

Index with NWCB	Correlation (r)	Correlation month
SOI	−0.52	Apr.
GEN	0.61	Apr.
DMI	−0.60	July
MJO	−0.48	July
Nino12	0.66	Apr.

The duration of a NWCB over Australia varies but can sometimes be up to two weeks (e.g., the NWCB that occurred in 1998, which appeared on 30 March and lasted until 12 April; see Figure 4). If the synoptic conditions are baroclinic, NWCBs can appear consecutively for more than a week. If the baroclinicity is reduced, a cloudband may still form but will usually dissipate after 24 hours. The frequency of NWCB days over Australia between April and October 1998 is shown in Figure 4. The synoptic conditions in April, which favoured cloudband activity, resulted in fourteen days of cloudband activity, followed by more cloudband activity less than a week later. Towards the end of the NWCB season, conditions were unfavourable for cloudband activity and cloudbands tended to dissipate after a day.

An inverse relationship between NWCB activity and the MJO is shown in Figure 7(a). The MJO index was created using OLR anomalies in the region of 90–100°E, 5°S–5°N. The MJO mainly occurs to the NW of the NWCB formation region, but also crosses into the NWCB formation region [14]. An increase in MJO activity was linked to a decrease in NWCB activity over Australia and more cloudiness (hence lower OLR) in the eastern Indian Ocean. The relationship between NWCB activity and the MJO was strongest in July (correlation of −0.48; see Table 2).

The NWCB formation region overlaps with the eastern part of the Indian Ocean dipole region. An inverse relationship was found between NWCB activity and the Indian Ocean dipole. The dipole mode index measures the difference in SST between the West and East Indian Ocean and is used to measure the Indian Ocean dipole. The Indian Ocean dipole event of 1994 coincided with one of the lowest recorded NWCB activity seasons. Low NWCB activity also coincided with a peak in warm SSTs in the Western Indian Ocean and cool SSTs in the Eastern Indian Ocean in 1982; this year also corresponded with the 1982-1983 El Niño event. The relationship between NWCB activity and the Indian Ocean dipole was strongest in July (correlation of −0.60; see Table 2).

3.4. Australian Rainfall and Northwest Cloudbands

NWCBs produce high rainfall amounts in Australia between April and October. In some inland regions where the annual rainfall is low (e.g., <200 mm/year), the contribution of >50 mm of rainfall from a cloudband can be significant. The interactions of NWCBs with other features, such as troughs, fronts, and convective activity, can also generate rainfall.

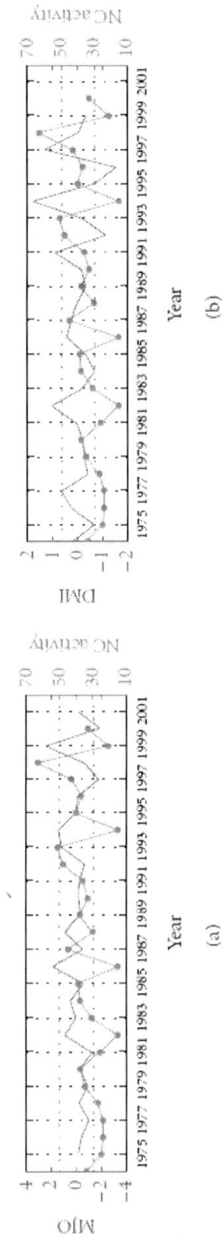

Figure 7. (a) The mean MJO anomaly and seasonal (April–October) NWCB frequency for 1975–1999 (MJO in W/m^2, based on OLR values) and (b) the mean dipole mode index anomaly and seasonal (April–October) NWCB frequency for 1975–1999. The dipole mode index represents the temperature anomaly difference (°C).

Two years of extreme cloudband activity, 1982 and 1998, were chosen for analysis. Cloudband activity was low in 1982, with only 15 NWCB days recorded. The cloudbands were mostly limited to the west and south coasts, with few NWCBS recorded in Central Australia (Figure 8(a)). Cloudband-related rainfall was also mostly limited to coastal areas, especially the west coast (Figure 8(b)). Most of the inland regions received less than 30 mm of cloudband-related rainfall between April and October.

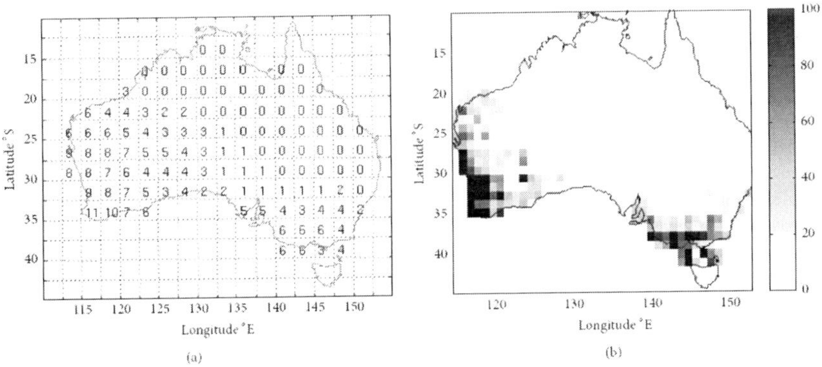

Figure 8. The (a) frequency of NWCB activity and (b) NWCB-related rainfall for 1982 (a low cloudband activity year).

Cloudband activity was high in 1998, with 63 NWCB days recorded. The cloudband coverage was more widespread in 1998 (Figure 9(a)) than it was in 1982. The cloudband contribution to inland rainfall, especially over Western Australia, is shown in Figure 9(b). Some of the inland rainfall stations received 65% of their April–October rainfall from cloudband-related activity (see next section). Most of the rainfall stations in Western Australia received >100 mm of cloudband-related rainfall between April and October, which is a significant amount for the inland stations.

3.5. Case Study: Southwest Australian Rainfall

Winter rainfall in SW Australia has declined more than 25% since the 1950s [25]. The decline has affected farming and other industries in coastal and inland regions. Complex interactions between atmospheric and oceanic circulations affect rainfall in the SW, with cold fronts producing most of the winter rainfall. NWCBs and convective activity also generate winter rainfall, but to a lesser degree [26].

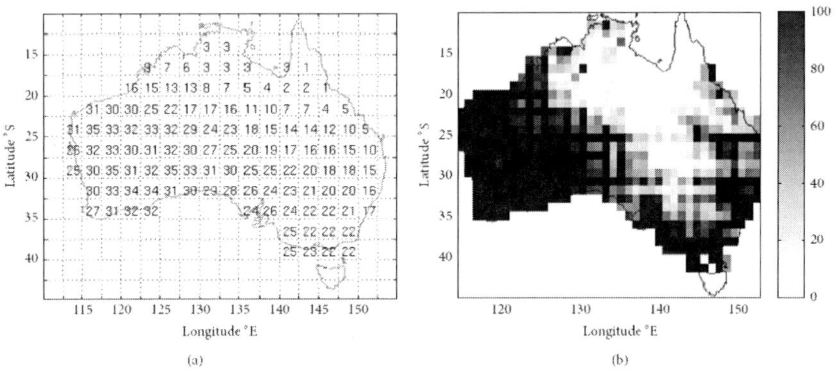

Figure 9. The (a) frequency of NWCB activity and (b) NWCB-related rainfall for 1998 (a high cloudband activity year).

Wright [19] and Kuhnel [18] discussed the importance of NWCBs as rainfall generators. Cloudbands can cause heavy rainfall, especially when they interact with an extratropical front (e.g., the cloudband activity in NW Victoria discussed in [19]); however, this heavy rainfall is irregular. A study of the rainfall gauge data associated with cloudband activity revealed this irregularity. The SW rainfall stations were chosen from the high-quality data set that Lavery et al. [27] described. Of these stations, rainfall gauges that were spread evenly over the SW and were 100% complete for the study period (1950–1999) were selected. The high-quality rainfall stations for the SW Australian region are shown in Figure 10, with the shaded circles denoting the stations chosen for cloudband analysis. These nine rainfall stations, along with their latitudes and longitudes, are listed in Table 3.

Table 3. Details of the SW rainfall stations.

Station number	Station name	Latitude (°S)	Longitude (°E)
8147	Yuna	28.3261	114.9572
9519	Cape Naturaliste	33.5381	115.0183
9557	Hopetoun P.O.	33.9511	120.1247
9564	King River	34.9444	117.9228
9616	Westbourne	34.0928	116.6603
10092	Merredin	31.4769	118.2778
10592	Lake Grace P.O.	33.1017	118.4608
10795	Avondale Research Station	32.1189	116.8664
12052	Menzies	29.6933	121.0286

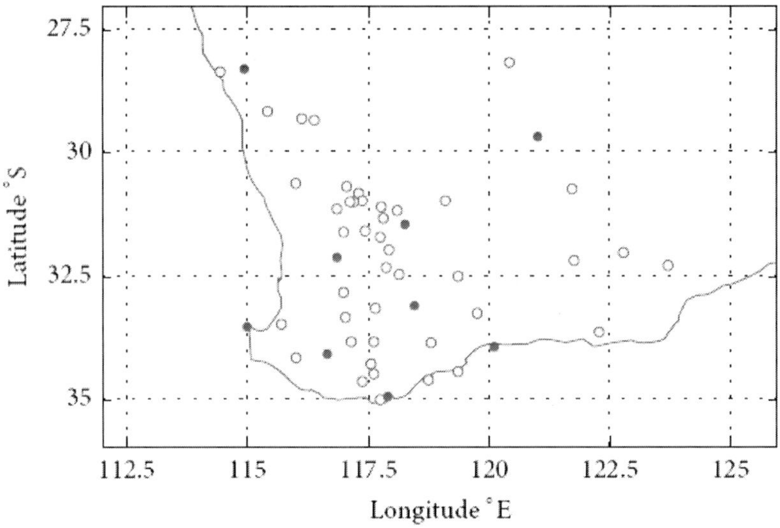

Figure 10. The locations of high-quality rainfall stations for the SW Australian region (courtesy of the Australian Bureau of Meteorology). The shaded circles denote the nine rainfall stations used in this study.

When NWCBs cross SW Australia and interact with frontal systems, heavy, widespread rainfall can occur. Cloudbands, and their interactions with cold fronts, account for up to 25% of the SW winter rainfall [1]. The results from this study showed cloudband-related rainfall was sometimes as high as 65%, as seen with station 12052 (Menzies) in 1998 (Figure 10). The cloudband-related rainfall, however, was highly variable. For Menzies, for example, the percentage of cloudband-related rainfall was sometimes less than 5% of the total April–October rainfall.

Over SW Australia, fronts provide a form of uplift for NWCBs. The rainfall produced as a result of cold fronts is much less than that produced from the interaction of cold fronts with NWCBs. Cloudband-related rainfall can exceed 75 mm a day, whereas frontal activity may produce only 25 mm of rainfall a day. Kuhnel [18] and Wright [1] showed the rain-producing capacity of a cloudband decreased towards the south and east. In SW Australia, the interaction of cold fronts with NWCBs produced most of the cloudband-related rainfall. In the northern regions (e.g., the Gascoyne-Murchison region), cloudband-related rainfall may occur without these interactions.

The seasonal (April–October) NWCB-related rainfall data for 1975–1999 for the nine stations listed in Table 3are shown in Figure 11. The coastal stations generally recorded higher rainfall than the inland stations. There was no corresponding upward trend in the rainfall compared with the upward trend in cloudband activity. Between 1975 and 1999, unusually heavy rainfall was recorded in 1980, 1988, and 1998, with 1988 and 1998 corresponding to La Niña years.

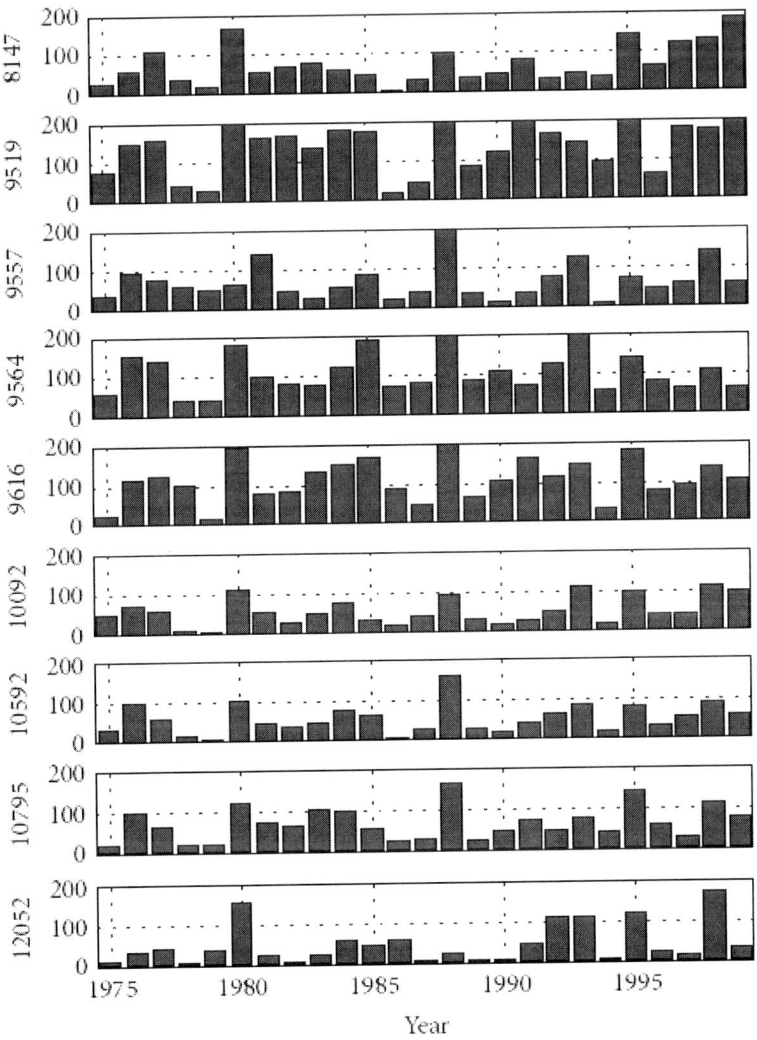

Figure 11. The seasonal (April–October) NWCB-related rainfall for 1975–1999 recorded at the nine SW rainfall stations.

The NWCB-related rainfall contributions to the total April–October rainfall varied seasonally and spatially (Figure 12). Years that seem to have no cloudband contribution to rainfall (i.e., 0%) were due to missing seasonal data. Most of the rainfall stations that recorded high rainfall amounts were located near the coast (e.g., station 9557 in 1980). Cloudbands and their interactions with meteorological features contributed to 15–30% of the April–October rainfall. In an extreme year (e.g., 1980), up to 75% of the seasonal rainfall could be attributed to cloudbands and their interactions with meteorological features.

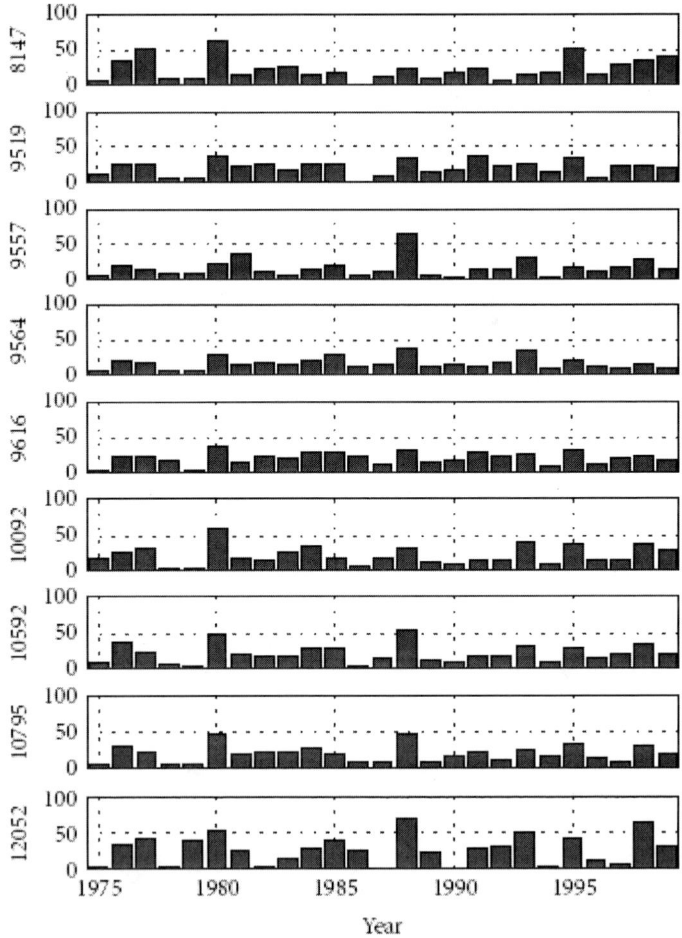

Figure 12. Percentages of NWCB-related rainfall contributions to the total April–October rainfall for 1975–1999 for the nine SW rainfall stations.

4. CONCLUSIONS

This study examined NWCB activity over Australia from 1975 to 1999 and an index representing cloudband activity over Australia between 1974 and 1999 was created and dissected to create regional indices, which showed the NWCB activity had a NW-SE bias. Atmospheric and oceanic influences on cloudband activity were examined and revealed that NWCB activity was related to the MJO, Indian Ocean dipole, El Niño southern oscillation, and SSTs in the NWCB formation region. These influences were not independent; thus NWCB activity over Australia was hard to predict.

The NWCB-related rainfall data were collated from the regional indices to show the cloudband-related rainfall for each 2.5° grid point across Australia. Most cloudband-related rainfall occurred in coastal regions, where the interaction of NWCBs with other features, such as frontal disturbances, increased the rainfall. Cloudband-related rainfall was also found to be spatially and temporally variable.

ACKNOWLEDGMENTS

The Gascoyne-Murchison Strategy and the Climate Variability in Agriculture Program funded this research. The authors would like to thank the Climate Diagnostics Centre, Boulder, for kindly providing the Reynolds reconstructed SST and reanalysis data, and the Australian Bureau of Meteorology for their rainfall data sets and processed GMS-5 satellite images (courtesy of the Japan Meteorological Agency). They would also like to thank the following people for their advice and guidance with this research: Dr. Neville Nicholls and Dr. Wasyl Drosdowsky, Australian Bureau of Meteorology Research Centre; Dr. William Wright, NCC; John Cramb and Joseph Courtney, Australian Bureau of Meteorology; Dr Gary Meyers, CSIRO Marine Research; Dr. Ian Smith, CSIRO Division of Atmospheric Research; Professor Greg Ivey, UWA; and Alisa Krasnostein, UWA.

REFERENCES

1. W. J. Wright, "Tropical-extratropical cloudbands and Australian rainfall: I. Climatology," International Journal of Climatology, vol. 17, no. 8, pp. 807–829, 1997.

2. W. K. Downey, T. Tsuchiya, and A. J. Schreiner, "Some aspects of a northwestern Australian cloudband,"Australian Meteorological Magazine, vol. 29, no. 3, pp. 99–113, 1981.

3. N. C. G. Hart, C. J. C. Reason, and N. Fauchereau, "Cloud bands over southern Africa: seasonality, contribution to rainfall variability and modulation by the MJO," Climate Dynamics, vol. 41, no. 5-6, pp. 1199–1212, 2013.

4. I. Kuhnel, "Tropical-extratropical cloudband climatology based on satellite data," International Journal of Climatology, vol. 9, no. 5, pp. 441–463, 1989.

5. R. G. Tapp and S. L. Barrell, "The north-west Australian cloud band: climatology, characteristics and factors associated with development," Journal of Climatology, vol. 4, no. 4, pp. 411–424, 1984.

6. B. Geerts and E. Linacre, "Northwest cloud bands," University of Wyoming, 1998, http://www-das.uwyo. edu/~geerts/cwx/notes/chap08/nw_cloud.html.

7. J. S. Frederiksen and C. S. Frederiksen, "Twentieth century winter changes in Southern Hemisphere synoptic weather modes," Advances in Meteorology, vol. 2011, Article ID 353829, 16 pages, 2011.

8. C. C. Ummenhofer, M. H. England, P. C. McIntosh et al., "What causes southeast Australia's worst droughts?" Geophysical Research Letters, vol. 36, no. 4, Article ID L04706, 2009.

9. A. Hoell, M. Barlow, and R. Saini, "Intraseasonal and seasonal-to-interannual indian ocean convection and hemispheric teleconnections," Journal of Climate, vol. 26, no. 22, pp. 8850–8867, 2013.

10. G. A. Meehl, "The annual cycle and interannual variability in the tropical Pacific and Indian Ocean regions," Monthly Weather Review, vol. 115, no. 1, pp. 27–50, 1987.

11. P. J. Webster, V. O. Magaña, T. N. Palmer et al., "Monsoons: processes, predictability, and the prospects for prediction," Journal of Geophysical Research C: Oceans, vol. 103, no. 7, pp. 14451–14510, 1998.

12. R. A. Madden and P. R. Julian, "Observations of the 40–50-day tropical oscillation—a review," Monthly Weather Review, vol. 122, no. 5, pp. 814–837, 1994.

13. A. C. Subramanian, M. Jochum, A. J. Miller, R. Murtugudde, R. B. Neale, and D. E. Waliser, "The Madden-Julian oscillation in CCSM4," Journal of Climate, vol. 24, no. 24, pp. 6261–6282, 2011.

14. P. J. Webster, "The Asian-Australian monsoon system: a prospectus for a research program. Draft recommendations from the St Michaels, Maryland, monsoon workshop July 28-30, 1998," U.S. Climate Variability and Predictability Research Program (CLIVAR) Working Group Report, 1998.

15. N. H. Saji, B. N. Goswami, P. N. Vinayachandran, and T. Yamagata, "A dipole mode in the tropical Indian ocean," Nature, vol. 401, no. 6751, pp. 360–363, 1999.

16. S. Sankar, M. R. R. Kumar, and C. Reason, "On the relative roles of El Nino and Indian Ocean Dipole events on the monsoon onset over Kerala," Theoretical and Applied Climatology, vol. 103, no. 3-4, pp. 359–374, 2011.

17. I. D. Bell, Extratropical cloud systems: satellite interpretation and three-dimensional airflows [Ph.D. thesis], Monash University, Melbourne, Australia, 1982.

18. I. Kuhnel, "Tropical-extratropical cloudbands in the Australian region," International Journal of Climatology, vol. 10, no. 4, pp. 341–364, 1990.

19. W. J. Wright, Synoptic and climatological factors influencing winter rainfall variability in Victoria [Ph.D. thesis], The University of Melbourne, Melbourne, Australia, 1987.

20. J. S. Godfrey, "The effect of the Indonesian throughflow on ocean circulation and heat exchange with the atmosphere: a review," Journal of Geophysical Research C: Oceans, vol. 101, no. 5, pp. 12217–12237, 1996.

21. A. C. Hirst and J. S. Godfrey, "The response to a sudden change in Indonesian throughflow in a global ocean GCM," Journal of Physical Oceanography, vol. 24, no. 9, pp. 1895–1910, 1994.

22. G. Meyers, "Variation of Indonesian throughflow and the El Niño-Southern Oscillation," Journal of Geophysical Research C: Oceans, vol. 101, no. 5, pp. 12255–12263, 1996.

23. G. A. Meehl, "A coupled air-sea biennial mechanism in the tropical Indian and Pacific regions: role of the ocean," Journal of Climate, vol. 6, no. 1, pp. 31–41, 1993.

24. G. A. Meehl, "Coupled land-ocean-atmosphere processes and South Asian monsoon variability," Science, vol. 266, no. 5183, pp. 263–267, 1994.

25. N. Nicholls, L. Chambers, M. Haylock, C. Frederiksen, D. Jones, and W. Drosdowsky, Climate Variability and Predictability for South-West Western Australia: Phase 1 Report to the Indian Ocean Climate Initiative, Bureau of Meteorology Research Centre, Melbourne, Australia, 1999.

26. J. Gentilli, Australian Climate Patterns, Thomas Nelson, Melbourne, Australia, 1972.

27. B. Lavery, G. Joung, and N. Nicholls, "An extended high-quality historical rainfall dataset for Australia,"Australian Meteorological Magazine, vol. 46, no. 1, pp. 27–38, 1997.

Chapter 7

On the Differences in the Intraseasonal Rainfall Variability between Western and Eastern Central Africa: Case of 10–25-Day Oscillations

Alain Tchakoutio Sandjon,[1,2] Armand Nzeukou,[1] Clément Tchawoua,[3] and Tengeleng Siddi[1,4]

[1]Laboratory of Industrial Systems and Environmental Engineering, Fotso Victor Technology Institute, University of Dschang, Cameroon

[2]Laboratory for Environmental Modeling and Atmospheric Physics, Department of Physics, Faculty of Sciences, University of Yaoundé 1, P.O. Box 812, Yaoundé, Cameroon

[3]Laboratory of Mechanics, Department of Physics, Faculty of Science, University of Yaoundé 1, Cameroon

[4]Higher Institute of Sahel, University of Maroua, Cameroon

ABSTRACT

In this paper, we analyze the space-time structures of the 10–25 day intraseasonal variability of rainfall over Central Africa (CA) using 1DD GPCP rainfall product for the period 1996–2009, with an emphasis on the comparison between the western Central Africa (WCA) and the eastern Central Africa (ECA) with different climate features. The results of Empirical Orthogonal Functions (EOFs) analysis have shown that the amount of variance explained by the leading EOFs is greater in ECA than WCA (40.6% and 48.1%, for WCA and ECA, resp.). For the two subregions, the power spectra of the principal components (PCs) peak around 15 days, indicating a biweekly signal. The lagged cross-correlations computed between WCA and ECA PCs time series showed that most of the WCA PCs lead ECA PCs time series with a time scale of 5–8 days. The variations of Intraseasonal Oscillations (ISO) activity are weak in WCA, when compared with ECA where the signal exhibits large annual and interannual variations. Globally, the correlation coefficients computed between ECA and WCA annual mean ISO power time series are weak, revealing that the processes driving the interannual modulation of ISO signal should be different in nature or magnitude in the two subregions.

1. INTRODUCTION

The monitoring and prediction of climate in the tropics remain a crucial problem in the scientific community. It is well-known that many regions in the tropics are vulnerable to climate change because their resources are highly rainfall dependent. Amongst these regions, Central Africa (CA) is particularly vulnerable because the majority of its population is rural and practice rain-fed agriculture [1]. A strong rainfall intensity can result in devastating floods, whereas a weak rainfall is usually associated with droughts, thus affecting living conditions and the economy of the densely populated region in many sectors such as agriculture, livestock, and energy [2,3].

The CA extends from 15°S to 15°N and 0–50°E mainly over the land and part of Atlantic and Indian Oceans on its edges (Figure 1). The topography of the region is quite various, including highlands, mountains, and Plateaus. The western part (15°S–15°N; 0–30°E) is consisting of the zones of intense precipitation, especially over the Congo Basin [4]. Some of the highest rainfall totals are reported over Mountain Cameroon, at its western edge, where mean annual rainfall exceeds 10 meters. The Congo basin was proven to experience one of the world's most intense thunderstorms and highest frequency of lightning flashes [5, 6]. The western central Africa is almost covered by the Congo forest, which keep this region quite wet within the year. The Eastern part (15°S–15°N; 30–50°E) is characterized by widely diverse climates ranging from desert to forest over relatively small areas. The complex topography of the region is an important contributing factor to climate because it leads to the orographically induced rainfall and affects local climates [7]. This region has suffered both excessive and deficient rainfall in recent years [8, 9]. In particular, the frequency of anomalously strong rainfall causing floods has increased. The difference between the east and west boundaries of CA is of approximately 3000 m in surface elevation and 3000 mm in annual mean rainfall. Then the two subregions (ECA and WCA), separated from each other by the Rift Valley, are very different in terms of topography, surface conditions, and precipitation (Table 1).

Table 1. Comparison of some geographical features between ECA and WCA.

	Western Central Africa (west of Rift Valley)	Eastern Central Africa (east of Rift Valley)
Average topography	Lower than 700 m	Greater than 1500 m
Vegetation	Almost covered by the Congo forest	From desert to small forests over relatively small areas
Mean annual rainfall	Greater than 1600 mm	Lower than 700 mm
Borders	Rift Valley-Atlantic Ocean	Rift Valley-Indian Ocean

Rainfall variability in the tropics is complex, including different timescales ranging from few hours to a few months. Before the 1970s, the lack of high temporal resolution datasets leads many authors to the study of the rainfall fluctuations in the tropics in terms of annual cycle and interannual variability. The processes that induce rainfall variability within the season could not be investigated because it requires a high temporal resolution data. But since the beginning of the 1970s, the emergence of satellite data produced capability to estimate regional or global rainfall fields at higher resolution, from the satellite measurements. Therefore, the notion of intraseasonal variability (ISV) arose in the scientific community and many authors used different rainfall or rainfall proxy datasets to study ISV in the tropics [10, 11].

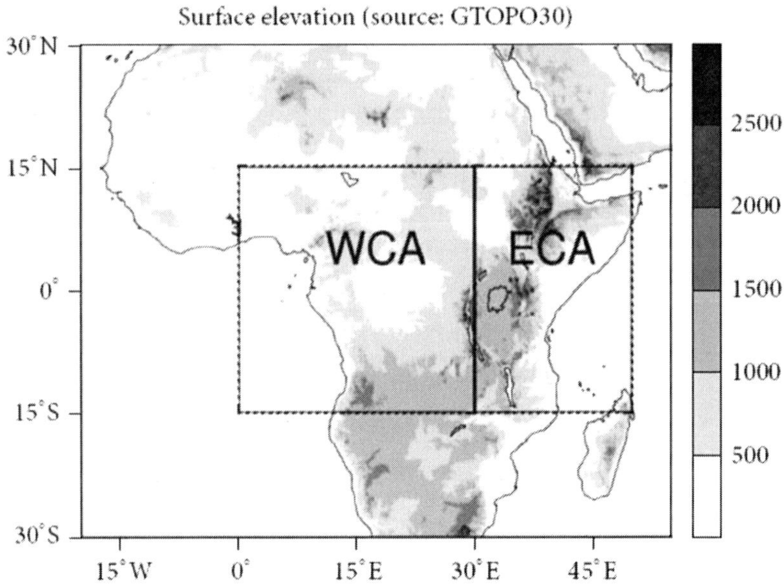

Figure 1. Surface elevation over the study area based on 30 min topographic data (m) from Digital Elevation Model (DEM) of the US Geological Survey. The study domain (15°S–15°N 0–50°E) and the division line between WCA and ECA are shown on the plot.

Some of these authors have studied the ISV in precipitation for different geographical regions of Africa. For example, significant spectral peaks around 15- and 40-day period were found for Sahel precipitation by [12]. Analysis was made for Africa and signals over 10–25- and 25–60-day period were found in convection and precipitation in the western region [12–14]. [15] using outgoing long-wave radiation data developed climatology for tropical intraseasonal convective anomalies. [16] investigated the influence of the Madden-Julian oscillation on East African rainfall and showed that MJO phases leading to wet spells in the eastern (coastal) region are often those associated with overall suppressed deep convection in the Africa/Indian Ocean region. [17] have investigated the spatial and temporal structure of intraseasonal oscillations over Central Africa using 21 years (1980–2000) of pentad (5 day) rainfall and low level wind and they found three dominant modes for intraseasonal rainfall with the period of around 40–50 days. [18] used principal components analysis and Satellite-Derived

Rainfall data to show that major cycles of intraseasonal rainfall events over Southern Africa are around 20 and 40 days.

Our understanding of the climate processes in CA is limited. Indeed, the literature on African climate variability is strongly biased towards West Africa (including the Sahel), East Africa, and Southern Africa. Thus, the CA region represents a notable gap in the study of the tropical climate system. Significantly, the structure of the intraseasonal oscillations (ISO) in Central Africa is only discussed by few authors [4, 17, 19, 20]. More recently, [19] used one-Degree Daily Global Precipitation Climatology Project (1 DD GPCP) higher resolution rainfall product to study the ISO patterns in CA, but they considered the whole area (15°S–15°N, 0–50°N) and did not take into account the contrast patterns between western and eastern parts of the region with different climate features.

Reference [4] also considered the whole area, and they showed that the East African modes (Ethiopian and Tanzanian modes) exhibit strong interannual variations, when compared with West African mode (Congo mode). Then the major issue of this study is the comparison of intraseasonal oscillations (ISO) patterns between western Central Africa (WCA) and eastern Central Africa (ECA). From the difference between physical features of ECA and WCA, one central question we may ask is "what could be the difference in the ISO patterns between WCA and ECA?" Some of the previous studies revealed two dominant frequency bands (10–25 and 25–60 days) at the intraseasonal timescale in tropical Africa [12–14], but for this study, we focused on the case of 10–25-day band. We use simple mathematical tools to assess the detailed spatial and temporal patterns of the leading modes of ISV of rainfall over the region, both throughout the year and from year to another.

In the next section, the data and method of analysis will be described. Section 3 will present the main results obtained and their analysis. Finally, Section 4 is devoted to a few discussions and conclusions.

2. DATA AND METHODS

For this study, we used the 1DD GPCP rainfall data. The GPCP algorithm combines precipitation estimates from several sources, including infrared (IR) and passive microwave (PM) rain estimates and rain gauge observations [22, 23]. The IR data came mainly from the different

Geostationary Meteorological Satellites but data from polar-orbiting satellites were also used to fill in the gaps at higher latitudes. The IR based estimates used the Geostationary Operational Environmental Satellite (GOES) precipitation index (GPI) described by [24]. The microwave data come mainly from the Special Sensor Microwave Imager (SSM/I) onboard the Defense Meteorological Satellite Program. The PM estimates were used to adjust the GPI estimate. Then, the multisatellite estimate was adjusted towards the large-scale gauge average for each grid box. The gauge-adjusted multisatellite estimates were then combined with gauge analysis using a weighted average, where the weights are the inverse error variances of the respective estimates. The current products include a monthly analysis at 2.5° × 2.5° longitude-latitude grids, a 5-day (pentad) analysis at the same spatial resolution, and a daily product at a special resolution of one degree. The one-degree daily (1DD) product does not use PM rain estimates and gauge measurements directly [22]. SSM/I data were used within the framework of the threshold-matched precipitation index (TMPI) to delineate rain areas in the IR data. Gauge data were involved indirectly when the 1DD product was scaled so that monthly accumulations of 1DD matched the monthly GPCP product. The monthly and pentad analyses extend from 1979 to current, while the daily product is available starting from October 1996. The daily products are made available 2 to 3 months after the end of each month. The product used in this paper is the one-degree daily precipitation data available on the NOAA website http://www.esrl.noaa.gov/psd/data/gridded/data.gpcp.html. After some global and regional validations [25–27], the GPCP rainfall estimates are actually being used widely in place of gauge observations or to supplement gauge observations in the climate variability studies.

In the atmospheric sciences, climate is affected by the many phenomena with different spatial and temporal scales. To study the climate variability within a given scale, we generally use a digital filter to extract only the desired frequencies in a data time series. The Lanczos filtering [28, 29] used in this study is one of Fourier methods of filtering digital data. Its principal feature is the reduction of the amplitudes of Gibbs oscillation. Fourier coefficients for the smoothed response function are determined by multiplying the original weight function by a function that Lanczos called "sigma factors" (1):

$$\overline{w}_k = \frac{\sin 2\pi k f_c}{\pi k} \cdot \frac{\sin \pi k/n}{\pi k/n}, \tag{1}$$

where the first fraction is the original weight function and the second is Lanczos factor.

The wavelet analysis (WA) is a time series analysis method that has increasingly been applied in geophysics during the last three decades. It is becoming a common tool for analyzing temporal variations of power within a time series. The transformation in time series from the time space into the time-frequency space show that the WA is able to determine both the dominant timescales of variability and how they vary with time. WA has several attractive advantages over the traditional Fourier analysis, especially when dealing with time series with time-varying amplitudes. In contrast, to Fourier transform, that generate values of amplitudes and phases averaged over the entire time series for each frequency component or harmonic, wavelet transform provide a localized instantaneous estimate of the amplitude and phase for each spectral component of the series. This gives WA an advantage in the analysis of nonstationary data in which the amplitude and phase of the harmonic components may change rapidly in time or space. While Fourier transform of the nonstationary time series would smear out any detailed information on the changing features, the WA keeps track of the evolution of the signal characteristics throughout the time. Further details on wavelet analysis can be found in [30].

Empirical Orthogonal Functions (EOF) analysis is a powerful tool for identifying coherent patterns that explains the largest fraction of the total variance of a field and it has been used extensively in studies of intraseasonal variations in tropical convection [31–33]. The EOF analysis is used in this study to define space-time structures of the dominant modes of intraseasonal variability in our study area. In the EOF analysis, the time-dependent deviations from the long-term mean are decomposed into a sum of products of fixed spatial patterns and time-dependent amplitudes (the principal components) as follows:

$$f(x_i, t) = \sum p_k(x_i)\,\alpha_k(t)\,. \qquad (2)$$

In (2), the principal components, , are uncorrelated to one another and described subsequently a maximum of variance in the original anomalies field, . Under these conditions, EOFs, are eigenvectors of the covariance matrix of.

3. RESULTS AND ANALYSIS

3.1. Spectral Analysis

For every day, we averaged the data of all grid points in each of the two subregions (WCA and ECA) of the study area to have a mean value. The spectral analysis was then performed separately on the two daily area mean time series obtained. In Fourier decomposition, the importance of a given frequency in a time series can be quantified by the spectral variance, defined as the amplitude of the harmonic corresponding to the this frequency. The power spectra in the two subregions reveal the importance of the intraseasonal variability in Central African climate (Figure 2). However, one should note that the dominance is less pronounced in WCA when many significant peaks are observed at different frequencies in synoptic scale. The fraction of total variance occurring in the 35- to 80-day range is 0.33 for WCA and 0.48 for ECA. Wavelet analysis clearly revealed that the processes inducing rainfall variability in the CA vary from synoptic (below 10 days) to longer timescales (Figures 3(a) and 3(b)). From the spectra, it is easy to identify the two bands of high wavelet powers in the ECA spectrum (10–25 days and 25–70 days), while in the WCA the relative importance of synoptic scale (below 10 days) does not allow to easily highlight the two intraseasonal periods bands. This shows that in WCA, the mechanisms that induce are much rainfall varied. At the synoptic scale, the spectral power peaks near 5 days in the two subregions. Nevertheless, the signal is predominantly contained within 10 and 70 days with two dominant timescales (10–25 days and 25–70 days), much highlighted in the ECA spectrum (Figure3(b)). This result is consistent to that found by other authors [14, 34] with daily OLR datasets over West Africa. The seasonality of wavelet power is clearly defined, showing the maximum power during the beginning and end of the year. In fact, this seasonality can be explained by the location of study area along the equator [35]. The peak of ITCZ precipitation belt shifts from north to south of the equator during November-December and returns to north during March-April.

3.2. Intraseasonal Oscillations

The spectral analysis performed above leads us to filter the rainfall anomalies either between 10 and 25 days to derive the space-time structures of the leading EOFs of 10–25-day intraseasonal variability of rainfall over CA. All these analyses are performed separately in ECA and WCA. The leading modes are retained according to the Scree test [36] and the North criteria [37], and spatial loadings associated with a given mode are obtained by computing the correlation coefficients between the corresponding PC's time series and the filtered data field.

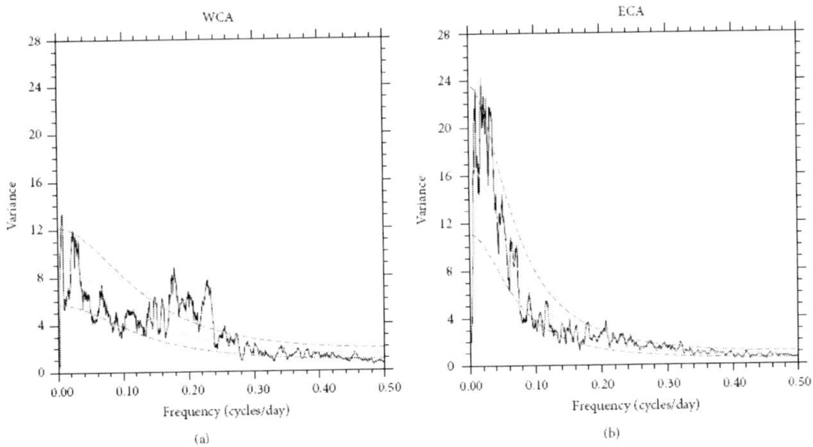

Figure 2. Calculated spectrum (solid black), red noise curve (green solid), and the curves indicating the upper (red dashed) and lower (blue dashed) bounds of 95% confidence level, for WCA (a) and ECA (b) subregions. The red noise spectrum is computed using the Markov processes [21].

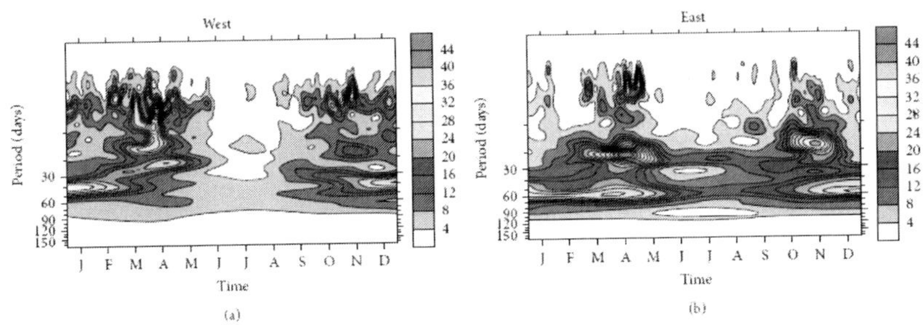

Figure 3. Annual mean wavelet power of the rainfall anomalies averaged over WCA (a) and ECA (b). A 120-day cutoff high pass filter was applied to long term anomalies in order to remove low frequencies variability such as interseasonal and interannual variability.

The leading eigenvector and the next two are accounting for, respectively, 15.8%, 13.8%, and 11.3% of the variance for WCA and 18.2%, 15.2%, and 14.7% for ECA (Figure 4). For WCA, all three leading EOFs have positive loadings (Figures 4(a)–4(c)). The first two modes are centered over the Atlantic Ocean near the equator and extend to cover a part of the land and the Atlantic Ocean, while the third is centered over the land and extends over the Congo Forest. Over ECA, the spatial loadings of first and third EOFs are negative, while the second is positive (Figures 4(d)–4(f)). The amount of the total variance explained by the leadings EOFs is smaller in WCA (40.6%) than ECA (48.1%), suggesting that rainfall time series in WCA have more degree of freedom. These results are consistent with the spectral analysis of Section 3.1.

The power spectra of the three principal components (PCs) peak around 15 days both in WCA and ECA (Figure 5). The amplitude time series reveals a difference between ECA and WCA 10–25-day rainfall oscillations. WCA experiences intraseasonal oscillations (ISO) almost throughout the year and the contrast between the seasons is less pronounced, but there are relative bursts of signal amplitude during the October–April season (Figures 6(a)–6(c)). However, the seasonality of oscillations is well defined in ECA (Figures6(d)–6(f)), as the oscillation amplitudes maximize during October–April and there are weakened or no oscillation during the rest of the year.

It is well known that the slow eastward propagation at an average speed of 5 $m \cdot s^{-1}$ [31] is one of the most fundamental features that distinguishes the MJO from other phenomena in the tropical atmosphere, especially, convectively coupled Kelvin waves, which propagate eastward at greater speeds of 15–17 $m\ s^{-1}$. This eastward propagation could potentially be the relationship between WCA and ECA modes of intraseasonal variability. In Figure 7, the lagged correlations are displayed between the WCA and ECA PCs time series. This figure clearly reveals that the observed PC's time series exhibit a characteristic lead/lag structure. The relatively high maximum positive correlation indicates that most of the WCA PC's lead ECA PCs time series with a time scale of 5–8 days. This result suggests a zonal propagation of EOF modes found above. Averaging filtered data between 15°N and 15°S, and plotting lagged correlation as a function of longitude, succinctly captures ISO propagation (Figure 8). Regression of the ISO indices [4], with filtered GPCP and 850 hPa zonal wind, are performed using the data over Central Africa (15°N–15°S; 0–50°E).

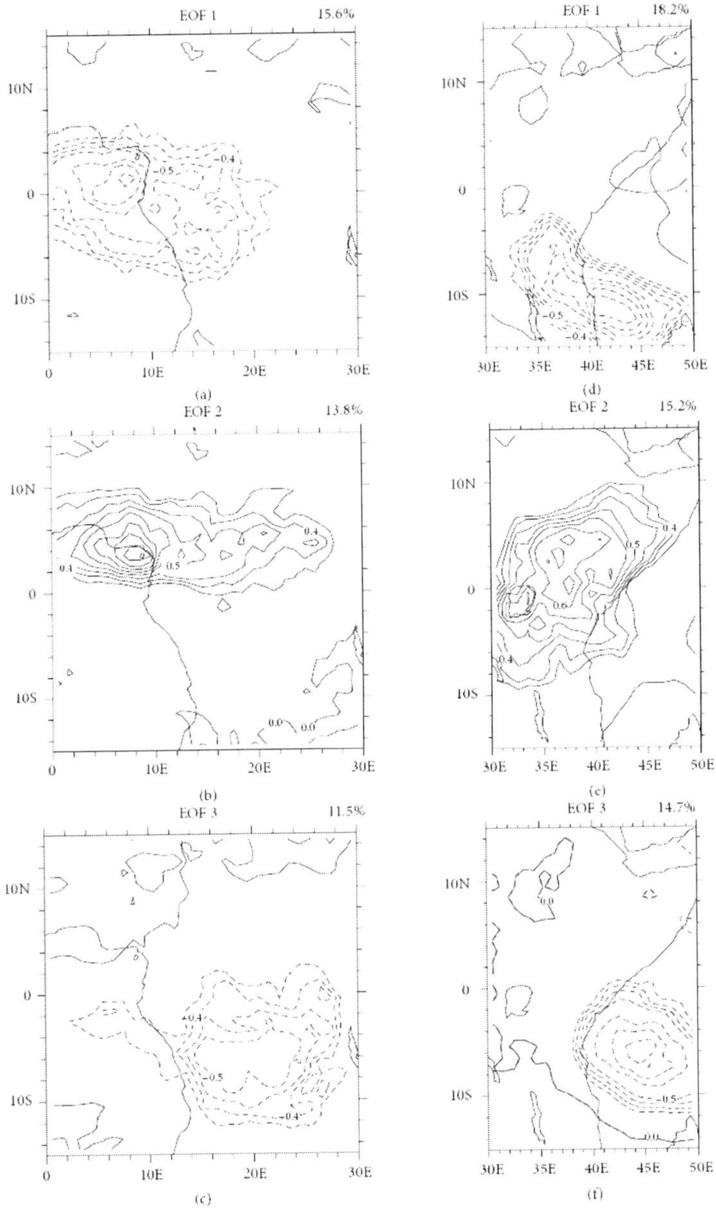

Figure 4. Correlation coefficient map (loadings) between rainfall anomalies field and PCs for WCA (a, b, c) and ECA (d, e, f). PCs were computed using the whole time series of daily rainfall for the period 1996–2009. Contour intervals are plotted every 0.05 and correlations in between −0.2 and 0.2 are omitted. The negative correlations are represented as dashed line and the positive correlation the solid lines.

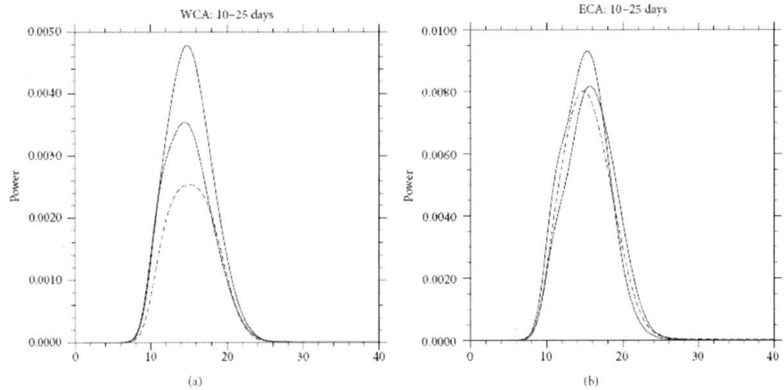

Figure 5. Power spectra of PC1 (Bold solid), PC2 (thin solid), and PC3 (thin dashed), respectively, over WCA (left top and left bottom) and ECA (right top and right bottom). The scale of the -axis is in days for all plots and the power dimensionless because the anomalies field is normalized.

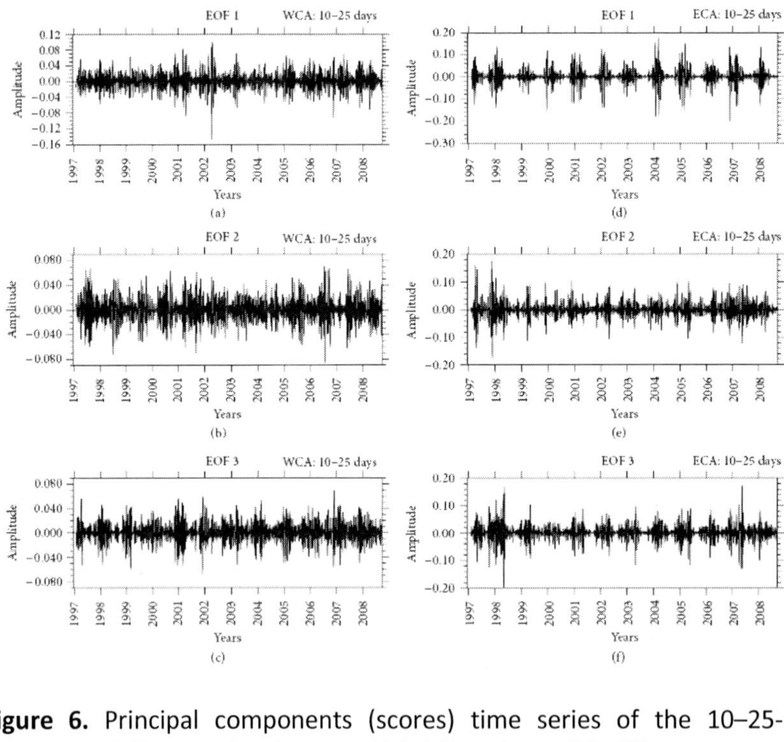

Figure 6. Principal components (scores) time series of the 10–25-day filtered rainfall anomalies for WCA ((a)–(c)) and ECA ((d)–(f)).

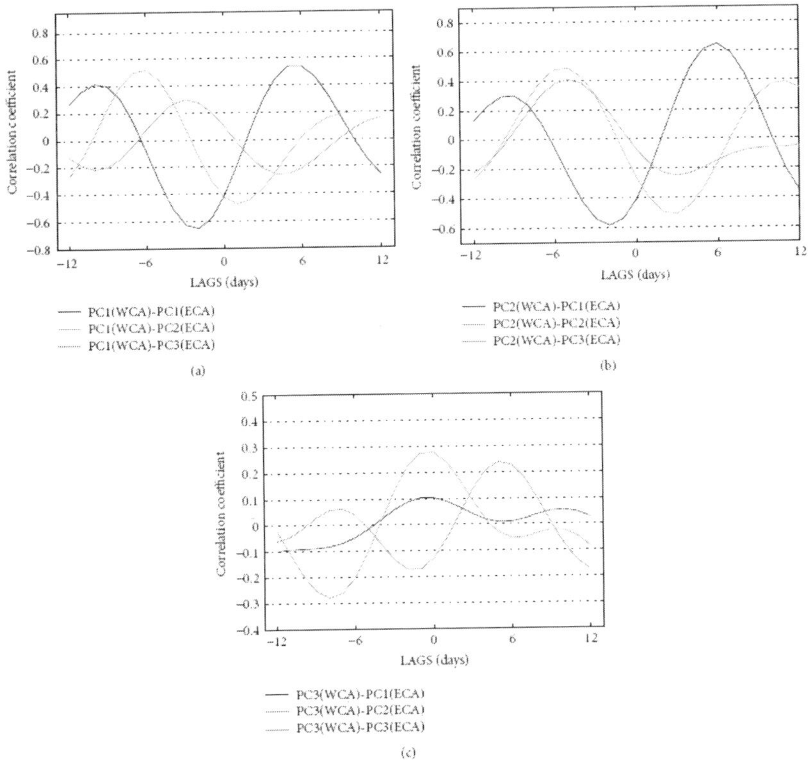

Figure 7. Lead-lag correlation between ECA and WCA 10–25-day PC's time series. Positive correlations at positive time lags indicate that convection over WCA leads that over the ECA.

The eastward propagation of the ISO is associated with low-level convergence. The 850 hPa convergence anomalies lead the convection, showing enhanced easterlies (westerlies) ahead (behind) enhanced rainfall as the ISO propagates through the global tropics. The average duration of an intraseasonal event to move from 0 to 45°E is approximately 7 days.

3.3. Interannual Variations in the ISO Amplitude

In the previous section, it was shown that the ISO amplitude exhibits large interannual variations in ECA, when compared with WCA. Moreover, it was proved by many previous studies that an important characteristic of the

MJO (ISO) is its irregularity from year to another [4, 38, 39]. Figure 9 shows the interannual variations in the ISO power over WCA (Figure 9(a)) and ECA (Figure 9(b)). For every year, the ISO strength is accessed by taking the mean squared amplitude of each PCs time series. The interannual variations of ISO strength are coherent for all the three PCs in each of the two subregions, as can be seen by the high values of the correlation coefficients (Table 2). But the correlation coefficients computed between the WCA and ECA ISO power time series are relatively low (Table 3), showing that the processes driving the interannual modulation of ISO signal are different in nature or magnitude in the two subregions. It can also be seen that the ISO strength experienced the highest fluctuations during the end of the 1990s and the beginning of the 2000s. For the two subregions, the variations of ISO activity during the end of the 1990s and early 2000s are similar. This period is marked by above normal ISO activity in the year 1997, followed by the below normal ISO activity during 1997–2001. The difference clearly appears toward the end of the study period. For the two timescales, ECA is characterized by below normal ISO activity during the years 2005–2008, while during the same period, the WCA experienced above normal ISO activity in many years. Further investigations are required to identify the causation of the coherent interannual variations of ISO signal during this period.

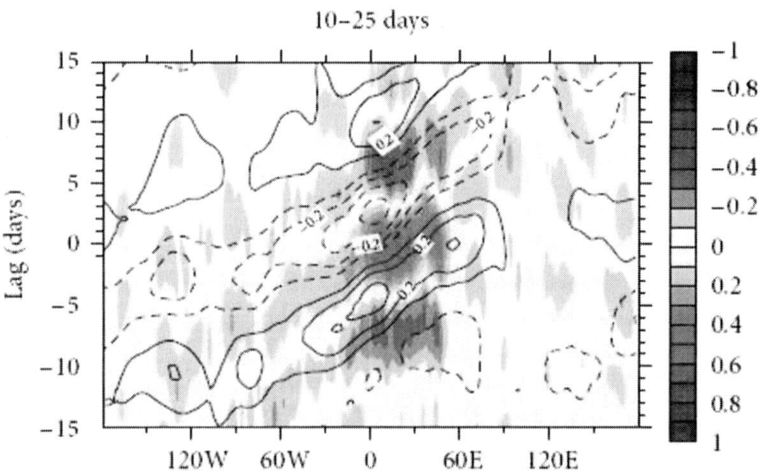

Figure 8. ISO composite based on the correlation between bandpass (10–25 days) filtered 850-hPa zonal wind (contours) and precipitation (colors) upon ISO index time series. The reference time series is Central Africa precipitation time series. The ISO index was computed using the average wavelet power.

Table 2. Correlation coefficients between WCA and ECA annual mean ISO power.

	WCA	ECA
PC1-PC2	0.43	0.51
PC1–PC3	0.61	0.12
PC2-PC3	0.73	0.47

Table 3. Cross-correlation coefficients between WCA and ECA annual mean ISO power.

WCA		ECA	
	PC1	PC2	PC3
PC1	−0.09	0.06	−0.05
PC2	−0.12	0.15	0.13
PC3	0.1	−0.11	−0.01

4. SUMMARY AND CONCLUSIONS

ISV of rainfall over Central Africa was examined using 1DD GPCP rainfall product for the period 1996–2009, with an emphasis on the comparison between western Central Africa (WCA) and eastern Central Africa (ECA), with different climate features. The classical Fourier analysis revealed the importance of ISV in Central African climate, but the dominance of intraseasonal band is more highlighted in ECA, when compared with WCA. A wavelet analysis showed that intraseasonal precipitation variability in CA is dominated by two distinct timescales (10–25 days and 25–70 days) with significant spectral peaks centered, respectively, around periods of 15 days and 50 days, in consistency with the classical Fourier analysis. The seasonality of the ISO power is evident, showing the maximum power at the beginning and end of the year. But this seasonality is well defined in ECA, unlike WCA where the signal persists almost throughout the year.

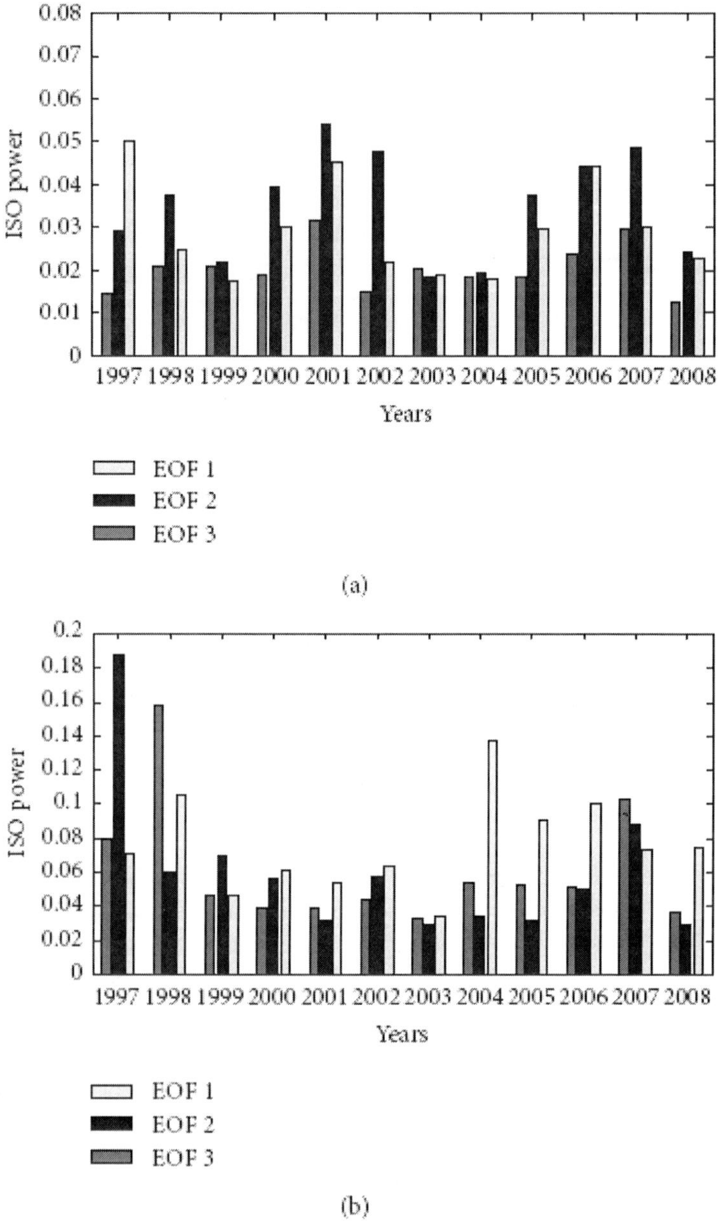

Figure 9. Interannual variations of ISO indices in WCA (a) and ECA (b), within the period 1997–2008, and for the three EOFs retained.

For 10–25-day band, the result of EOF analysis of the filtered data has shown that the leading eigenvectors vectors explain about 40.6% of total intraseasonal rainfall variability for WCA and 48.1% for ECA, suggesting that filtered rainfall time series in WCA have more degree of freedom than ECA. The spatial loadings are coherent, showing either positive or negative correlation coefficient between each eigenvector and rainfall. The power spectra of PCs peak around 15 days both in WCA and ECA.

The amplitude time series confirmed that rainfall variability in ECA is highly seasonal, revealing significant oscillations (wet/dry spells) with greater amplitudes during the beginning and end of the year (October–April) and weakened or no oscillation during the rest of the year. Once again, this seasonality is less pronounced in WCA PC's time series, but nevertheless the ISO amplitudes are relatively higher during October–April than other months.

An ISO strength index was built by averaging the wavelet power corresponding to each PC and for the 10–25-day timescales to form a daily indices time series. The plots of interannual variations of ISO strength clearly show that the signal over ECA exhibits larger interannual variations than WCA. Further investigations are needed to explain the causation of these patterns found.

The picture that emerges from these results is that the 10–25-day intraseasonal rainfall variability detected in 1DD GPCP rainfall shows some different patterns in ECA and WCA. The main difference between WCA and ECA modes is the variations of ISO strength. The variations in WCA are weak, compared to ECA where the signal exhibits large annual and interannual variations.

ACKNOWLEDGMENTS

Most of the scripts used in this study were produced from the NCL website http://www.ncl.ucar.edu/. The GPCP data was obtained from the NOAA website, http://www.esrl.noaa.gov/. All the administrator members of these websites are gratefully acknowledged for maintaining the updated data. The authors also thank the senior researchers of LEMAP and LISIE for helpful comments.

REFERENCES

1. UNEP DEWA/PAN-AFRICAN START, "Vulnerability of water resources to environmental change in Africa," River Basin Approach, UNEP, 2004.

2. T. James and P. Karl, "Economic losses and poverty effects of droughts and floods in Malawi," OER Africa, 2009.

3. P. Dierk, M. Charles, and H. Stefan, "Diagnosing the droughts and floods in equatorial east Africa during boreal autumn 2005–08," Journal of Climate, vol. 23, no. 3, pp. 813–817, 2010.

4. A. S. Tchakoutio, A. Nzeukou, and C. Tchawoua, "Intraseasonal atmospheric variability and its interannual modulation in Central Africa," Meteorology and Atmospheric Physics, vol. 117, no. 3-4, pp. 167–179, 2012.

5. E. J. Zipser, D. J. Cecil, C. Liu, S. W. Nesbitt, and D. P. Yorty, "Where are the most intense thunderstorms on earth?" Bulletin of the American Meteorological Society, vol. 87, no. 8, pp. 1057–1071, 2006.

6. G. Hong and G. Heygster, "Intense tropical thunderstorms detected by the special sensor microwave imager/sounder," IEEE Transactions on Geoscience and Remote Sensing, vol. 46, no. 4, pp. 996–1005, 2008.

7. M. Thomas, C. H. C. John, G. Alexander, and J. C. Nicolas, "Temporal precipitation variability versus altitude on a tropical high mountain: observations and mesoscale atmospheric modelling," Quarterly Journal of the Royal Meteorological Society, vol. 135, no. 643, pp. 1439–1455, 2009.

8. A. M. Moore, J. P. Loschnigg, P. J. Webster, and R. R. Leben, "Coupled ocean-atmosphere dynamics in the Indian Ocean during 1997-98," Nature, vol. 401, no. 6751, pp. 356–360, 1999.

9. C. Mutai, S. Hastenrath, and D. Polzin, "Diagnosing the 2005 drought in equatorial East Africa," Journal of Climate, vol. 20, no. 18, pp. 4628–4637, 2007.

10. R. A. Madden and P. R. Julian, "Description of global-scale circulation cells in the tropics with a 40–50 day period," Journal of the Atmospheric Sciences, vol. 29, pp. 1109–1123, 1972.

11. T. Yamagata and Y. Hayachi, "A simple diagnostic model for the 30–50 days oscillation in the tropics,"Journal of the Meteorological Society of Japan, vol. 62, no. 5, pp. 709–717, 1984.

12. B. Sultan and S. Janicot, "Intra-seasonal modulation of convection in the West African monsoon,"Geophysical Research Letters, vol. 28, no. 3, pp. 523–526, 2001.

13. B. Sultan and S. Janicot, "The West African monsoon dynamics—part II: the pre-onset and onset of the summer monsoon," Journal of Climate, vol. 16, pp. 3407–3427, 2003.

14. S. Janicot and F. Mounier, "Evidence of two independent modes of convection at intraseasonal timescale in the West African summer monsoon," Geophysical Research Letters, vol. 31, no. 16, Article ID L16116, 2004.

15. R. W. Higgins, D. E. Waliser, J.-K. E. Schemm, C. Jones, and L. M. V. Carvalho, "Climatology of tropical intraseasonal convective anomalies: 1979–2002," Journal of Climate, vol. 17, no. 3, pp. 523–539, 2004.

16. P. Camberlin and P. Benjamin, "Influence of the Madden-Julian oscillation on East African rainfall—I: intraseasonal variability and regional dependency," Quarterly Journal of the Royal Meteorological Society, vol. 132, no. 621, pp. 2521–2539, 2006.

17. L. Tazalika and M. R. Jury, "Intra-seasonal rainfall oscillations over central Africa: space-time character and evolution," Theoretical and Applied Climatology, vol. 94, no. 1-2, pp. 67–80, 2008.

18. M. R. Jury and C. Hector, "Intraseasonal variability of satellite-derived rainfall and vegetation over Southern Africa," Earth Interactions, vol. 14, no. 3, pp. 1–26, 2010.

19. A. S. Tchakoutio, A. Nzeukou, C. Tchawoua, F. M. Kamga, and D. Vondou, "A comparative analysis of intraseasonal variability in OLR and 1DD GPCP data over central Africa," Theoretical and Applied Climatology, vol. 116, no. 1-2, pp. 37–49, 2014.

20. A. S. Tchakoutio, A. Nzeukou, C. Tchawoua, B. Sonfack, and T. Siddi, "Comparing the patterns of 20–70 days intraseasonal oscillations over Central Africa during the last three decades," Theoretical and Applied Climatology, 2013.

21. D. L. Gilman, F. J. Fuglister, and J. M. Mitchell, "On the power spectrum of 'red noise'," Journal of the Atmospheric Sciences, vol. 20, no. 2, pp. 182–184, 1963.

22. G. J. Huffman, M. Morrissey, D. Bolvin et al., "Global precipitation at one degree daily resolution from multisatellite observations," Journal of Hydrometeorology, vol. 2, pp. 36–50, 2001.

23. R. F. Adler, "The version-2 Global Precipitation Climatology Project (GPCP) monthly precipitation analysis (1979-present)," Journal of Hydrometeorology, vol. 4, pp. 1147–1167, 2003.

24. P. A. Arkin and B. N. Meisner, "The relationship between large-scale convective rainfall and cold cloud over the Western Hemisphere during 1982–84," Monthly Weather Review, vol. 115, no. 1, pp. 51–74, 1987.

25. J. McCollum, E. Nelkin, D. Klotter et al., "Validation of TRMM and other rainfall estimates with a high-density gauge dataset for West Africa—part I: validation of GPCC rainfall product and pre-TRMM satellite and blended products," Journal of Applied Meteorology, vol. 42, no. 10, pp. 1337–1354, 2003.

26. T. Dinku, P. Ceccato, E. Grover-Kopec, M. Lemma, S. J. Connor, and C. F. Ropelewski, "Validation of satellite rainfall products over East Africa's complex topography," International Journal of Remote Sensing, vol. 28, no. 7, pp. 1503–1526, 2007.

27. F. Ruiz, T. Dinku, S. J. Connor, and P. Ceccato, "Validation and intercomparison of satellite rainfall estimates over Colombia," Journal of Applied Meteorology and Climatology, vol. 49, no. 5, pp. 1004–1014, 2010.

28. C. Lanczos, Applied Analysis, Prentice-Hall, 1956.

29. C. E. Duchon, "Lanczos filtering in one and two dimensions," Journal of Applied Meteorology, vol. 18, no. 8, pp. 1016–1022, 1979.

30. C. Torrence and G. P. Compo, "A practical guide to wavelet analysis," Bulletin of the American Meteorological Society, vol. 79, no. 1, pp. 61–78, 1998.

31. K. M. Weickmann and T. R. Knuston, "30–60 day atmospheric oscillations: composite life cycle of convection and circulation

anomalies," Monthly Weather Review, vol. 115, no. 7, pp. 1407–1436, 1987.

32. H. H. Harry and Z. Chindong, "Propagating and standing components of the intraseasonal oscillation in tropical convection," Journal of the Atmospheric Sciences, vol. 54, pp. 741–751, 1996.

33. A. J. Matthews, "Intraseasonal variability over tropical Africa during northern summer," Journal of Climate, vol. 17, no. 12, pp. 2427–2440, 2004.

34. E. D. Maloney and S. Jeffrey, "Intraseasonal variability of the West African monsoon and Atlantic ITCZ,"Journal of Climate, vol. 21, no. 12, pp. 2898–2918, 2008.

35. H. Rui and B. Wang, "Development characteristics and dynamic structure of tropical intraseasonal convection anomalies," Journal of the Atmospheric Sciences, vol. 47, no. 3, pp. 357–379, 1990.

36. R. B. Cattell, "The scree test for the number of factors," Multivariate Behavioral Research, vol. 1, no. 2, pp. 245–276, 1966.

37. G. R. North, T. L. Bell, R. F. Cahalan, and F. J. Moeng, "Sampling errors in the estimation of empirical orthogonal functions," Monthly Weather Review, vol. 110, no. 7, pp. 699–6706, 1982.

38. H. H. Hendon, C. Zhang, and J. D. Glick, "Interannual variation of the Madden-Julian oscillation during austral summer," Journal of Climate, vol. 12, no. 8, pp. 2538–2550, 1999.

39. J. M. Slingo, D. P. Rowell, K. R. Sperber, and F. Nortley, "On the predictability of the interannual behaviour of the Madden-Julian oscillation and its relationship with El Niño," Quarterly Journal of the Royal Meteorological Society, vol. 125, no. 554, pp. 583–609, 1999.

Chapter 8

Variability in Severe Coastal Flooding, Associated Storms, and Death Tolls in Southeastern Australia since the Mid–Nineteenth Century

Scott B. Power[1], Jeff Callaghan[2]*

[1]Bureau of Meteorology, Melbourne, Victoria, Australia

[2]Bureau of Meteorology, Queensland Office, Brisbane, Queensland, Australia

ABSTRACT

The variability in the number of severe floods that occurred in coastal catchments in southeastern Australia since the mid–nineteenth century, along with the variability in both the frequency of the weather types that triggered the floods and the associated death tolls, is analyzed. Previous research has shown that all of the severe floods identified were associated with one of two major weather types: east coast lows (ECLs) and tropical interactions (TIs). El Niño–Southern Oscillation (ENSO) is shown to strongly modulate the frequency of severe coastal flooding, weather types, and the number of associated deaths. The analysis presented herein, which examines links over more than a century, provides one of very few known statistically significant links between ENSO and death tolls anywhere in the world. Over the period 1876/77–2013/14 the average numbers of coastal floods, ECLs, TIs, and deaths associated with freshwater drowning in La Niña years are 92%, 55%, 150%, and 220% higher than the corresponding averages in El Niño years. The average number of deaths per flood in La Niña years is 3.2, which is 66% higher than the average in El Niño years. Death tolls of 10 or more occurred in only 5% of El Niño years, but in 27% of La Niña years. The interdecadal Pacific oscillation also modulates the frequency of severe floods, weather types, and death tolls. The results of this study are consistent with earlier research over shorter periods and broader regions, using less-complete datasets.

KEYWORDS

Australia; Synoptic climatology; Climate variability; Flood events

1. INTRODUCTION

Severe floods in southeastern Australia frequently isolate towns, often causing death, major disruptions to road and rail links, the evacuation of many houses and business premises, and the widespread flooding of farmland. Severe flooding can cause death. For example, there were at least 90 known deaths caused by floods in and around Gundagai in June 1852, and 25 deaths during the Brisbane River catchment floods in January 2011 (Callaghan and Power 2014).

Callaghan and Power (2014) recently developed a record of the frequency of severe floods that have occurred in coastal catchments in eastern Australia during 1860–2012, from Brisbane (Queensland) in the north, through Sydney and to Eden in New South Wales (NSW), 1500 km farther south. Their study region, which we adopt, is depicted in Fig. 1.

Callaghan and Power (2014) considered a flood to be major, which we refer to as severe in this paper to better describe their nature, if (i) it caused inundation of a river within approximately 50 km of the coast or (ii) there was nonriverine flooding overland near the coast, from the active part of a weather system, that extends at least 20 km along the coast. If severe flooding was caused in more than one catchment, it is regarded as a single severe flooding event.

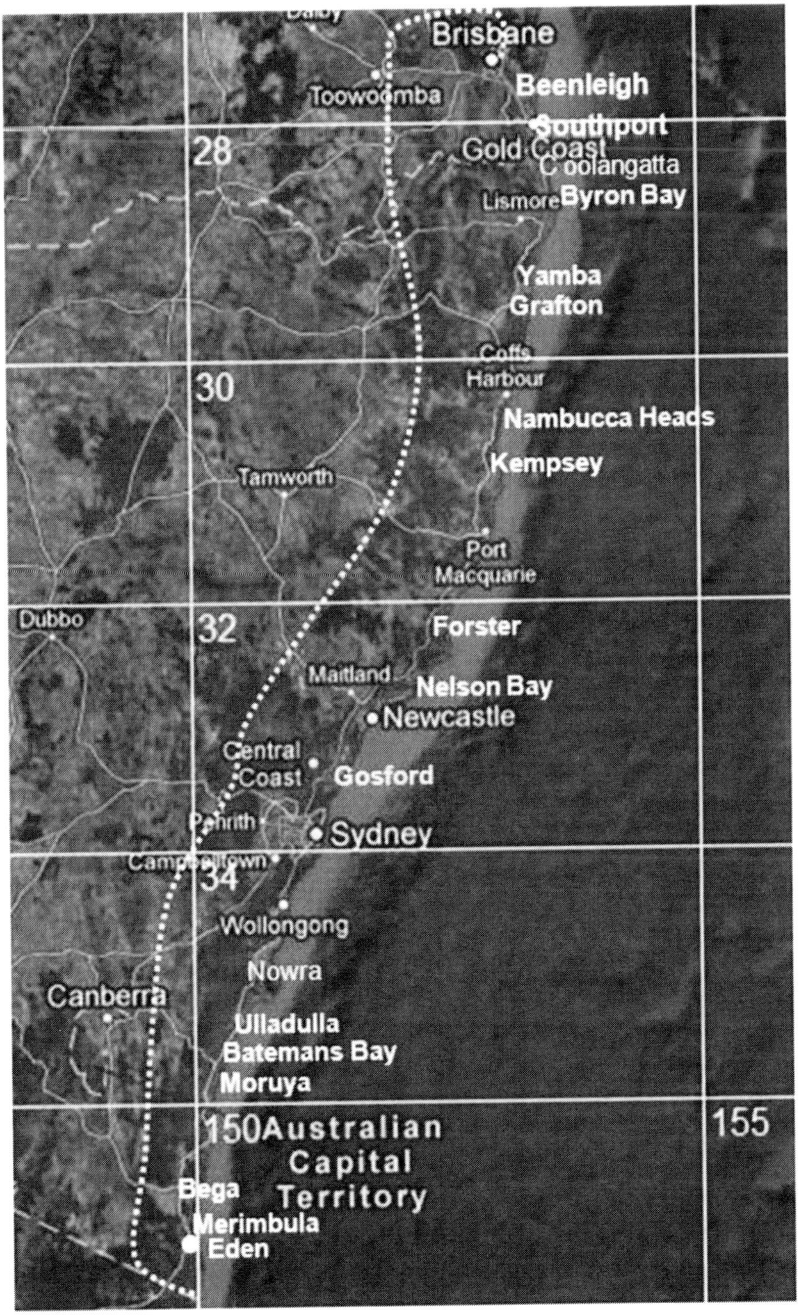

Figure 1. Study region is shown on the left side enclosed by the dotted white line.

In an attempt to produce reliable (or homogeneous) records of the frequency of severe flooding and weather types, Callaghan and Power (2014) restricted attention to the period from 1860, when the region (i) is extensively populated, (ii) has extensive coverage of meteorological stations, (iii) is extensively connected by telecommunication, and (iv) has busy coastal shipping traffic offshore. They concluded that all of the floods were linked to either an east coast low (ECL) or a tropical interaction (TI). Examples of a TI included a tropical cyclone (TC) that moves into the study region retaining its TC characteristics and extratropical transitions, whereby a TC or tropical low is reorganized or transformed into a more ECL-like structure (while interacting with a deep layered trough) as it moves into the study region. The importance of ECLs and other east coast low pressure systems to heavy rainfall or flooding is discussed by, for example, Pepler et al. (2014), Hopkins and Holland (1997), and G. McKay (2011, unpublished conference paper titled "The life blood of floods: Flood producing weather systems").

Previous research indicates that there is a great deal of variability in eastern Australian rainfall, streamflow, flooding and flood risk on interannual (e.g., McBride and Nicholls 1983; Chiew et al. 1998; Power et al. 1998, 1999a; Kiem and Franks 2001; Kiem et al. 2003; Verdon et al. 2004; Speer 2008; Speer et al. 2009, 2011; King et al. 2014), and decadal–multidecadal time scales (Power et al. 1999a,b; McKeon et al. 2004; Kiem et al. 2003). Much of the interannual variability is due to El Niño–Southern Oscillation (ENSO; McBride and Nicholls 1983; Chiew et al. 1998; Power et al. 1998,1999a; Kiem and Franks 2001; Kiem et al. 2003; Verdon et al. 2004; Pui et al. 2011; Speer 2008;Speer et al. 2011). Much of the decadal/multidecadal variability is due to the interdecadal Pacific oscillation (IPO; Power et al. 1999a; Folland et al. 1999; Parker et al. 2007; Christensen et al. 2013;Henley et al. 2015; Power et al. 1999a,b; McKeon et al. 2004; Verdon et al. 2004; Cai and van Rensch 2012). These links also extend to variability in heavy and extreme rainfall, and flood frequency in eastern Australia (Franks and Kuczera 2002; Braganza et al. 2015, chapter 4; Micevski et al. 2006;Kiem et al. 2003; Kiem and Verdon-Kidd 2013; King et al. 2013).

Here, we determine if ENSO and the IPO modulate the frequency of the severe coastal floods, and the frequency of the associated weather types and death tolls, in our much longer record. We will also quantify the magnitude of any modulation evident. An investigation of this kind, in which we examine variability in storm types, major flooding, and death tolls, has not been conducted previously for this region.

The data and methods used in this investigation are described in section 2. Variability in severe floods, ECLs, TIs, and their associated death tolls, as well as the roles of ENSO and the IPO in causing the variability, are described in section 3. Results are summarized and briefly discussed in section 4.

2. DATA AND METHODS

Callaghan and Power (2014) included a description of severe coastal floods in the same study region from January 1860 to December 2012. An extension of this list to include the severe floods that occurred during the period January 2013–December 2014 is given in appendix A. The date of occurrence, the type of weather system that triggered the flood, and the death toll are provided.

In this paper, we use the term ENSO years, defined here as the period from May in one year to April in the following calendar year. This choice reflects the typical life cycle of ENSO events (i.e., both El Niño and La Niña events), which tend to grow, mature, erode, and end within this period (see, e.g.,Power and Smith 2007). We use the first year in the period to refer to that period, so that 1860, for example, is used to refer to the period April 1860–May 1861. The analysis described covers the period 1860–2013. The Southern Oscillation index (SOI) averaged over the period June–December is used as the ENSO index. This index is also consistent with the life cycle of ENSO and has been used successfully in previous studies (e.g., Power et al. 1995; Power and Smith 2007; Callaghan and Power 2011; Power and Kociuba 2011; Chung et al. 2014; Salinger et al. 2014). Note that the SOI becomes available from 1876, and so correlations are performed on data from this year.

Correlation coefficients between time series of severe floods, ECLs, TIs, and associated death tolls, as well as either the SOI or an index for the IPO, are calculated. The statistical significance of correlation coefficients is tested using the method described by Power et al. (1998), a method that takes temporal persistence in the time series into account. We will refer to this as the P98 method (seeappendix B). Monte Carlo methods involving randomly generated simulated data (also described inappendix B) are then used as a double check on the statistical significance of correlations involving the SOI. We will find that the Monte Carlo method gives similar results to the P98 method.

The significance of the results is expressed either as a p value, or as a percentage value for the corresponding statistical significance level. The p value is equal to the estimated probability that the magnitude of the observed correlation coefficient can be exceeded in magnitude by chance, under the assumption that the data can be approximated as a red-noise process (Power et al. 1998).

3. VARIABILITY AND ITS LINKS TO ENSO AND THE IPO

3.1 Variability

The time series of the number of severe floods (F), ECLs, TIs, and total deaths from freshwater flooding per year are presented in Fig. 2. Time series of ECLs and TIs are exhibited in Fig. 2a, and their sum equates to the number of floods. Deaths associated with ECLs and TIs are given in Fig. 2b, and their sum equates to the number of deaths. Substantial variability is evident, ranging between 0 and 6 floods per year, 0 and 5 ECLs per year, 0 and 6 TIs per year, and 0 and 46 deaths per year, about long-term averages of 1.6 floods per year, 0.92 ECLs per year, 0.73 TIs per year, and 4.0 deaths per year. These, and other basic statistics, are presented in Table 1.

The highest number of deaths occurred during the 12 months from May 1892. Most of these deaths occurred when a TI triggered flooding in the Brisbane–Ipswich region. This remains the highest flood reliably recorded on the Brisbane River.

In the last 50 yr, the single event causing the highest death toll (25) was the TI that caused major flooding in southeast Queensland. This event was dominated by extreme flash flooding in the Lockyer valley and on the Toowoomba Range.

3.2 Links to ENSO

Some of this variability in severe floods, ECLs, TIs, and total deaths from freshwater flooding per year in the study region of coastal southeastern Australia is also linked to ENSO. This can be illustrated in several ways. For example, the correlation coefficient between the variables and the SOI (over the period 1876–2013, the period when SOI data are available) are all statistically significant (see Table 2).

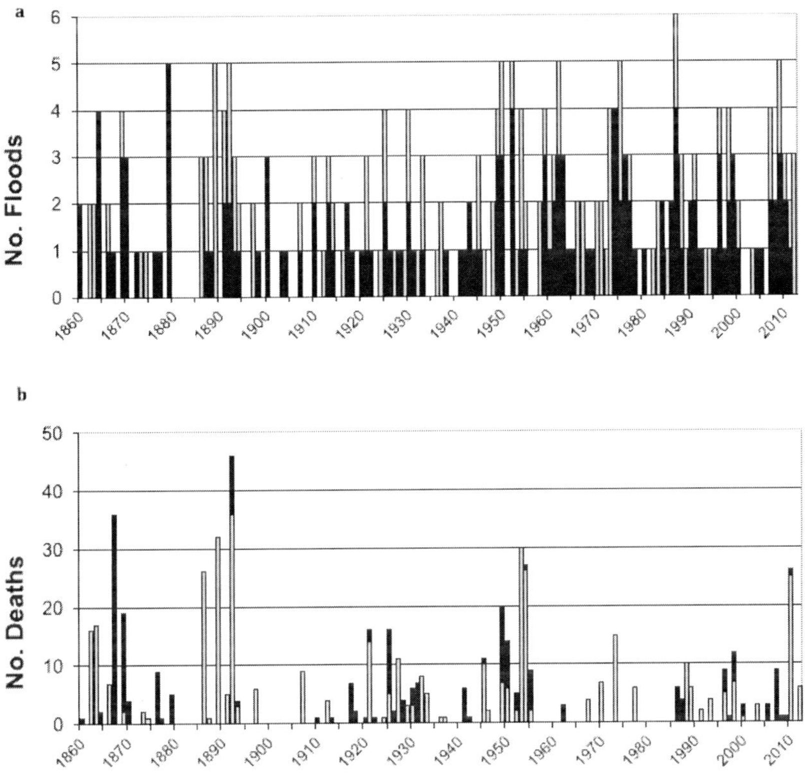

Figure 2. Number of (a) severe floods and (b) deaths, 1860–2013. The floods triggered by ECLs are represented with black bars and the floods triggered by TIs are represented with by gray bars in (a). The death tolls during ECL-triggered floods are represented with black bars and the death tolls during TI-triggered floods are represented by gray bars in (b). Statistics for the data presented are summarized in Table 1.

Table 1. Basic statistics include averages in all years, El Niño years, and La Niña years, as well as the differences (%) between the averages in La Niña and El Niño years.

Event	Avg (per year)			Change between La Niña and El Niño years (%)
	All years	El Niño years	La Niña years	
Major flood	1.6	1.2	2.3	92
ECL	0.92	0.77	1.2	55
TI	0.73	0.48	1.2	150
Deaths per year	4	2.28	7.3	220
Deaths per flood	2.5	1.93	3.2	66
Percentage of years with major floods	74	63	82	30
Percentage of years with ECLs	53	46	64	39
Percentage of years with TIs	45	37	56	51
Percentage of years with deaths	47	37	59	59
Years with >10 deaths (%)	12	5	27	440

Table 2. Correlation coefficients between the number of floods, ECLs, TIs, and deaths with the SOI and IPO index. The p values are given in parentheses.

Variable	Correlation coefficient between variable and the SOI (with *p* values)	Correlation coefficient between variable and the IPO index (with *p* value)
Major flood frequency	0.3 (<0.01, 0.0)	−0.5 (0.15)
ECL frequency	0.2 (0.04, 0.05)	−0.4 (0.2)
TI frequency	0.3 (0.001, 0.001)	−0.5 (0.2)
Death toll	0.3 (0.001, 0.0001)	−0.4 (0.27)

The long-term average number of events experienced in the study region differs between El Niño and La Niña years. This is illustrated in Fig. 3, which shows that the average number of floods, ECLs, TIs, and deaths in La Niña years are all much higher than their corresponding averages in El Niño years. The percentage increase is especially marked for death tolls (220%). This large increase occurs because the frequency of both ECLs and TIs increases, and the increase in the deadliest systems—the TIs (Callaghan and Power 2014)—is also large (150%). Death tolls of 10 or more occurred in 27% of La Niña years, but only 5% of El Niño years. The average number of deaths per flood in La Niña years is 3.2, which is 66% higher than the average in El Niño years. See Table 1 for further details.

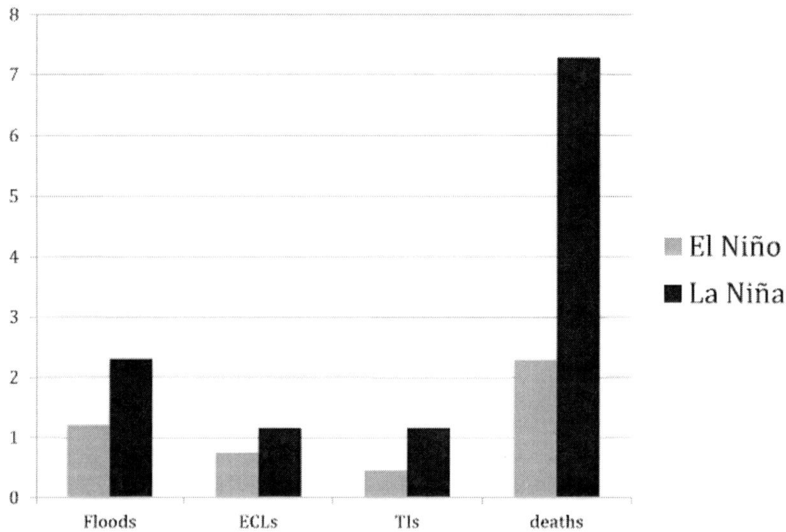

Figure 3. Average number of severe floods, ECLs, TIs, and deaths for both El Niño (gray bars) and La Niña (black bars) years.

Approximately 75% of all years have at least one flooding event. Approximately 63% of El Niño years and 82% of La Niña years have at least one flooding event.

Further illustration of the impact of ENSO on the frequency of the events is evident in the proportion of years (%) in which floods, ECLs, TIs, and deaths occur. In every case, the events occurred more frequently during La Niña years than they did during El Niño years. The proportion of La Niña years that witnessed each event type is greater than the proportion of El Niño years in all cases. See Table 1 for further details.

This analysis clearly indicates that ENSO has an impact on the number of events. This occurs despite the fact that the impact of ENSO on seasonal rainfall in coastal southeastern Australia at certain times of the year is weak (Pepler et al. 2014). This earlier result should therefore not be interpreted as meaning that ENSO's impact on coastal southeastern Australia is unimportant.

3.3 Decadal variability and its links to the IPO

Ten-year running sums of the number of floods, ECLs, TIs (all in Fig. 4a), deaths (Fig. 4b), and deaths per flood (Fig. 4c) are presented. All variables show pronounced variations in the range 7–26 floods per decade, 2–15 ECLs per decade, 0–18 TIs per decade, 3–114 deaths per decade, and 0–9 deaths per flood per decade. Death tolls have been relatively low since 1960, despite an increase in the number of floods.

A time series for the IPO, using the method described by Parker et al. (2007), is made available by the Met Office. The correlation coefficients between 10-yr running averages in the IPO time series and 10-yr running averages of the time series of the frequency of severe floods, ECLs, TIs, and deaths are −0.53, −0.38, −0.46, and −0.38, respectively. None of the correlation coefficients are statistically significant (see Table 2). This occurs despite the larger correlation coefficients because the number of degrees of freedom is markedly reduced when taking decadal averages (see, e.g., Power et al. 1998).

The signs of the correlation coefficients are nevertheless consistent with previous research (Power et al. 1999b; Verdon et al. 2004; Kiem et al. 2003), the relationships evident on interannual time scales described in the previous subsection, and what is seen in climate models (Walland et al. 2000; Arblaster et al. 2002; Power et al. 1995). The statistical relationships therefore very likely reflect genuine physical relationships.

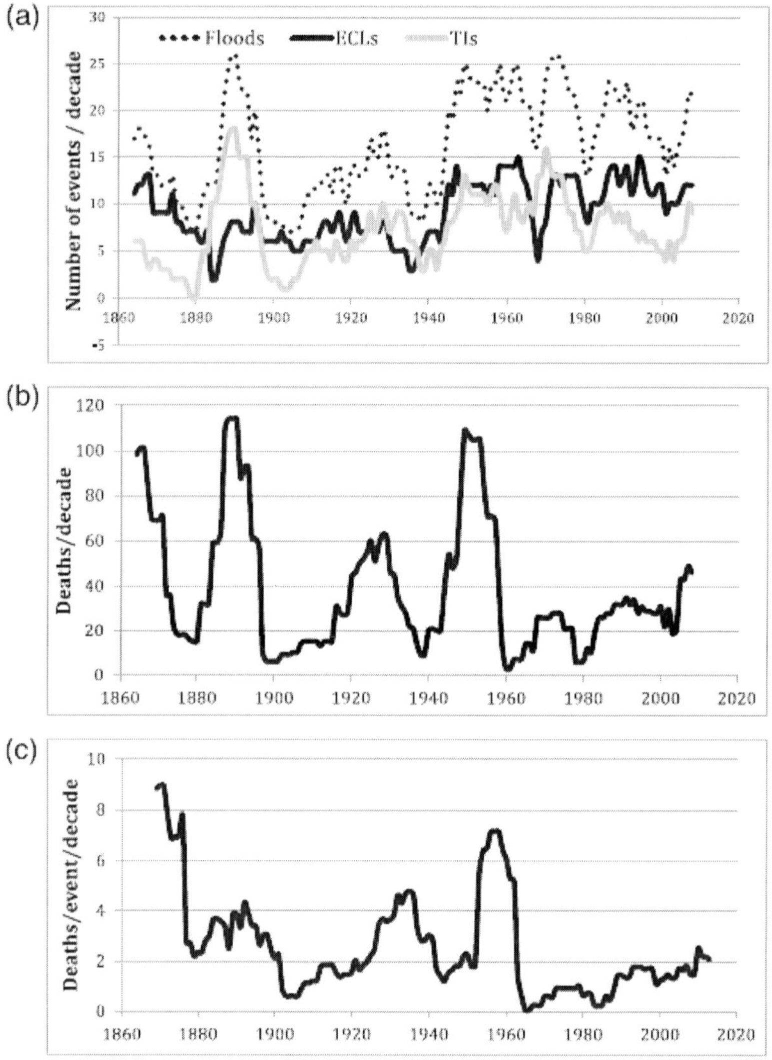

Figure 4. Decadal variability (10-yr running averages) in the frequency of (a) severe floods (dotted line), ECLs (black line), and TIs (gray line); (b) deaths; and (c) deaths per severe flood.

The average numbers of severe floods, ECLs, TIs, and deaths in years when the IPO time series was negative are 2.0, 1.0, 1.0, and 5.4, respectively. These figures are 67%, 39%, 110%, and 225% larger than the corresponding figures during years in which the IPO time series was positive. If we instead use decadal variability in the IPO time series to stratify the data rather than the annual IPO data, the last set of figures above become 54%, 33%, 85%, and 93%. Death tolls were 10 or more in 14 yr when the IPO time series was negative, but in only 1 yr when the IPO time series was positive. Similarly, death tolls of 10 or more occurred in 12% of years during negative decadal phases of the IPO, but in only 6% of years during positive decadal phases. Key statistics are summarized in Table 2.

There is also pronounced interdecadal variability (Fig. 4), which is partly due to the modulating effect of the IPO discussed above. The number of floods is generally greater after 1945 than before, consistent with the findings of Franks and Kuczera (2002) and Kiem et al. (2003), who examined flood frequencies over NSW more broadly. After 1960, 90% of years experienced at least one major flood, which represents a 50% increase in the same statistic calculated over the early part of the record (1860–1935).

There are two decadal peaks in the number of floods: one from the mid-1880s to the mid-1890s and another from the late 1960s to the late 1970s. Both periods experienced decades with 26 major floods. Every year during 1887–95 witnessed at least one major flood. There were 18 TIs, and 6 of these were TCs. There have not been this many tropical cyclones affecting the study region in any decade either before or since this time. This high level of storm activity caused the splitting of Stradbroke Island (Callaghan and Helman 2008). The highest death toll on record (46) discussed above also occurred during this period.

The later period experiencing an equal record high level of major flooding, triggered by larger than average numbers of both ECLs and TIs. In fact nearly the entire period between the early 1940s and the mid-1970s experienced a high frequency of major flooding. During 1967, for example, there were three major floods, two of which were triggered by tropical cyclones. Tropical cyclones also occurred offshore during 1967 that did not trigger major flooding. Together these offshore systems and the systems that did trigger major flooding caused the worst beach erosion on record at the Gold Coast and in northeastern NSW (e.g., Callaghan and Helman 2008). This record erosion event marked the beginning of persistent erosion issues in this region during the following decade.

Deaths have been relatively few since the late 1950s (Fig. 4b), and the number of deaths per event has been relatively low since the early 1960s (Fig. 4c). This reduction occurred even though the number of floods increases toward the end of the record (Fig. 2). Deaths per event were greatest during the early part of the record when warning services were largely unavailable, many settlers lived on fertile areas on the flood plains of rivers, and an awareness of the risks posed by flooding was not as wide (Callaghan and Power 2014).

4. SUMMARY AND DISCUSSION

We examined interannual and decadal variability in the frequency of severe floods, the storms that triggered them, and the resulting death tolls from freshwater flooding in an important coastal region in southeastern Australia. The region contains two of Australia's largest cities (Sydney and Brisbane) as well as some of most densely populated coastal zones in Australia, and the population is growing rapidly. Variability in this region is therefore of major interest. The new dataset also allows us to examine variability in this region over a longer period than has been possible previously.

ENSO is shown to strongly modulate the frequency of severe flooding, associated storms, and death tolls. Over the period 1876–2013 the average numbers of floods, ECLs, TIs, and deaths in La Niña years are 2.3, 1.2, 1.2, and 3.7 per year, respectively, which are all much higher than their El Niño year counterparts (Table 1). Elevated risk during La Niña years is also evident in related quantities. For example, while death tolls of 10 or more occurred in 27% of La Niña years, such tolls only occurred in 5% of El Niño years. The average number of deaths per flood in La Niña years (3.2) is also higher (by 66%) than the average in El Niño years.

We also identified a statistically significant link between death tolls from freshwater drowning in our coastal study region, and a widely used ENSO index (the SOI). This adds to what seems to be a very limited list of known statistically significant links between ENSO and deaths, anywhere in the world. For example, a review of the impact of ENSO on human health (Kovats et al. 2003) identified only two studies in which statistically significant links had been established between ENSO and human mortality rates. The first study (Bouma and van der Kaay 1996) described statistically significant increases in malarial deaths in the Punjab in India and in Sri Lanka, arising from La Niña–driven increases in monsoonal rainfall. The second study (Kuhnel and Coates 2000) examined the impact of ENSO on the total number of deaths linked to natural hazards, over a much broader region of eastern Australia than we examine. Their study region encompasses all of the eastern mainland states (i.e., Queensland, New South Wales, and Victoria), of which our study region is a small but important subset. They found that La Niña elevated the risk of death from flooding over this much broader region, consistent with our results for the coastal zone alone. Interestingly, Kuhnel and Coates (2000)found that most deaths in eastern Australia linked to ENSO actually tend to occur during El Niño years, in association with heat waves. Whether or not this is true in our (coastal) study region is unclear.

It is well established that ENSO variability can be predicted to an extent on seasonal time scales (e.g., Latif et al. 1994). This, together with the strong links we have identified between ENSO and severe flooding, and associated death tolls, highlights the possibility that flood risk and death tolls might also be partially predictable. Whether or not this is the case, we can infer from the data presented in this study that there may be benefits in raising awareness of the risks posed to life and safety from flooding and from the severe storms identified, as part of seasonal prediction and weather warning services. Thus, for example, the increased risk from severe flooding might also be part of the seasonal outlook leading in to La Niña years, or when a severe storm of the type identified is forecast. The risks seem to be widely recognized in the broader community in association with tropical cyclones. There might also be an increasing awareness in the risks linked to east coast lows, but this does not seem to be the case for the other weather systems identified.

Marked decadal (and interdecadal) variability is evident in all variables, within ranges of 7–26 floods per decade, 2–15 ECLs per decade, 0–18 TIs per decade, 3–114 deaths per decade, and 0.0–8.9 deaths per flood per decade. It is reassuring to note that both death tolls and deaths per flood have been relatively low since 1960. Factors that have contributed to this trend include very substantial improvements in the quality of severe weather forecasts and emergency services, as well as greater access to forecasts. During the nineteenth and early twentieth centuries forecasts were largely unavailable, many settlers lived on fertile areas on the flood plains of rivers, and an awareness of the risks posed by flooding was not as widely appreciated (Callaghan and Power 2014). The reduction in deaths per event might also reflect a generally greater awareness of the risks posed by severe weather in the community. These factors have been offset by an increase in the frequency of severe floods and by a very marked increase in population density (Callaghan and Power 2014). For example, Australian Bureau of Statistics figures indicate that Sydney's population was 2953 in 1796, but thereafter grew rapidly to 629 503 in 1911, 3 204 696 in

1981, and 4 627 345 in 2011. An inadequate understanding of the risks posed by flood in some sections of the community might also have contributed to some of the recent deaths (Callaghan and Power 2014), highlighting the value of ongoing educational programs.

Evidence presented in section 3c, together with consistent information presented previously using observational data (Power et al. 1999a,b; Verdon et al. 2004; Kiem et al. 2003; Micevski et al. 2006;Kiem and Verdon-Kidd 2013) and climate model output (Walland et al. 2000; Arblaster et al. 2002;Power et al. 1995), indicates that the IPO modulates the frequency of severe coastal flooding on interdecadal time scales. The average number of severe floods, ECLs, TIs, and deaths during negative decadal phases of the IPO is higher in each case, than during positive decadal phases. Death tolls of 10 or more occurred in 12% of years during negative decadal phases, but in only 6% of years during positive decadal phases. The extent to which this decadal modulation reflects predictability beyond ENSO time scales is unclear (Power et al. 1995; Westra et al. 2015).

Note that in this study, we only considered deaths associated with freshwater flooding. Deaths arose from other factors linked to the same severe weather events, including death at sea. A large number of deaths at sea associated with severe weather off southeastern Australia occurred during the nineteenth century. For example, 121 lives were lost when The Dunbah was wrecked on Sydney Heads on 20 August 1857, in what has become known as the Dunbah storm. A storm on 12 July 1866 caused the loss of many vessels including the 552-ton Cawarra on Oyster Bank near Newcastle, with 62 lives lost. The total death toll at sea from this event was approximately 100. The SS Dandenong lost 40 lives during a storm on 10–12 September 1876 that became known as the Dandenong gale, and the Maitland gale on 5–7 May 1898 caused more than 50 fatalities in 15 vessels in NSW coastal waters, with 24 lost from the SS Maitland alone. We hope to provide additional information on deaths at sea linked to severe storms in southeastern Australia in a future study.

Note that in this study we restricted our attention to examining the impact of ENSO and the IPO on the variability. Future work could examine the influence of other modes of variability that are known to influence Australia (see, e.g., Risbey et al. 2009; Braganza et al. 2015, chapter 4).

APPENDIX A

Description of Additional Severe Floods

Callaghan and Power (2014) included a description of events from January 1860 to December 2012. In this appendix we extend this list up to December 2014. The event reference number, the type of weather event that triggered the flood [see Callaghan and Power (2014) for more details], and the death toll from freshwater flooding arising from the flood are provided. Callaghan and Power (2014)identified 253 severe floods, so the first additional severe flood identified here is assigned a reference number of 254. Only three additional severe floods occurred during January 2013–December 2014. (See Table A1.)

Table A1. Severe floods during January 2013–December 2014.

Event reference No.	Location	Date	Type	Fatalities
254	Brisbane Creeks, Lockyer Valley, and Grafton	27–28 Jan 2013	TI3	3
255	Mid-North Coast NSW	23–24 Feb 2013	TI3	2
256	Hunter	1–3 Mar 2013	TI2	0

APPENDIX B

Methods Used to Assess the Statistical Significance of the Correlation Coefficients

The statistical significance of the correlations is estimated using both the P98 method [as defined insection 2, i.e., the method described by Power et al. (1998)] and the Monte Carlo methods described below. The P98 method is ideally suited to red, normally distributed data. The Monte Carlo methods are suitable for non-normally distributed data. Results indicate that both methods give similar results for the data used in this investigation.

a. Correlations with the SOI

The statistical significance of correlations with the SOI is determined using both the P98 method and a Monte Carlo method described below.

The statistical significance of correlation coefficients can be affected by serial correlation. This is only true, however, if there is serial correlation present in both of the time series being correlated [see, e.g., formulas given by Power et al. (1998)]. As the lag(1yr) autocorrelation of the SOI is close to zero over the periods studied, we do not need to incorporate this affect into the assessment of statistical significance of correlation coefficients involving the SOI. This simplifies the Monte Carlo experiments that are required.

The synthetic data used when assessing the statistical significance of correlation coefficients between the SOI and each of severe floods, ECLs, TIs, and death tolls result in a randomly generated number derived from the 1860–2013 relative frequency distribution of these variables. These randomly generated time series are then correlated with the actual SOI time series. Three thousand synthetic time series are generated and correlated with the observed SOI time series. The proportion of correlation coefficients (expressed as a decimal, e.g., 0.01 for 1% of the simulations) that have magnitudes greater than or equal to the magnitude of the observed correlation coefficient provides the estimate of the pvalue in each case.

b. Correlations between 10-yr running averages and sums

Persistence obviously needs to be taken into account when assessing the statistical significance of correlation coefficients between two 10-yr running averages. However, use of 10-yr running values gives rise to statistics that are much more closely matched by normally distributed data (consistent with the central limit theorem), and so in this case the P98 method alone is used to assess the statistical significance.

ACKNOWLEDGMENTS

We thank Blair Trewin, Rob Smalley, and Gordon McKay for helpful discussions. We also thank the Met Office for the IPO time series and the National Library of Australia for making their excellent newspaper archive available online (https://trove.nla.gov.au/newspaper). Thanks are also given to Anthony Kiem, Ailie Gallant, Andrew King, and the anonymous reviewers for helpful comments on earlier drafts.

REFERENCES

1. Arblaster, J., G. Meehl, and A. Moore, 2002: Interdecadal modulation of Australian rainfall. Climate Dyn., 18, 519–531

2. Bouma, M. J., and H. J. van der Kaay, 1996: The El Niño–Southern Oscillation and the historic malaria epidemics of the Indian subcontinent and Sri Lanka: An early warning system for future epidemics? Trop. Med. Int. Health, 1, 86–96

3. Braganza, K., B. Murphy, B. Timbal, P. Hope, A. Dowdy, K. Hennessy, J. Bhend, and D. Kirono, 2015: Climate Change in Australia Technical Report: Information for Australia's Natural Resource Management Regions. CSIRO and Bureau of Meteorology, 216 pp. [Available online at http://www.climatechangeinaustralia.gov.au/media/ccia/2.1.5/cms_page_media/168/CCIA_2015_NRM_TechnicalReport_WEB.pdf.]

4. Cai, W., and P. van Rensch, 2012: The 2011 southeast Queensland extreme summer rainfall: A confirmation of a negative Pacific decadal oscillation phase? Geophys. Res. Lett., 39, L08702

5. Callaghan, J., and P. Helman, 2008: Severe Storms on the East Coast of Australia 1770–2008. Griffith Centre for Coastal Management, Griffith University, 240

6. Callaghan, J., and S. B. Power, 2011: Variability and decline in the number of severe tropical cyclones making land-fall over eastern Australia since the late nineteenth century. Climate Dyn., 37, 647–662

7. Callaghan, J., and S. B. Power, 2014: Major coastal flooding in southeastern Australia 1860–2012, associated deaths and weather systems. Aust. Meteor. Oceanogr. J., 64, 183–213.

8. Chiew, F. H. S., T. C. Piechota, J. A. Dracup, and T. A. McMahon, 1998: ENSO and Australian rainfall, streamflow and drought: Links and potential for forecasting. J. Hydrol., 204, 138–149

9. Christensen, J. H., and Coauthors, 2013: Climate phenomena and their relevance for future regional climate change. Climate Change 2013: The Physical Science Basis, T. F. Stocker et al., Eds., Cambridge University Press, 1217–1308.

10. Chung, C. T. Y., S. B. Power, J. M. Arblaster, H. A. Rashid, and G. L. Roff, 2014: Nonlinear precipitation response to El Niño and global warming in the Indo-Pacific. Climate Dyn., 42, 1837–1856

11. Folland, C. K., D. E. Parker, A. W. Colman, and R. Washington, 1999: Large scale modes of ocean surface temperature since the late nineteenth century. Beyond El Niño: Decadal and Interdecadal Climate Variability, A. Navarra, Ed., Springer-Verlag, 73–102.

12. Franks, S. W., and G. Kuczera, 2002: Flood frequency analysis: Evidence and implications of secular climate variability, New South Wales. Water Resour. Res., 38

13. Henley, B. J., J. Gergis, D. J. Karoly, S. B. Power, J. Kennedy, and C. K. Folland, 2015: A tripole index for the Interdecadal Pacific Oscillation. Climate Dyn., 45, 3077–3090

14. Hopkins, L. C., and G. J. Holland, 1997: Australian heavy-rain days and associated east coast cyclones: 1958–92. J. Climate, 10, 621–635

15. Kiem, A. S., and S. W. Franks, 2001: On the identification of ENSO-induced rainfall and runoff variability: A comparison of methods and indices. Hydrol. Sci. J., 46, 715–728

16. Kiem, A. S., and D. C. Verdon-Kidd, 2013: The importance of understanding drivers of hydroclimatic variability for robust flood risk planning in the coastal zone. Aust. J. Water Resour., 17, 126–134

17. Kiem, A. S., S. W. Franks, and G. Kuczera, 2003: Multi-decadal variability of flood risk. Geophys. Res. Lett., 30, 1035

18. King, A. D., L. V. Alexander, and M. G. Donat, 2013: Asymmetry in the response of eastern Australia extreme rainfall to low-frequency Pacific variability. Geophys. Res. Lett., 40, 2271–2277

19. King, A. D., N. P. Klingaman, L. V. Alexander, M. G. Donat, N. C. Jourdain, and P. Maher, 2014: Extreme rainfall variability in Australia: Patterns, drivers, and predictability. J. Climate, 27, 6035–6050

20. Kovats, R. S., M. J. Bouma, S. Hajat, E. Worrall, and A. Haines, 2003: El Niño and health. Lancet, 362, 1481–1489

21. Kuhnel, I., and L. Coates, 2000: El Niño–Southern Oscillation: Related probabilities of fatalities from natural perils in Australia. Nat. Hazards, 22, 117–138

22. Latif, M., T. P. Barnett, M. A. Cane, M. Flügel, N. E. Graham, H. von Storch, J.-S. Xu, and S. E. Zebiak, 1994: A review of ENSO prediction studies. Climate Dyn., 9, 167–179

23. McBride, J. L., and N. Nicholls, 1983: Seasonal relationships between Australian rainfall and the Southern Oscillation. Mon. Wea. Rev., 111, 1998–2004

24. McKeon, G. M., and Coauthors, 2004: Historical degradation episodes in Australia: Global climate and economic forces and their interaction with rangeland grazing systems. Pasture Degradation and Recovery in Australia's Rangelands, G. M. McKeon et al., Eds., Natural Resource Sciences, 27–86.

25. Micevski, T., S. W. Franks, and G. Kuczera, 2006: Multidecadal variability in coastal eastern Australian flood data. J. Hydrol., 327, 219–225

26. Parker, D., C. Folland, A. Scaife, J. Knight, A. Colman, P. Baines, and B. Dong, 2007: Decadal to multidecadal variability and the climate change background. J. Geophys. Res., 112, D18115

27. Pepler, A., A. Coutts-Smith, and B. Timbal, 2014: The role of east coast lows on rainfall patterns and inter-annual variability across the east coast of Australia. Int. J. Climatol., 34, 1011–1021

28. Power, S. B., and I. N. Smith, 2007: Weakening of the Walker circulation and apparent dominance of El Niño both reach record levels, but has ENSO really changed? Geophys. Res. Lett., 34, L18702

29. Power, S. B., and G. Kociuba, 2011: Impact of global warming on the SOI. Climate Dyn., 37, 1745–1754

30. Power, S. B., M. Haylock, R. Colman, and X. Wang, 1995: The predictability of interdecadal changes in ENSO and ENSO teleconnections. J. Climate, 8, 2161–2180

31. Power, S. B., F. Tseitkin, S. Torok, B. Lavery, and B. McAvaney, 1998: Australian temperature, Australian rainfall, and the Southern Oscillation, 1910–1996: Coherent variability and recent changes. Aust. Meteor. Mag., 47, 85–101.

32. Power, S. B., C. Folland, A. Colman, and V. Mehta, 1999a: Inter-decadal modulation of the impact of ENSO on Australia. Climate Dyn., 15, 319–324

33. Power, S. B., F. Tseitkin, V. Mehta, S. Torok, and B. Lavery, 1999b: Decadal climate variability in Australia during the 20th century. Int. J. Climatol., 19, 169–184

34. Pui, A., A. Lala, and A. Sharma, 2011: How does the interdecadal Pacific oscillation affect design floods in Australia? Water Resour. Res., 47, W05554

35. Risbey, J. S., M. J. Pook, P. McIntosh, M. C. Wheeler, and H. H. Hendon, 2009: On the remote drivers of rainfall variability in Australia. Mon. Wea. Rev., 137, 3233–3253

36. Salinger, M. J., S. McGree, F. Beucher, S. B. Power, and F. Delage, 2014: A new index for variations in the position of the South Pacific convergence zone 1910/11–2011/2012. Climate Dyn., 43, 881–892

37. Speer, M. S., 2008: On the late twentieth century decrease in Australian east coast rainfall extremes. Atmos. Sci. Lett., 9, 160–170

38. Speer, M. S., P. Wiles, and A. Pepler, 2009: Low pressure systems off the New South Wales coast and associated hazardous weather: Establishment of a database. Aust. Meteor. Mag., 58, 29–39.

39. Speer, M. S., J. Phillips, and B. N. Hanstrum, 2011: Meteorological aspects of the 31 March 2009 Coffs Harbour flash flood. Aust. J. Meteor. Oceanogr., 61, 201–210.

40. Verdon, D. C., A. M. Wyatt, A. S. Kiem, and S. W. Franks, 2004: Multidecadal variability of rainfall and streamflow: Eastern Australia. Water Resour. Res., 40, W10201

41. Walland, D., S. Power, and A. Hirst, 2000: Decadal climate variability in the CSIRO coupled GCM. Climate Dyn., 16, 201–211,

42. Westra, S., B. Renard, and M. Thyer, 2015: The ENSO–precipitation teleconnection and its modulation by the interdecadal Pacific oscillation. J. Climate, 28, 4753–4773,

Chapter 9

Climatic Variation at Thumba Equatorial Rocket Launching Station, India

K. V. S. Namboodiri, P. K. Dileep, and Koshy Mammen

Meteorology Facility, Thumba Equatorial Rocket Launching Station (TERLS), Vikram Sarabhai Space Centre (VSSC), Indian Space Research Organization, Thiruvananthapuram 695 022, India

ABSTRACT

Long-term (45 years) diversified surface meteorological records from Thumba Equatorial Rocket Launching Station (TERLS), India, were collected and analysed to study the long-term changes in the overall climatology, climatology pertained to a particular observational time, mean daily climatology in temperature, inter-annual variability in temperature, interannual variability in surface pressure, and rainfall for the main Indian seasons—South West and North East monsoons and inter-annual mean monthly anomaly structure in temperature. Results on various analyses show strong and vivid features contributed by climate change for this South Peninsular Indian Arabian Sea Coastal Station, and this paper may be a first time venture which discusses climate change imparted perturbations in several meteorological parameters in different time domains, like a specific time, daily, monthly, and interannually over a station. Being a coastal rocket launching station, climatic change information is crucial for long-term planning of its facilities as well as for various rocket range operational demands.

1. INTRODUCTION

Climate change refers to a statistically strong and significant variation in either the average state of the climate or in its variability, persisting for an extended period, typically in decades or longer. Short-period oscillations are statistically insignificant in the scenario of long-term climate change context. Climate change may be due to natural internal processes on earth (atmospheric, seismic, volcanic, and oceanic), external forces (variation in solar activity, rotation and revolution of earth), and more recently anthropogenic activities. It is now well established that anthropogenic activities cause intensification of the greenhouse effect and thereby contribute to the climate change. The paper by Oreskes [1] highlighted the listings in the ISI database on climate change from several abstracts published in refereed scientific journals between 1993 and 2003.

In India, the first time document on the concern of greenhouse effect and possible mitigation measures was published in the Science Reporter, CSIR, by Ayyar [2]. Several papers were published in a special session in Current Science (volume 90, 2006) focussing the concerns over the socioeconomic and scientific aspects attributed to climate change on Indian scenario. To develop high-resolution climate change simulations over the Indian region, Providing Regional Climates for Impact Studies (PRECIS) modelling system was applied [3], and it was observed that the warming is monotonously widespread over India with substantial spatial differences in the projected rainfall changes.

Different data sets have been used in order to examine the monotonic change in the level of global temperature by various investigators [4–7]. From the diversified methodologies adopted, it is observed that the global surface temperature has risen by 0.74°C with an error of ±0.18°C when estimated by a linear trend over the last 100 years (1906–2005) and 0.07°C ± 0.02°C per decade. An executive summary is made available by Solomon et al. [8] on climate change in the fourth assessment report of the Intergovernmental Panel on Climate Change (IPCC). Climate change features are local or regional highly influenced by agricultural practices, urbanisation, heat island effect, changing irrigation, aerosol modifications, desertification, deforestation, and decay processes of organic materials [9–

14]. Long-term changes in maximum temperature, minimum temperature, and their range have been analyzed over two stations in Cyprus by Price et al. [15] who established a reduction in range and thereby a possibility in increased cloud cover and tropospheric aerosols. Recently seasonal variability in mean sea level pressure extremes over the 11 zones of India and adjoining sea area has been studied by Singh et al. [16].

In the present paper, features on surface climate change over Thumba Equatorial Rocket Launching Station (TERLS), a coastal station near the South tip of Peninsular India, are critically examined. TERLS was established as an international rocket launching station dedicated to the United Nations for conducting meteorological and sounding rocket experiments on the geomagnetic equator. Meteorology group, TERLS has one of the heaviest responsibilities in the range to cater (i) weather forecasting and data needs of sounding rocket launch operations (in planning, pad operations and launching), (ii) climatological briefing to working crews including scientists to plan experiments, (iii) Organise various met. Observations, (iv) processing, analyses, storage and retrieval of various met. data (Class I surface observatory, autoweather stations, sonic anemometers, 50 m micro meteorological flux tower facility, meteorological balloon launches, rocket chaff derived wind data) and provide the data for users demand. For the planning of a scientific mission, the variability of climate is very crucial and thereby the importance of the present study on the climate change perspective. Importance of weather factors affecting rocket operations can be found elsewhere in detail [17, 18]. Steady increase in temperature anomaly at different levels in the troposphere over TERLS was investigated and documented in the Space research in India [19]. Joseph and Simon [20] have investigated weakening trend of the South West monsoon current through Peninsular India. Dash et al. [21] reported the decreasing frequency of monsoon depressions over the Indian region, and Krishnakumar et al. [22] have analysed significant decrease in South West monsoon rainfall while increase in post monsoon season over the whole Kerala State (also a meteorological subdivision), India. From the voluminous data bank, the present study analyses daily, monthly, seasonal, and interannual variability on different meteorological parameters based on both eye reading instruments and non-instrumental practices in cloud and visibility observations as per World Meteorological Organization (WMO) practices. Till the inception of long-term uninterrupted data collection through instrument-based cloud and visibility measurements, data collection from such non-instrumental means solely vested on conventional methods in surface meteorological observations adopted worldwide is to be accepted for a reasonable climatology extraction. The present work examines the imprints made by the climatic change on several meteorological parameters in different time domains, like a specific time, daily, monthly, and interannually unlike the climate change associated variations on a single meteorological parameter like temperature or rainfall alone have dealt in most of the climatic change studies.

2. DATA AND METHODOLOGY

A class I surface meteorological observatory was established at TERLS (8°32″N/76° 52″E, 6.7 amsl) by India Meteorological Department (IMD) approximately 70 m away from the coastline of the Arabian Sea coast (Figure 1). TERLS is situated hardly 8 km West North West of Thiruvananthapuram, the capital city of the Kerala state of India. Various meteorological data from 0830 IST (Indian Standard Time, IST = GMT + 0530 hr) observations for a period of 45 years (1964–2008), except for pressure where it is hourly data (1973–2008), are used in order to investigate changes encountered in meteorological parameters due to climate change. The data used are obtained from the following measurements at 0830 IST (1964–2008) namely, (i) maximum temperature (past 24 hr reported at 0830 IST), (ii) minimum temperature (past 24 hr reported at 0830 IST), (iii) past 24 hr rainfall (reported at 0830 IST), (iv) dry-bulb temperature at 0830 IST, (v) wet-bulb temperature at 0830 IST, (vi) relative humidity at 0830 IST, (vii) soil temperature at 0830 IST, (viii) non-instrumental eye observations on clouds and visibility at 0830 IST, (ix) wind speed and direction at 0830 IST, and (x) hourly surface pressure (1973–2008). Table 1 provides details on various instrumentation, exposure, maintenance, data span and the lowest and the highest in maximum absolute error values among various monthly statistics from 1964 to 2008.

Table 1. Details on instrumentation, data used, and the observed variation in maximum absolute error in monthly mean used in the discussion of climatology at 0830 IST.

S. no	Parameter	Data used	Instrument	Height (agl) (cm)	Exposure	Lowest and highest maximum absolute error	Make	Maintenance and calibration protocol
1	Maximum temperature (°C)	1964–2008	Maximum thermometer (Mercury-in-glass thermometer)	122	Stevenson screen	0.087 to 0.132 (°C)	IMD	
2	Minimum temperature (°C)	1964–2008	Minimum thermometer (spirit-in-glass)	122	Stevenson screen	0.061 to 0.145 (°C)	IMD	
3	Dry-bulb temperature (°C)	1964–2008	Drybulb thermometer (mercury-in-glass thermometer)	122	Stevenson screen	0.088 to 0.125 (°C)	IMD	
4	Rainfall (mm)		Nonrecording rain gauge of collector area 20 square cm	0	Ground			Daily, weekly, monthly, half yearly, annual, and periodic
4	Wet-bulb temperature (°C)	1964–2008	Wet-bulb thermometer	122	Stevenson screen	0.06 to 0.146 (°C)	IMD	
5	Relative humidity (%)	1964–2008	Drybulb thermometer, wetbulb thermometer, and psychrometer table.	122	Stevenson screen	0.512 to 0.916 (%)	IMD	
6	Soil temperature (°C)	1964–2008	Soil thermometer in metallic tube	−5		0.201 to 0.261 (°C)	IMD	
7	Pressure (mb)	1974–2008	Aneroid barograph (diaphragm type)	6.70 m amsl			IMD	
8	Wind speed	1964–2008	Anemometer	300		0.043 to 0.24 (m/s)	IMD/Dynalab	
9	Wind direction	1964–2008	Wind vane	300		Within 1 degree	IMD/Dynalab	

Figure 1. Location shows the class-I surface meteorological observatory at Thumba.

The above mentioned 45 years of data have partitioned into two climatological slots of 30 years each, namely, 1964 to 1993 and from 1979 to 2008, respectively, called as past data for the former and present data for the later. Maximum absolute error values in the total data set (1962–2008) are the summation of errors [23] of past and present data sets. Individual error is computed as the ratio of standard deviation to the number of observations of a particular data set. Trenberth [24] has established the climate shift as a prominent upward trend in temperature after 1976. In order to include this observation influence, half portion (15 years) period from 1979 to 1993 is made common in the latter half of past data and in the former part of present data. The difference obtained among present data and past data has been represented as Δ in the discussions throughout. To see genuineness in findings attributed to climate change, Δ values are compared with absolute errors so as to see that Δ do not fall within absolute errors of corresponding meteorological parameters.

3. RESULTS AND DISCUSSION

3.1. Δ Climatology of the Past 24 hr Observations

Maximum temperature, minimum temperature, and 24 hr accumulated rainfall are the only 3 meteorological parameters registered at 0830 IST having effect of the past 24 hr variability behaviour. Climatology for various parameters has been derived for past data and present data and presented in Tables 2, 3, 4, and 5 in which months that constitute the different seasons are mentioned. Derived Δ values are also incorporated in the above tables. Comparison between past data and present data maximum temperature climatology shows marked increase in present data climatology in relation to the past data counterpart, and thereby, positive values alone are observed in Δ. Δ values vary from +0.01°C in the North East monsoon (NE monsoon) month November to +0.43°C in the pre- South West monsoon (SW monsoon) month May. The NE monsoon season comprises October and November, whereas the four months from June to September designate SW monsoon season, which are the main two Indian seasons. In minimum temperature, all Δ values are positive with maximum value in the winter month February with a value +0.53°C. The observations among maximum and minimum temperature climatology derived in past data and present data along with Δ values can infer that the minimum daily temperatures have generally increased at a larger rate than the maximum daily temperatures. Karl et al. [25] studied the possible cause of increased urbanisation effects and Rebetez, and Beniston [26] justified the result of increased cloud cover in the large increase in minimum temperature and small increase in maximum temperatures.

Table 2. Climatology and Δ of maximum temperature, minimum temperature, mean daily temperature, mean range in temperature, and monthly rainfall.

Month	Maximum temperature °C			Minimum temperature °C			Mean daily temperature °C			Mean range in temperature °C			Monthly rainfall mm		
	1964–1993	1979–2008	Δ	1964–1993	1979–2008	Δ	1964–1993	1979–2008	Δ	1964–1993	1979–2008	Δ	1964–1993	1979–2008	Δ
JAN	32.59	32.76	+0.17	21.30	21.58	+0.28	26.94	27.16	+0.22	11.29	11.17	−0.12	11.80	11.10	−0.70
FEB	33.12	33.32	+0.20	21.89	22.42	+0.53	27.51	27.87	+0.36	11.24	10.91	−0.33	16.20	16.40	+0.20
MAR	33.84	34.07	+0.23	23.75	24.13	+0.38	28.78	29.09	+0.31	10.06	9.92	−0.14	33.20	34.80	+1.60
APR	34.25	34.41	+0.16	25.24	25.51	+0.27	29.76	29.95	+0.19	9.05	8.88	−0.17	104.10	99.40	−4.70
MAY	33.03	33.46	+0.43	25.14	25.50	+0.36	29.12	29.48	+0.36	7.97	7.96	−0.01	179.20	165.80	−13.40
JUN	30.36	30.71	+0.35	23.99	24.36	+0.37	27.20	27.53	+0.33	6.43	6.35	−0.08	292.70	272.20	−20.50
JUL	29.38	29.48	+0.10	23.36	23.79	+0.43	26.38	26.63	+0.25	6.03	5.68	−0.35	203.00	181.30	−21.70
AUG	29.27	29.60	+0.33	23.39	23.80	+0.41	26.35	26.70	+0.35	5.92	5.79	−0.13	141.50	120.90	−20.60
SEP	30.34	30.70	+0.36	23.68	24.00	+0.32	27.04	27.34	+0.30	6.71	6.67	−0.04	158.20	171.00	+12.80
OCT	31.19	31.53	+0.34	23.74	24.04	+0.30	27.50	27.76	+0.26	7.53	7.44	−0.09	204.90	241.40	+36.50
NOV	31.76	31.77	+0.01	23.21	23.57	+0.36	27.48	27.67	+0.19	8.53	8.20	−0.33	166.80	170.70	+3.90
DEC	32.30	32.56	+0.26	22.44	22.44	+0.00	27.35	27.50	+0.15	9.83	10.11	+0.28	66.60	53.70	−12.90

DEC, JAN, and FEB: winter; MAR, APR, and MAY: summer; JUN, JUL, AUG, and SEP: SW monsoon; OCT, NOV: NE monsoon.

Table 3. Climatology and Δ of dry-bulb temperature, wet bulb temperature, dew point temperature, relative humidity, and soil temperature.

Month	Dry-bulb temperature (0830 IST) °C			Wet-bulb temperature (0830 IST) °C			Dew point temperature (0830 IST) °C			Relative humidity (%) (0830 IST)			Soil temperature (0830 IST) °C		
	1964–1993	1979–2008	Δ	1964–1993	1979–2008	Δ	1964–1993	1979–2008	Δ	1964–1993	1979–2008	Δ	1964–1993	1979–2008	Δ
JAN	26.06	27.07	+1.01	22.33	22.90	+0.57	20.35	20.69	+0.34	71.36	68.85	−2.51	29.03	31.38	+2.35
FEB	26.77	27.71	+0.94	23.04	23.55	+0.51	21.12	21.43	+0.31	71.76	69.41	−2.35	30.41	32.54	+2.13
MAR	28.31	29.20	+0.89	24.58	24.94	+0.36	22.81	22.94	+0.13	72.55	69.53	−3.02	32.80	34.25	+1.45
APR	29.29	30.00	+0.71	26.09	26.30	+0.21	24.71	24.72	+0.01	76.56	73.61	−2.95	33.73	35.11	+1.38
MAY	28.70	29.43	+0.73	26.12	26.43	+0.31	25.02	25.17	+0.15	80.94	78.33	−2.61	32.46	34.16	+1.70
JUN	26.61	27.33	+0.72	25.20	25.57	+0.37	24.55	24.81	+0.26	88.51	86.43	−2.08	29.55	30.99	+1.44
JUL	25.76	26.36	+0.60	24.55	24.88	+0.33	23.95	24.21	+0.26	89.80	88.03	−1.77	29.01	30.00	+0.99
AUG	25.90	26.60	+0.70	24.60	24.96	+0.36	23.97	24.22	+0.25	89.40	86.98	−2.42	29.60	30.96	+1.36
SEP	26.62	27.37	+0.75	24.84	25.21	+0.37	24.02	24.25	+0.23	85.77	83.45	−2.32	31.10	32.23	+1.13
OCT	26.81	27.59	+0.78	24.90	25.33	+0.43	24.02	24.32	+0.30	85.01	82.61	−2.40	31.40	32.45	+1.05
NOV	26.90	27.63	+0.73	24.40	24.86	+0.46	23.27	23.58	+0.31	80.84	79.35	−1.49	30.99	32.02	+1.03
DEC	26.70	27.59	+0.89	23.35	23.72	+0.37	21.61	21.74	+0.13	74.48	71.39	−3.09	30.52	32.09	+1.57

DEC, JAN, and FEB: winter; MAR, APR, and MAY: summer; JUN, JUL, AUG, and SEP: SW monsoon; OCT, NOV: NE monsoon.

Table 4. Climatology and Δ of visibility code, low cloud, high cloud, and total cloud.

Month	Visibility code (0830 IST)			Low cloud (0830 IST)			Middle cloud (0830 IST)			High cloud (0830 IST)			Total cloud (0830 IST)		
	1964–1993	1979–2008	Δ	1964–1993	1979–2008	Δ	1964–1993	1979–2008	Δ	1964–1993	1979–2008	Δ	1964–1993	1979–2008	Δ
JAN	96.79	96.98	+0.19	0.55	0.47	−0.08	0.81	0.78	−0.03	1.18	1.35	+0.17	2.51	2.53	+0.02
FEB	96.75	96.91	+0.16	0.48	0.47	−0.01	0.66	0.58	−0.08	1.08	1.16	+0.08	2.21	2.17	−0.04
MAR	96.61	96.88	+0.27	0.62	0.71	+0.09	0.54	0.37	−0.17	1.06	1.21	+0.15	2.20	2.24	+0.04
APR	96.77	96.86	+0.09	1.25	1.57	+0.32	1.00	0.87	−0.13	1.72	1.79	+0.07	3.91	3.94	+0.03
MAY	96.70	96.68	−0.02	2.02	2.08	+0.06	1.45	1.43	−0.02	1.56	1.64	+0.08	4.87	4.68	−0.19
JUN	96.26	96.43	+0.17	2.86	2.84	−0.02	2.31	2.37	+0.06	1.16	1.49	+0.33	5.96	5.90	−0.06
JUL	96.10	96.37	+0.27	2.74	2.74	+0.00	2.50	2.82	+0.32	1.26	1.67	+0.41	5.96	6.06	+0.10
AUG	96.41	96.58	+0.17	2.27	2.38	+0.11	2.31	2.28	−0.03	1.31	1.86	+0.55	5.63	5.68	+0.05
SEP	96.48	96.63	+0.15	2.16	2.15	−0.01	1.63	1.62	−0.01	1.34	1.62	+0.28	4.93	4.90	−0.03
OCT	96.76	96.70	−0.06	1.75	1.96	+0.21	2.07	2.14	+0.07	1.29	1.73	+0.44	4.92	5.06	+0.14
NOV	96.68	96.65	−0.03	1.44	1.55	+0.11	1.86	1.94	+0.08	1.26	1.72	+0.46	4.40	4.65	+0.25
DEC	96.81	96.79	−0.02	1.01	0.91	−0.10	1.40	1.15	−0.25	1.20	1.40	+0.20	3.56	3.25	−0.31

DEC, JAN, and FEB: winter; MAR, APR, and MAY: summer; JUN, JUL, AUG, and SEP: SW monsoon; OCT, NOV: NE monsoon.

Table 5. Climatology and Δ of wind speed (scalar), wind speed (resultant), and wind steadiness factor.

Month	Wind speed (scalar) (0830 IST) ms^{-1}			Wind speed (resultant) (0830 IST) ms^{-1}			Wind steadiness factor (0830 IST) (%)		
	1964–1993	1979–2008	Δ		1979–2008	Δ	1964–1993	1979–2008	Δ
JAN	1.04	0.75	−0.29	0.71	0.33	−0.38	68.89	44.48	−24.41
FEB	1.07	0.81	−0.26	0.73	0.38	−0.35	67.95	47.36	−20.59
MAR	1.21	1.00	−0.21	0.83	0.56	−0.27	68.49	56.15	−12.34
APR	1.29	1.00	−0.29	0.92	0.63	−0.29	71.46	63.15	−8.31
MAY	1.45	1.04	−0.41	1.14	0.75	−0.39	78.40	72.44	−5.96
JUN	1.49	1.09	−0.40	1.03	0.8	−0.23	68.97	73.65	+4.68
JUL	1.68	1.35	−0.33	1.15	0.94	−0.21	68.08	69.60	+1.52
AUG	1.61	1.31	−0.30	1.36	1.05	−0.31	84.68	79.99	−4.69
SEP	1.57	1.10	−0.47	1.17	0.81	−0.36	74.57	74.06	−0.51
OCT	1.10	0.82	−0.28	0.64	0.40	−0.24	58.69	49.26	−9.43
NOV	0.97	0.74	−0.23	0.49	0.23	−0.26	50.29	31.17	−19.12
DEC	1.01	0.75	−0.26	0.67	0.37	−0.30	66.44	48.88	−17.56

DEC, JAN, and FEB: winter; MAR, APR, and MAY: summer; JUN, JUL, AUG, and SEP: SW monsoon; OCT, NOV: NE monsoon.

From each individual daily maximum and minimum temperature, mean daily temperatures ((maximum temperature + minimum temperature)/2) are computed and further derive the past data and present data climatologies of mean temperatures. Only positive Δ values are observed throughout the year, which clearly indicate that the temperature over this tropical coastal station is increasing. Temperature climatologies have shown the tropical monsoonal characteristic of two maxima and two minima in monthly variation.

The diurnal temperature range is calculated from individual maximum and minimum temperature value and calculated past data and present data temperature range climatologies. Range climatologies are featured by annual variation. The prominent feature of negative Δ values in range is seen, which shows that the difference between maximum temperature and minimum temperature in a day is decreasing. Also, the larger rate of increase in minimum temperatures than maximum temperatures results in these negative Δ values in the diurnal temperature range.

The rainfall Δ shows decrease in rainfall in present data for the summer months April and May. Well-marked negative Δ values of about 21 mm are observed in the SW monsoon months (June–August). An increase in rainfall activity is detected in the SW monsoon month September and in the NE monsoon months October and November (debted to monsoonal climate) with maximum value of +36.50 mm in October. The station weather is dominated by convective mesoscale systems during the NE monsoon. Probably this investigation poses the doubt about whether such meso-scale weather phenomena like thunderstorm formations are on increasing trend during NE Monsoon and thereby an increase in Δ.

3.2. Climatology of a Particular Observation at 0830 IST

Analyses are carried out on different parameters collected at 0830 IST. Δ in dry bulb temperature, wet bulb temperature, and dew point temperature obtained from the two climatologies show only positive values. It ranges from +0.60°C in July to +1.01°C in January for dry bulb temperature. Variations in Δ among the SW monsoon months are less in general for all the above three variables, even though a maximum value in Δ is observed in dry bulb temperature. The less variation of wet bulb and dew point temperatures may be as a result of moisture laden atmosphere during the period which imparts corresponding variations in the moisture parameters like wet bulb temperature and dew point temperature.

Station relative humidity values at 0830 IST show more than 70% throughout the year with maximum in SW monsoon of the order 89%. Low negative values in Δ relative humidity are detected as the air becomes drier in present data compared to past. As the air is getting dried, both dew point and the wet bulb temperatures are supposed to show negative Δ values, but the feature is not detected. This may be inferred as the dry bulb temperature is the key parameter in elevating Δ values of both dew point and wet bulb temperatures, and the meagre relative humidity decrease in present data than past data is insignificant to impart Δ changes in dew point and wet bulb temperatures. Climatologies of RH also have shown two maxima and two minima, whereas the secondary minima around the NE monsoon is shallow in character.

Significant positive Δ values are observed among the two climatologies in soil temperature with a low value of +0.99°C to a high of +2.35°C in July and January, respectively. Perhaps the Δ increase observed in the winter months January and February may contribute reduction in radiational cooling which leads to a reduction in fog formation probability over the station. Soil temperature climatologies also have shown tropical monsoonal type with two maxima and minima.

Wind speed (both scalar and resultant) is decreased in present climatologies. Mathematically resultant wind speed shall be determined as the square root of the sum of the squares of zonal and meridional wind components. The encountered Δ values that rise to a maximum value of the order 0.5 m/s are due to decrease in wind speeds in the present data climatology in comparison with past data climatology. Wind steadiness factor values at 0830 IST are computed as the ratio of mean resultant wind speed to mean scalar wind (average value of scalar wind speeds) expressed in %, and the two slot climatologies are made. Wind steadiness factor is a general meteorological technique in order to derive the quantitative estimate of steadiness of wind direction [27]. The feature of losing controllability of blowing winds from a specific direction (computed from wind steadiness) as the demand of a particular season might be justified as a result of heat island effect of the very fast developing urbanisation hardly 8 km East South East of the observatory and the increasing vehicle populations and thereby drastic modifications in at 0830 IST observations which belong to land breeze prevalence over the station in the non-SW monsoonal months (up to a maximum of 20% decrease). Throughout the day, strong seabreeze is in prevalence in the SW monsoon months. Even in the SW monsoon which is dominantly associated with unidirectional wind flow, slight deteriorations in the steadiness of the wind directions have observed. Both the above observations brought out the phenomenal changes in the steadiness of offshore winds (land breeze) at 0830 IST during the non-SW monsoonal months at greater extend and onshore wind (sea breeze) at 0830 IST during the SW monsoon steadiness in smaller grades. The feature of very prominent decrease in direction steadiness in preset data to the past is clearly detected. As negative Δ dominance exists in the non-SW monsoon months where change over of land-sea breezes are so pronounced for the station, this observation may be further studied in order to check any possible modifications in the meso-scale meteorological phenomena like sea/land breeze, thunderstorms due to associated thermal and moisture environment changes.

Visibility and cloud climatologies are also studied. As these observational procedures are non-instrumental, the data is too subjective and quality observation is solely vested on the experience of the weather observer. The observation shows, better visibility codes in the present data climatology. A visibility code of 96 and above signifies that a well-defined object (like a mobile phone tower) at least 1 km far can be distinctly differentiable. Among various cloud climatology, dominant positive Δ values are existed in high clouds through out the year which may be pointed out as one of the reason for subdued albedo effect which enhances greenhouse effect, and hereby surface temperature escalation. This may be the result of number of high cloud types (Ci and Cc) are less compared to the varieties of low and middle cloud types, which enables easy estimation for the observer while making counts on high clouds. The increase in high cloud climatology values in present data may be the result of increase in air traffic resulted contrails over the region which results an increase in cirrus cloud formations due to tropical dynamics unlike microphysical processes in mid-latitudes as in Minnis et al. [28], Zerefos et al. [29]. Eleftheratos et al. [30] have studied the variability of cirrus clouds over the tropics governed by the activity of El Nino Southern Oscillation, the Quasi Biennial Oscillation and solar activity. Possible radiation budget alterations, modifications at the time of occurrence of minimum and maximum temperatures are to be critically examined as a part of future investigations. Prominent two maxima (secondary maximum is very shallower compared to the primary) and two minima were seen in middle, high and total cloud climatology.

3.3. Mean Daily Climatology in Temperature

Mean of daily maximum, minimum, mean and range in temperature between daily maximum and minimum temperatures are computed and depicted for the 365 days in a year (Figures 2, 3, 4 and 5) along with their standard deviations. The increase in mean daily values in present data temperatures are quite obvious and standard deviation values are fallen within 2.7°C for all the four mean daily climatologies. A difference statistics for 365 days between present data to past data is made and is shown in Figure 6. Out of 365 days, more than 80% mean daily maximum, minimum and mean temperatures are distributed as greater than zero occurrences in the difference statistics. An elevated frequency of occurrence of more than 65% under less than zero class is a concrete feature which implies that the range among maximum and minimum temperature is alarmingly diminishing.

3.4. Interannual Variability

Inter annual variability has critically examined for annual mean of minimum, maximum and mean temperatures, annual rainfall in the SW and NE monsoon over the station and for the Kerala State and average annual surface pressure in the SW and NE monsoons for the station. Each inter annual variability data has undergone an 11 year moving average with end point extrapolation till the raw data availability to bring out long term trend and the trend line also has shown in each inter annual variability figures similar to Krishnakumar et al. [22] for the interannual variability of rainfall over Kerala. Interpolated values are obtained for each year from the trend and such a data set obtained, has been subjected to statistical significant test. interpolated data has not taken for a particular year, if the raw data point is missing for the corresponding year. This method has enabled no reduction in the degrees of freedom (df). To confirm the statistical significance on various 11 year moving averaged data and also to assess the strength of sample correlation coefficient (), linear regression t-test were carried out and the results are presented (Table 6). Generally strong and significant linear trend variations are observed. Even though linear trends of maximum and minimum temperatures have shown strong and significant statistical trends, the derived variable-temperature range has resulted a weak and insignificant test statistic. This may be statistically inferred as a variable derived out of two variables which favours strong and significant test statistic separately need not show the same test static of strong and significant.

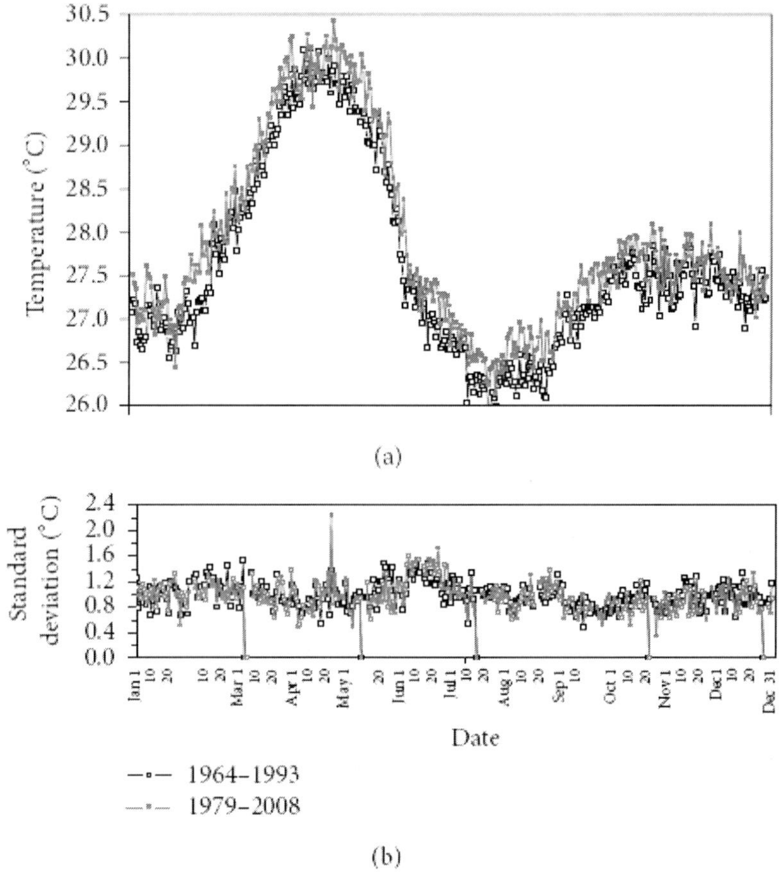

Figure 2. (a) Mean daily temperature climatologies in 1964–1993, 1979–2008, and their differences, along with (b) standard deviation of corresponding climatologies.

3.4.1. Interannual Variability in Temperature

To assess the interannual variability with temperature, annual mean minimum, mean maximum and mean of mean temperatures are found out for each year from 1964 to 2008. Also to ascertain deviation from the normal, each yearly mean value is deducted from the mean of the entire values available for present data and thereby extracted the anomaly in temperature. Mean and standard deviation of daily mean, maximum, minimum and daily temperature range are computed for a year and depicted in Figures 7, 8, 9 and 10. The fitted linear trend line slope (dy/dx) as well as correlation coefficient () values are annotated in graphs. It is seen that from both mean daily temperature as well as from the maximum temperature an abrupt prominent shift or a discernible upward trend is occurred after 1976 may be like the "climate shift" as stated by Trenberth [24]. The 45 year rate in mean daily, maximum and minimum are obtained as +0.585°C, +0.540°C and +0.630°C respectively or with decadal rates of (+)°C, (+)°C and (+)°C accordingly. In the obtained value for the global mean surface temperature as a whole (+)°C, surface temperatures over land have risen at about double the ocean rate after 1979 (more than (+)°C per decade versus (+)°C per decade) as per the Trenberth et al. [31]. From 1950 to 2004, the global land trend in diurnal temperature range was extracted as negative (−0.07°C per decade) although tropical areas are still underrepresented [31]. Kumar et al. [32] showed trends of increase in diurnal temperature range, significant increase in annual maximum temperature, and practically trendless annual minimum temperature over India in the last century. Our inter annual variability results coincide mostly with former studied data period which supports the decrease of diurnal temperature range, and the same is very vivid from Figure 6 discussed earlier. It infers that for this tropical coastal station the obtained decadal rate of (+)°C, which is in agreement with global ocean rate, may be due to the station being coastal in origin. The error value reduction by ±0.03°C for TERLS can be due to more stabilized single station data. Also rate of increase in trend for minimum temperature is high in comparison to maximum temperature. A decreasing trend in maximum temperature is noticed after 2005 with decreasing standard deviation values. This observation might be correct as a result of shifting the observatory since 1964 to hardly 30 m towards north in 2005 were smooth and free sea-breeze flow which enables abrupt reach of maximum temperature in the diurnal course around the time of sea breeze onset. The feature of attaining the maximum temperature of the day around the time of onset of sea breeze is known [33]. As the station is of coastal in origin, the occurrence of daily maximum temperature is very much linked to the setting and strong establishment of sea breeze front towards the land areas.

Mean monthly anomaly structures in minimum, maximum, and mean temperatures are presented in time section contour diagrams (Figures 11,

12 and 13). All contour depictions show well-marked temperature increase during the course of years and are so pronounced after 1976.

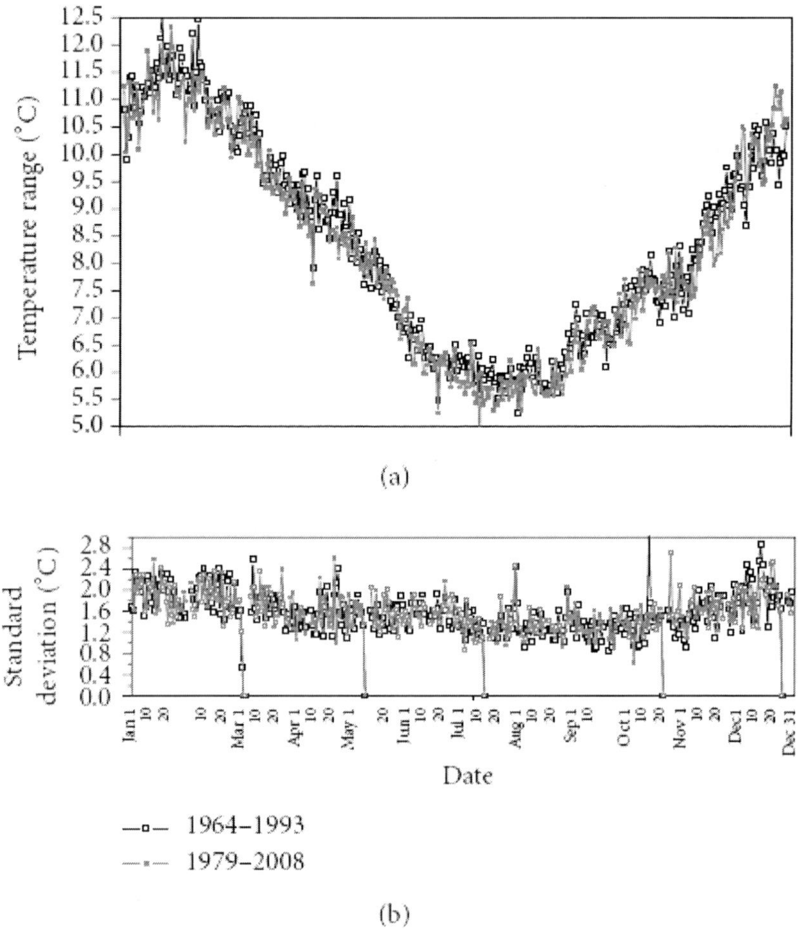

Figure 3. (a) Mean daily temperature range climatologies in 1964–1993, 1979–2008, and their differences, along with (b) standard deviation of corresponding climatologies.

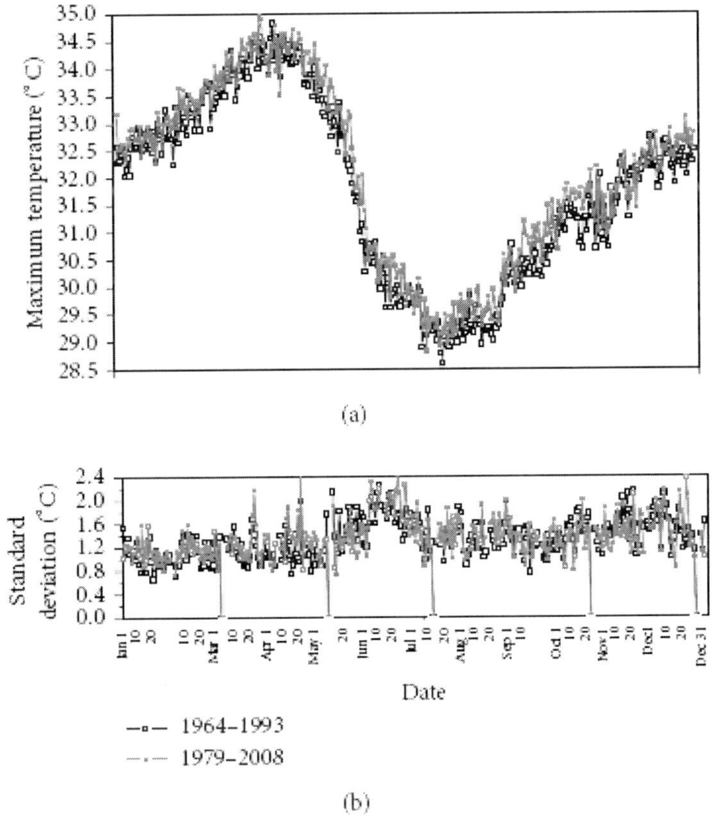

Figure 4. (a) Maximum daily temperature climatologies in 1964–1993, 1979–2008, and their differences, along with (b) standard deviation of corresponding climatologies.

3.4.2. Interannual Variability in SW and NE Monsoon Rainfall and Surface Pressure

Interannual variability in rainfall and surface pressure particularly for SW monsoon and NE monsoon are studied. A decrease in trend in SW monsoon and an increase in trend in NE monsoons are the observed features over the station with decadal variability of (−) mm and (+) mm, respectively (Figures 14 and 15). Analysis by considering initial rainfall values (at the beginning year on the fitted straight line) of South West monsoon, North East monsoon, corresponding slope of them as 817.822 mm, 296.535 mm,

−21.5 mm/decade and 39.3 mm/decade, respectively, shows the likelihood exceed of North East monsoon rainfall over South West monsoon over this region after the year 2050 if the current trend continues. Similar analyses are carried out on the slot 1965 to 2007 for the Kerala meteorological subdivision from the longest instrumental rainfall series of the Indian regions (1813–2006) ([34]; http://www.tropmet.res.in/). The actual rainfall amount over Kerala during 2007 is obtained from the quarterly journal the MAUSAM (July 2008 and October 2008 issues). Decadal variations show a decreasing trend in SW monsoon by (−) mm and increasing trend in NE monsoon by (+) mm (Figures 16 and 17).

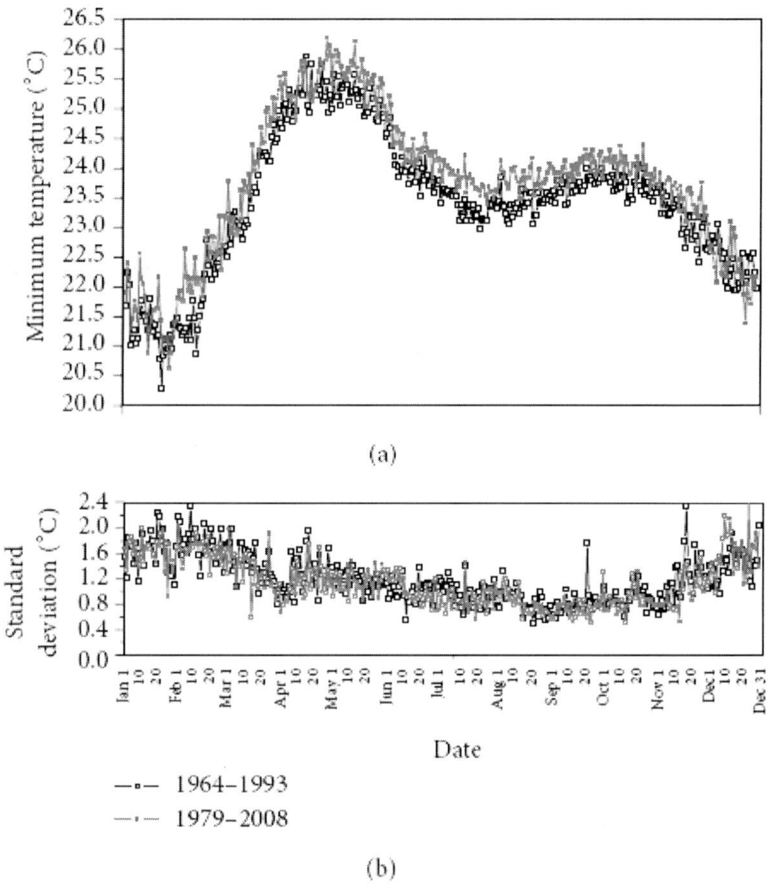

Figure 5. (a) Minimum daily temperature climatologies in 1964–1993, 1979–2008, and their differences, along with (b) standard deviation of corresponding climatologies.

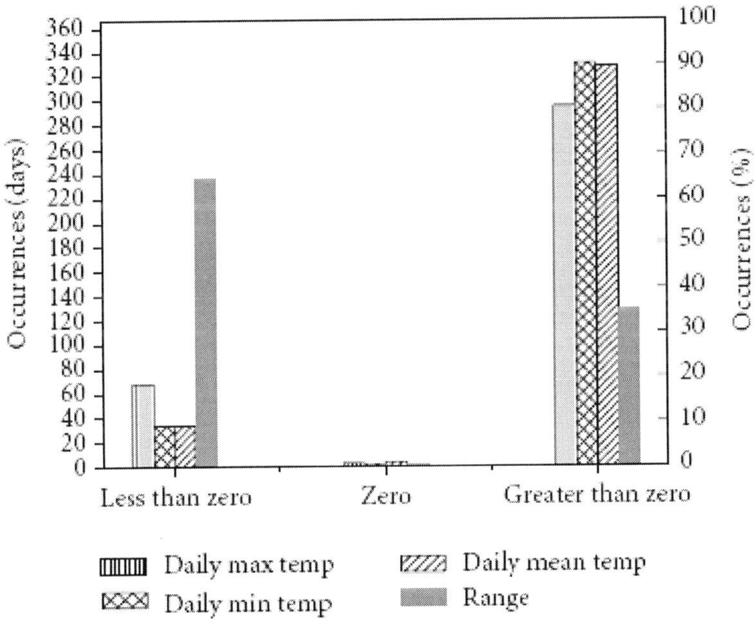

Figure 6. Difference statistics for 365 days between present data and past data on daily maximum, minimum, mean temperatures, and range in temperature within a day.

Interannual variability in surface pressure was derived and depicted for SW monsoon (Figure 18) and NE monsoon (Figure 19). For this, every hourly data available since 1973 was subjected to mean monthly value computation. Number of observations available ranges from 1100 to 2245 in SW monsoon and 500 to 1495 in NE monsoon, respectively, and provided along with the standard deviation values in the above figures. Since atmospheric pressure is a highly stabilized property over a region for a particular season, the decadal trend variability observed is (+) mb and (–) mb for SW monsoon and NE monsoon, respectively. The positive (increase) and negative (decrease) interannual trends observed in SW monsoon and NE monsoon, respectively, might have contributed possible implications in the development of associated synoptic scale meteorological systems (position and orientation of troughs/ridges and low pressures) and thereby corresponding decrease and increase in SW monsoon and NE monsoon rainfall trends.

Table 6. Linear trends of 11 year moving averaged data on various inter-annual variability. Level of significance was choosen at 0.05% to read out critical t values (t_c) corresponding to the degree of freedom (*df*). If $r > 0.5$ strong, $r < 0.5$ weak, $t > t_c$ significant, and $t < t_c$ not significant.

No.	Interannual variability in	r	Slope (dy/dx) ± error	Calculated t	*df*	Comments	Data missing
1	Mean temperature	0.891	0.013 ± 0.001	12.08154	38	Strong and significant. (Figures 7(a) and 7(b))	1974, 1975, 1976, 1982, and 1992
2	Mean maximum temperature	0.725	0.0121 ± 0.002	6.50392	38	Strong and significant. (Figures 8(a) and 8(b))	1974, 1975, 1976, 1982, and 1992
3	Mean minimum temperature	0.952	0.014 ± 0.000	19.29677	38	Strong and significant. (Figures 9(a) and 9(b))	1974, 1975, 1976, 1982, and 1992
4	Mean temperature range	−0.190	−0.002 ± 0.002	−1.19078	38	Weak and not significant (Figure 10(a)).	1974, 1975, 1976, 1982, and 1992
5	South West monsoon rainfall (Thumba)	−0.477	−2.145 ± 0.616	−3.483	41	Weak and significant (Figure 14)	1964, 1998
6	North East monsoon rainfall (Thumba)	0.848	3.928 ± 0.393	10.000	39	Strong and significant (Figure 15)	1964, 1977, 1992, and 1998
7	South West monsoon rainfall (Kerala)	−0.721	−6.149 ± 0.934	−6.587	40	Strong and significant (Figure 16)	1964, 2007, and 2008
8	North East monsoon rainfall (Kerala)	0.737	3.430 ± 0.491	6.986	41	Strong and significant (Figure 17)	1964, 2008
9	South West monsoon surface pressure	0.445	0.006 ± 0.002	2.899	34	Weak and significant (Figure 18)	1964 to 1972
10	North East monsoon surface pressure	−0.616	0.010 ± 0.002	−4.555	34	Strong and significant (Figure 19)	1964 to 1972

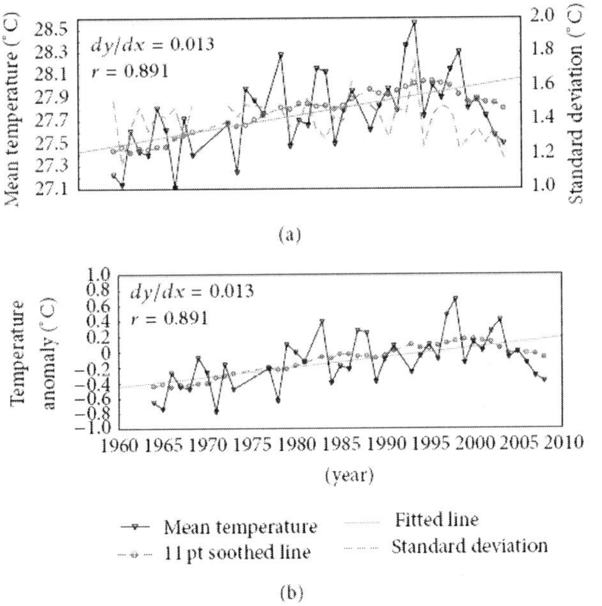

Figure 7. (a) Yearly mean temperature along with standard deviation and (b) anomaly.

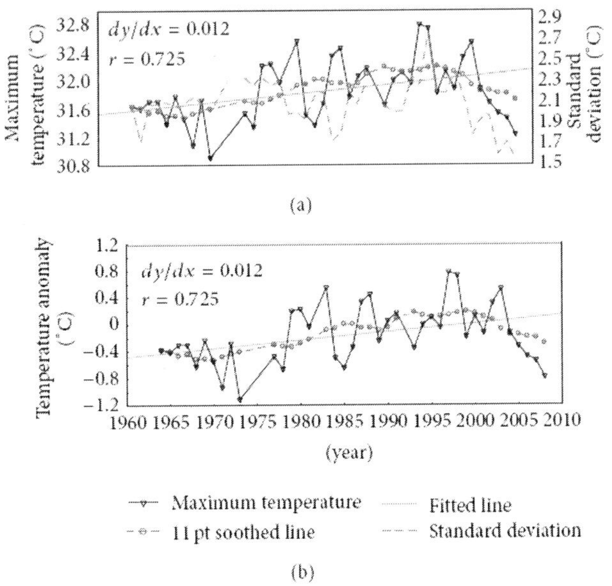

Figure 8. (a) Yearly mean maximum temperature along with standard deviation and (b) anomaly.

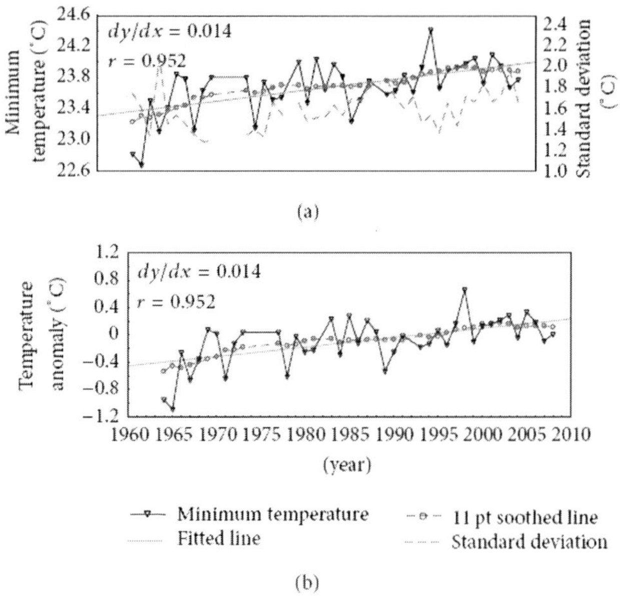

Figure 9. (a) Yearly mean minimum temperature along with standard deviation and (b) anomaly.

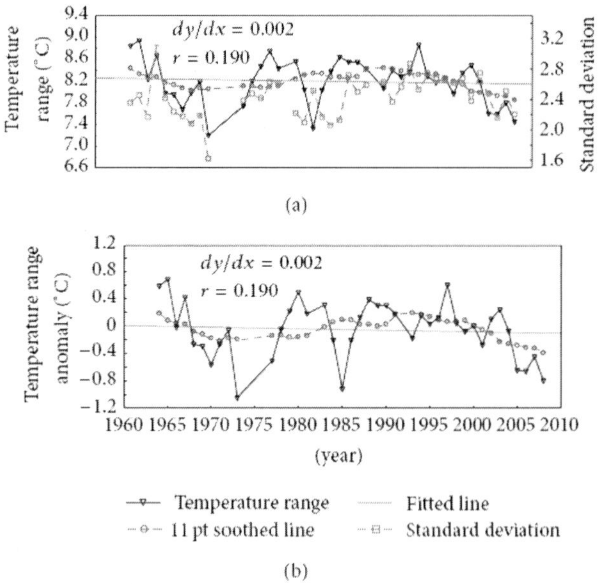

Figure 10. (a) Yearly mean temperature range along with standard deviation and (b) anomaly.

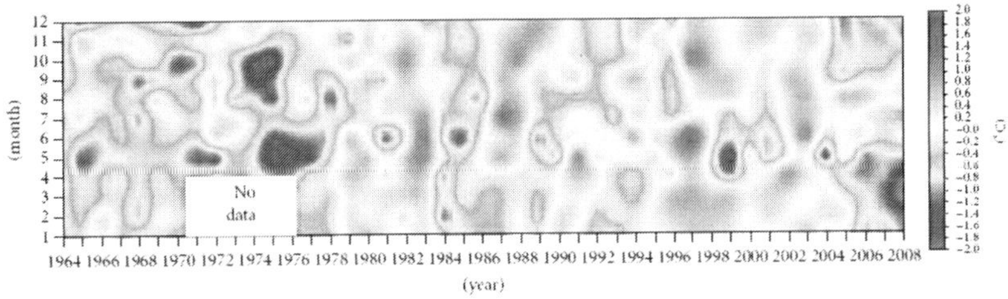

Figure 11. Maximum temperature anomaly.

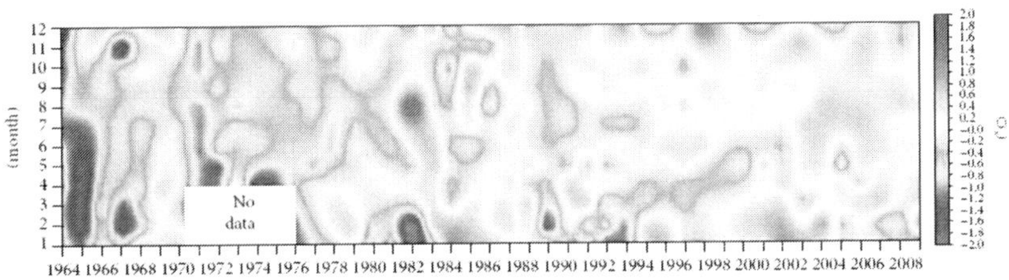

Figure 12. Minimum temperature anomaly.

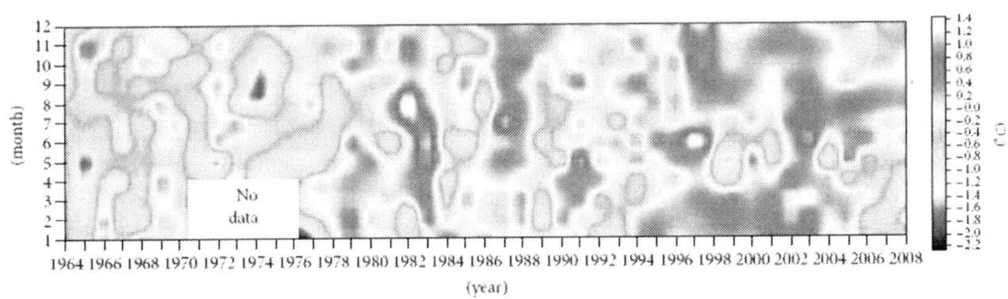

Figure 13. Mean temperature anomaly.

4. CONCLUSION

Long-term surface meteorological records from TERLS were collected and analysed to study the long-term changes in overall climatology, climatology pertained to a particular time of observation, mean daily climatology in temperature, interannual variability in temperature, interannual variability in SW and NE monsoon rainfall, and mean monthly anomaly structure in temperature. Results on various analyses show strong and vivid features contributed by climate change. The salient features observed contributed by climate change are as follows:(1)All the temperatures show increase in their behaviour with slight decrease in RH between the present climatology and past climatology.(2)Steadiness of wind decreases especially in non-SW monsoon months, which gives an indirect signature of possible modifications in meso-scale systems.(3)High cloud climatology increases and thereby probable modifications in radiation budget, time of occurrence of maximum and minimum temperatures.(4)Daily statistics for 365 days show more than 80% exceedance in present climatology in comparison to past in minimum, maximum, and mean temperature.(5)Interannual variability analysis in mean temperature for this tropical coastal station shows +1.3°C/100 years which is same as that of global ocean rate.(6)Observed interannual rainfall variability in SW monsoon and North East monsoon shows decrease in trend by –21.45 mm/decade and increase in trend by +39.3 mm/decade, respectively. Prognosis on this observation points out the likelihood exceedance of NE monsoon rainfall over the South West monsoon after the year 2050 for the station. But it is prejudicial to infer that the rainfall increase in NE monsoon can compensate the forecasted decrease in SW monsoon, since NE monsoon is mostly of heavy rainfall in nature as a result of meso-scale convective systems. Such rainfall runoff is very fast than the slow water infiltration feature in association with SW monsoon rainfall, which is essential for irrigation and agricultural needs.(7)Decadal variability in surface pressure has shown +0.06 mb and –0.10 mb for SW monsoon and NE monsoon, respectively.

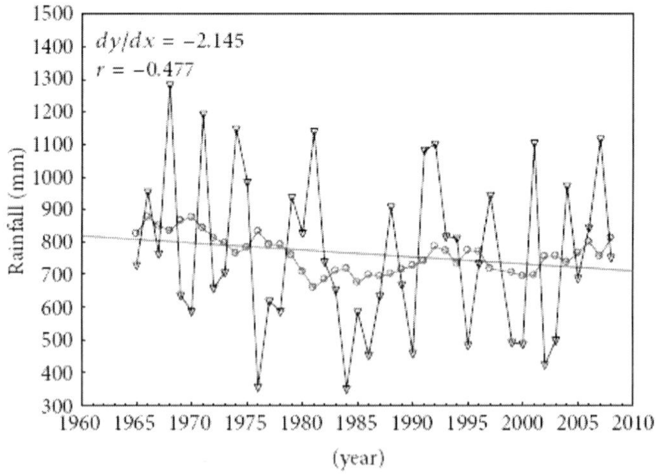

Figure 14. South West monsoon rainfall over Thumba.

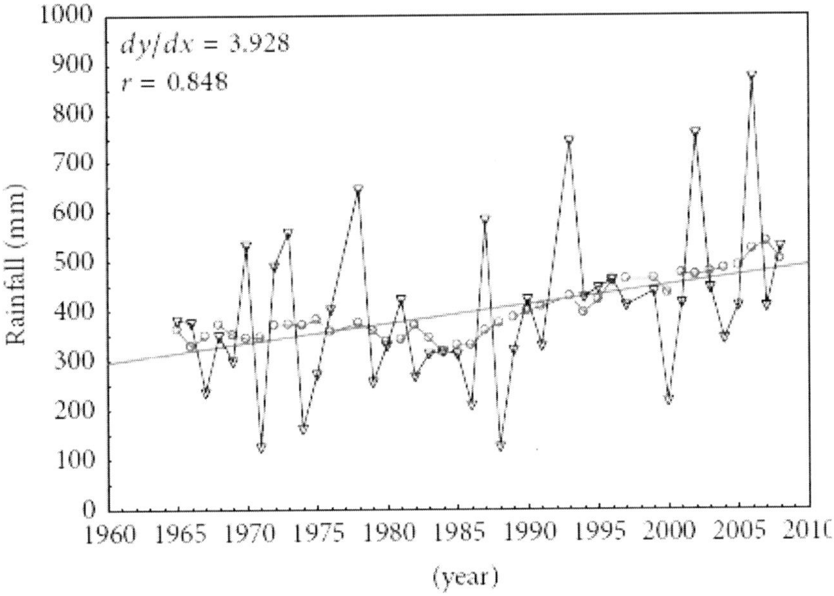

Figure 15. North East monsoon rainfall over Thumba.

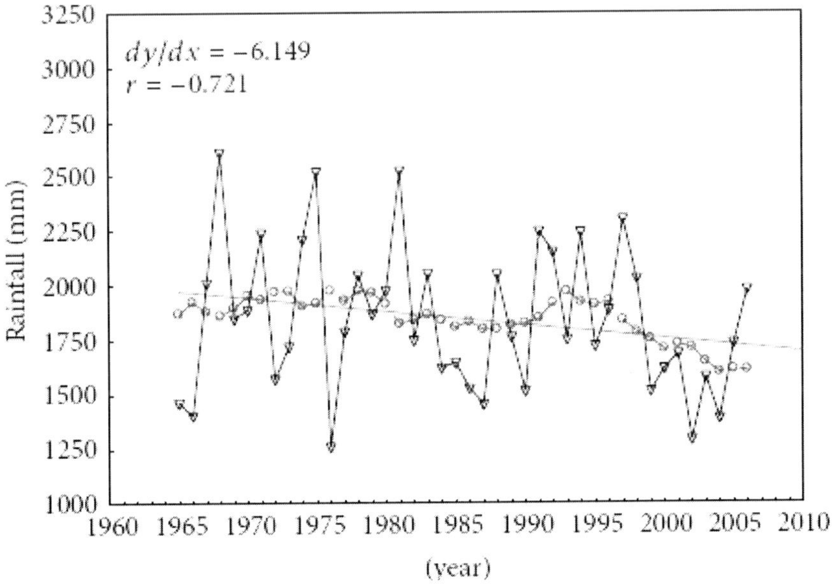

Figure 16. South West monsoon rainfall over Kerala.

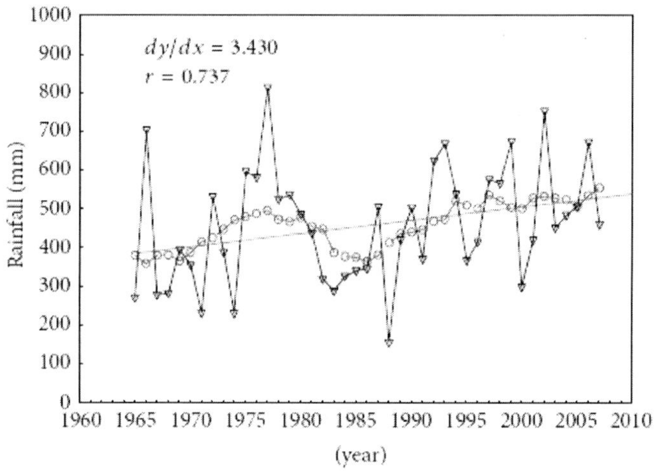

Figure 17. North East monsoon rainfall over Kerala.

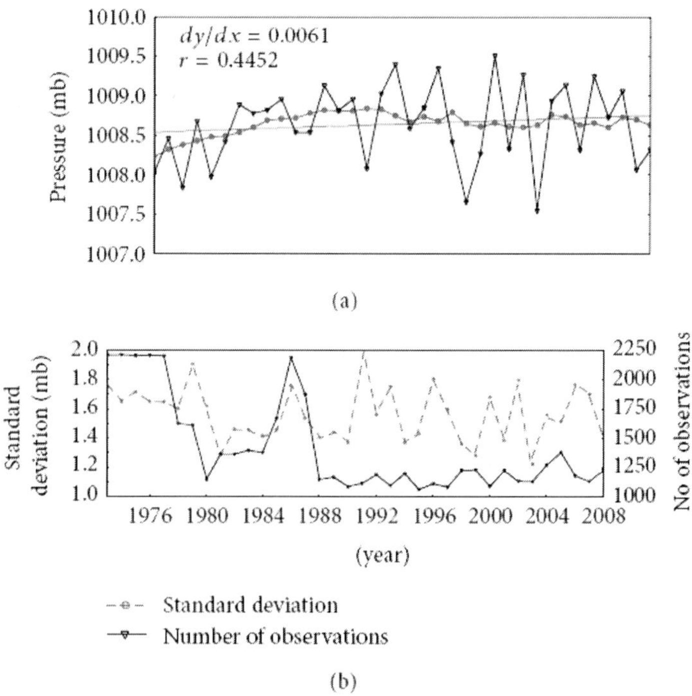

Figure 18. (a) Interannual variability in surface pressure for South West monsoon season along with (b) standard deviation and number of observations.

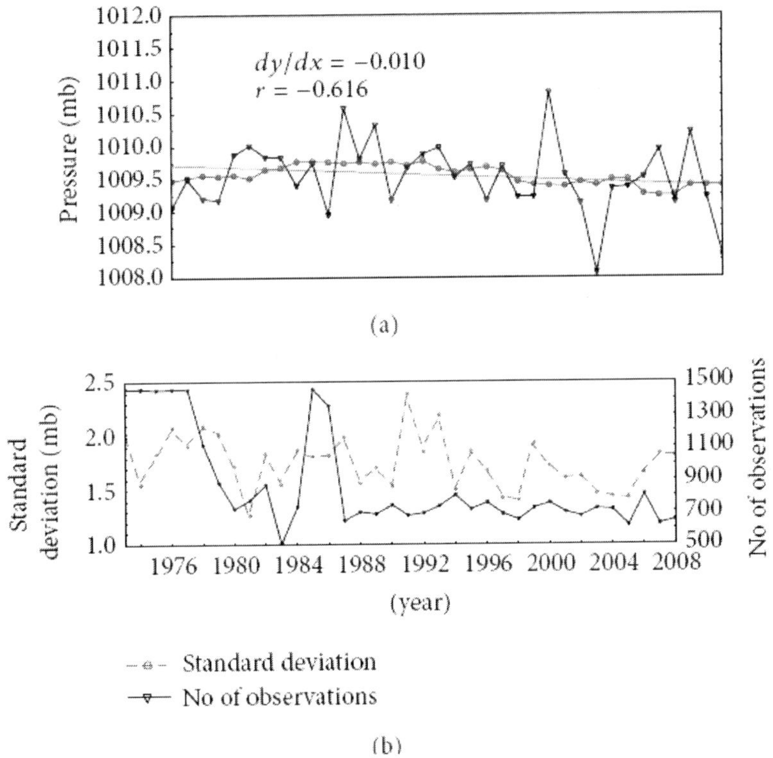

Figure 19. (a) Interannual variability in surface pressure for North East monsoon season along with (b) standard deviation and number of observations.

The imprints due to climate change on various meteorological parameters over this coastal station proved that the region also faces vulnerable situation due to the phenomena, and therefore, necessary engineering methods/socioeconomic steps are to be undertaken.

ACKNOWLEDGMENTS

The availability of this unique data for a continuous period of 45 years is the outcome of concerted and whole hearted efforts put in by the staffs of TERLS/India Meteorological Department. It is a pleasure to acknowledge with esteemed gratitude the team that worked since 1963 at the TERLS,

Thumba, whose efforts have yielded the voluminous unique data bank which is available for investigations. The authors are extremely grateful to Dr. K Krishnamoorthy, Director, Space Physics Laboratory, VSSC, and the reviewers for their valuable comments that helped to improve the paper to the final form.

REFERENCES

1. N. Oreskes, "The scientific consensus on climatic change," Science, vol. 306, p. 1686, 2004.

2. H. Ayyar, Effect of Pollution on Weather. Science Reporter. Council of Scientific and Industrial Research (CSIR), 1973.

3. K. Rupa Kumar, A. K. Sahai, K. Krishna Kumar et al., "High-resolution climate change scenarios for India for the 21st century," Current Science, vol. 90, no. 3, pp. 334–345, 2006.

4. P. Brohan, J. J. Kennedy, I. Harris, S. F. B. Tett, and P. D. Jones, "Uncertainty estimates in regional and global observed temperature changes: a new data set from 1850," Journal of Geophysical Research D, vol. 111, no. 12, Article ID D12106, 2006.

5. T. M. Smith and R. W. Reynolds, "A global merged land-air-sea surface temperature reconstruction based on historical observations (1880–1997)," Journal of Climate, vol. 18, no. 12, pp. 2021–2036, 2005.

6. K. M. Lugina, P. Y. Groisman, K. Y. Vinnikov, V. V. Koknaeva, and N. A. Speranskaya, "Monthly surface air temperature time series area-averaged over the 30-degree latitudinal belts of the globe, 1881-2004 In Trends: a compendium of data on global change," Carbon Dioxide Information Analysis Centre, Oak Ridge National Laboratory, US Department of Energy, Oak Ridge, Tenn, USA, 2005, http://cdiac. esd.ornl. gov/trends/ temp/lugina/lugina.html.

7. J. Hansen, R. Ruedy, M. Sato et al., "A closer look at United States and global surface temperature change," Journal of Geophysical Research D, vol. 106, no. 20, pp. 23947–23963, 2001.

8. S. Solomon, D. Quin, M. Manning, et al., "Contribution of working group I to the Fourth Assessment Report of the IPCC," Tech. Rep. IPCC-AR4, 2007.

9. R. C. Balling Jr. and S. W. Brazel, "Recent changes in Phoenix, Arizona summertime diurnal precipitation patterns," Theoretical and Applied Climatology, vol. 38, no. 1, pp. 50–54, 1987.

10. R. Bornstein and Q. Lin, "Urban heat islands and summertime convective thunderstorms in Atlanta: three case studies," Atmospheric Environment, vol. 34, no. 3, pp. 507–516, 2000.

11. P. G. Dixon and T. L. Mote, "Patterns and cause of Atlanda's urban heat island-initiated precipitation,"Journal of Applied Meteorology, vol. 42, pp. 1273–1284, 2003.

12. F. Fujibe, "Long-term surface wind changes in the Tokyo metropolitan area in the afternoon of sunny days in the warm season," Journal of the Meteorological Society of Japan, vol. 81, no. 1, pp. 141–149, 2003.

13. T. Inoue and F. Kimura, "Urban effects on low-level clouds around the Tokyo metropolitan area on clear summer days," Geophysical Research Letters, vol. 31, no. 5, Article ID L05103, 2004.

14. J. M. Shepherd, H. Pierce, and A. J. Negri, "Rainfall modification by major urban areas: observations from spaceborne rain radar on the TRMM satellite," Journal of Applied Meteorology, vol. 41, no. 7, pp. 689–701, 2002.

15. C. Price, S. Michaelides, S. Pashiardis, and P. Alpert, "Long term changes in diurnal temperature range in Cyprus," Atmospheric Research, vol. 51, no. 2, pp. 85–98, 1999.

16. H. N. Singh, S. D. Patil, S. D. Bansod, and N. Singh, "Seasonal variability in mean sea level pressure extremes over the Indian region," Atmospheric Research, vol. 101, no. 1-2, pp. 102–111, 2011.

17. J. Kingwell, J. Shimizu, K. Narita, H. Kawabata, and I. Shimizu, "Weather factors affecting rocket operations: a review and case history," American Meteorological Society, vol. 72, pp. 778–793, 1991.

18. R. E. Turner and C. K. Hill, Terrestrial Environment (Climatic) Guidlines for Use in Aerospace Vehicle Development, 1982 Revision, Technical Memor. 82473, National Aeronautics and Space Administration, 1982.

19. "Space Research in India," in Proceedings of the 38th COSPAR Meeting, pp. 199–204, Bremen, Germeny, 2010.

20. P. V. Joseph and A. Simon, "Weakening trend of the southwest monsoon current through peninsular India from 1950 to the present," Current Science, vol. 89, no. 4, pp. 687–694, 2005.

21. S. K. Dash, J. R. Kumar, and M. S. Shekhar, "On the decreasing frequency of monsoon depressions over the Indian region," Current Science, vol. 86, no. 10, pp. 1404–1411, 2004.

22. K. N. Krishnakumar, G. S. L. H. V. Prasada Rao, and C. S. Gopakumar, "Rainfall trends in twentieth century over Kerala, India," Atmospheric Environment, vol. 43, no. 11, pp. 1940–1944, 2009.

23. National Council of Educational Research and Training (NCERT), India, Text book of physics—part I for class 11, 2006.

24. K. E. Trenberth, "Recent observed interdecadal climate changes in the Northern Hemisphere," Bulletin of the American Meteorological Society, vol. 71, no. 7, pp. 988–993, 1990.

25. T. K. Karl, H. F. Diaz, and G. Kukla, "Urbanization: its detection and effect in the United States climate record," Journal of Climate, vol. 1, pp. 1099–1123, 1988.

26. M. Rebetez and M. Beniston, "Changes in sunshine duration are correlated with changes in daily temperature range this century: an analysis of Swiss climatological data," Geophysical Research Letters, vol. 25, no. 19, pp. 3611–3613, 1998.

27. K. V. S. Namboodiri, P. K. Dileep, and K. Mammen, "Wind steadiness up to 35 km and its variability before the South West Monsoon onset and the withdrawal," Mausam, vol. 63, pp. 275–282, 2012.

28. P. Minnis, A. J. Kirk, P. Rabindra, and P. Dung, "Contrails, cirrus trends, and climate," Journal of Climate, vol. 17, pp. 1671–1685, 2004.

29. C. S. Zerefos, K. Eleftheratos, D. S. Balis, P. Zanis, G. Tselioudis, and C. Meleti, "Evidence of impact of aviation on cirrus cloud formation," Atmospheric Chemistry and Physics, vol. 3, no. 5, pp. 1633–1644, 2003.

30. K. Eleftheratos, C. S. Zerefos, C. Varotsos, and I. Kapsomenakis, "Interannual variability of cirrus clouds in the tropics in el niño southern oscillation (ENSO) regions based on international satellite cloud climatology project (ISCCP) satellite data," International Journal of Remote Sensing, vol. 32, no. 21, pp. 6395–6405, 2011.

31. K. E. Trenberth, P. D. Jones, and P. Ambenje, "Observations: surface and climate change," in Climate Change 2007: The Physical Science Basis. Contribution of working group I to the Fourth Assessment Report of the Intergovernmental Panel on Climate Change, S. Solomon, D. Qin, and M. Manning, Eds., Cambridge University Press, Cambridge, UK, 2007.

32. K. R. Kumar, K. K. Kumar, and G. B. Pant, "Diurnal asymmetry of surface temperature trends over India," Geophysical Research Letters, vol. 21, no. 8, pp. 677–680, 1994.

33. T. R. Sivaramakrishnan and P. S. Prakash Rao, "Sea-breeze features over Sriharikota, India,"Meteorological Magazine, vol. 118, no. 1400, pp. 64–67, 1989.

34. N. A. Sontakke, N. Singh, and H. N. Singh, "Instrumental period rainfall series of the Indian region (AD 1813–2005): revised reconstruction, update and analysis," Holocene, vol. 18, no. 7, pp. 1055–1066, 2008.

Chapter 10

Observed trends and changes in daily temperature and precipitation extremes over the Koshi river basin 1975–2010

Arun B. Shrestha[a]*, Sagar R. Bajracharya[a], Aseem R. Sharma[a], Chu Duo[c], Ashwini Kulkarni[d]

[a] River Basins, International Centre for Integrated Mountain Development (ICIMOD), Kathmandu, Nepal

[b] Natural Resources and Environmental Studies Program, University of Northern British Columbia, Prince George, Canada

[c] Institute of Tibetan Plateau Atmospheric and Environmental Sciences, Tibet Autonomous Region, Lhasa, China

[d] Forecasting Research Division, Indian Institute of Tropical Meteorology, Pune, India

ABSTRACT

The Koshi river basin is a sub-basin of the Ganges shared among China, Nepal, and India. The river system has a high potential for investment in hydropower development and for irrigation in downstream areas. The upper part of the basin contains a substantial reserve of freshwater in the form of snow and glaciers. Climate variability, climate change, and climate extremes might impact on these reserves, and in turn impact on systems that support livelihoods, such as agriculture, biodiversity and related ecosystem services. Climatological variability and trends over the Koshi river basin were studied using RClimDex. Daily temperature data (20 stations) and precipitation data (50 stations) from 1975 to 2010 were used in the analysis. The results show that the frequency and intensity of weather extremes are increasing. The daily maximum temperature (TXx) increased by 0.1 °C decade^{-1} on average between 1975 and 2010 and the minimum (TNn) by 0.3 °C decade^{-1}. The number of warm nights increased at all stations. Most of the extreme temperature indices showed a consistently different pattern in the mountains than in the Indo-Gangetic plains, although not all results were statistically significant. The warm days (TX90p), warm nights (TN90p), warm spell duration (WSDI), and diurnal temperature range (DTR) increased at most of the mountain stations; whereas monthly maximum and minimum values of daily maximum temperature, TX90p, cool nights (TN10p), WSDI, cold spell duration indicator (CSDI), DTR decreased at the stations in the Indo-Gangetic plains, while the number of cold days increased. There was an increase in total annual rainfall and rainfall intensity, although no clear long-term linear trend, whereas the number of consecutive dry days increased at almost all stations. The results indicate that the risk of extreme climate events over the basin is increasing, which will increase people's vulnerability and has strong policy implications.

1 INTRODUCTION

For decades, global climate change studies using observational temperature and precipitation data focused only on changes in mean values and there were few studies of climate extremes. In recent decades, because the issue of climate change became more prominent, the research on past climate trends has progressed enormously, especially for precipitation and temperature (Manton *et al.*, **2001**; Frich *et al.*, **2002**; Alexander *et al.*, **2006**; Klein Tank *et al.*, **2006**; Donat *et al.*, **2013**; Whan *et al.*, **2013**). The Second Assessment Report of the Intergovernmental Panel on Climate Change (IPCC) highlighted the fact that insufficient studies were available on observed historical trends in climate extremes, and suggested that more studies would give a clearer indication of what changes might already have occurred in the global climate (Nicholls *et al.*, **1996**). At the time it was already envisaged that one of the effects of global warming might be an increase in weather and climate extremes. Studies of climate extremes are also useful for the verification of climate model results and thus increasing confidence in projections of future climate (Kruger, **2006**).

Very few studies have been made of climate extremes in Nepal or the adjacent countries in the Hindu Kush Himalayan region (especially not in a basin wise context). Hingane *et al.* (**1985**) made a detailed study of historical time series of temperature data from India and found a definite warming trend in mean annual all India temperature with an increase of 0.4 °C over the preceding century. Shrestha *et al.* (**1999**) analysed the trend in maximum temperature in Nepal and found an average annual warming of 0.06 °C $year^{-1}$ between 1971 and 1994. The increase in maximum temperature was more pronounced in the high-altitude regions. Some results indicate the importance of spatial variation. An analysis of climate changes in the Nepalese part of the Koshi river basin showed a significant increase in temperature and precipitation in a few of the areas within the basin, decreasing trends in a very few areas, but no trends at most stations (Sharma *et al.*, **2000**). A study by Islam (**2009**) showed that wet days over Nepal are increasing; consecutive dry days, monthly maximum 1-day precipitation, and monthly maximum consecutive 5-day precipitation are decreasing; maximum and minimum temperatures are increasing; the number of cool days and cool nights is decreasing, and the number of warm nights is increasing. Baidya *et al.* (**2008**) also observed an increasing trend in temperature and precipitation extremes over Nepal. A study done by Caesar *et al.* (**2011**) considering seven stations from Nepal (among those three stations from Koshi river basin) showed that warm extremes are increasing and cold extremes are decreasing. They found that trends in precipitation extremes are less spatially consistent across the region. You *et al.* (**2008**) assessed the changes in daily climate extremes in the eastern central Tibetan Plateau during 1961–2005 and found large proportion of stations showing statically significant trends for all temperature indices with decrease in extreme cold days/nights together with increase in extreme warm days/nights. While most precipitation indices show increasing trends in the southern and northern Tibetan Plateau and show decreasing trends in the central Tibetan Plateau. Mantonet *al.* (**2001**) assessed trends in extreme daily rainfall and temperature in Southeast Asia from 1961 to 1998 and found a significant increase in annual hot days and warm nights, a significant decrease in cool days and cold nights, a significant decrease in the number of rainy days across the region, and at many stations an increase in the proportion of annual rainfall from extreme events. Klein Tank *et al.* (**2006**) investigated the changes in daily temperature and precipitation extremes in central and south Asia from 1961 to 2000. They found a significant increase in the percentage of warm nights and days and a decrease in the percentage of cold nights and days at 70% of stations; the daytime trends in maximum temperature extremes were smaller than the night-time trends in minimum temperature extremes. There was no consistent pattern of change in precipitation extremes, but there was a significant increase in extreme precipitation indices. Singh *et al.* (**2014**) assessed extreme wet and dry spells over the core region during the South Asia summer monsoon season from 1951 to 2011 and found a significant increase in the frequency, but decrease in intensity, of dry spells, and a significant increase in the intensity of wet spells. Donat *et al.* (**2013**) analysed temperature and precipitation extreme indices from 1901 to 2010 on a global scale. Their main findings included widespread and significant warming trends related to temperature extremes, mostly stronger for indices based on daily minimum temperatures than for indices calculated from daily maximum temperatures. The changes in precipitation extremes were in general spatially more complex and locally less significant. However, on a global scale, there was a tendency towards wetter conditions for most precipitation indices, i.e. the intensity, frequency, and duration of extreme precipitation is increasing on average.

Changes in the frequency and intensity of extreme climate events are likely to have an immediate and intense impact on society and the environment, and climate change is likely to exacerbate any existing problems in a river basin. The Koshi river basin has been selected as a key river basin for studies to improve understanding of changes and trends in climate indices. In the following, we describe the results of a study of spatial and temporal differences and trends in climate extremes over the Koshi river basin using RClimDex software. The study provides a comprehensive analysis of observed temperature and precipitation extremes over the whole basin, including some stations in parts of the basin in China (Tibet Autonomous Region), Nepal and India; and includes data and information differentiated by elevation.

2 METHODS AND DATA

2.1 Study area

The Koshi river originates in the high-altitude Tibetan Plateau and flows through the Himalayan range, the headwater area for a number of major tributaries, and Nepal's mid-mountains to the Terai lowland region in Nepal and India, where it joins the Ganges (Figure 1). A vast alluvial fan has developed over the centuries along the plains reaches of the river. The basin drains an area of some 87 970 km^2; approximately 32% in China, 45% in Nepal, and 23% in India. The elevation range within the basin is extreme, extending from the highest point in the world, 8848 masl at the summit of Mount Everest, to just over 30 masl in the plains. The population density ranges from less than 4 people per square kilometre in the northern portion to more than 1000 people per square kilometre in the plains.

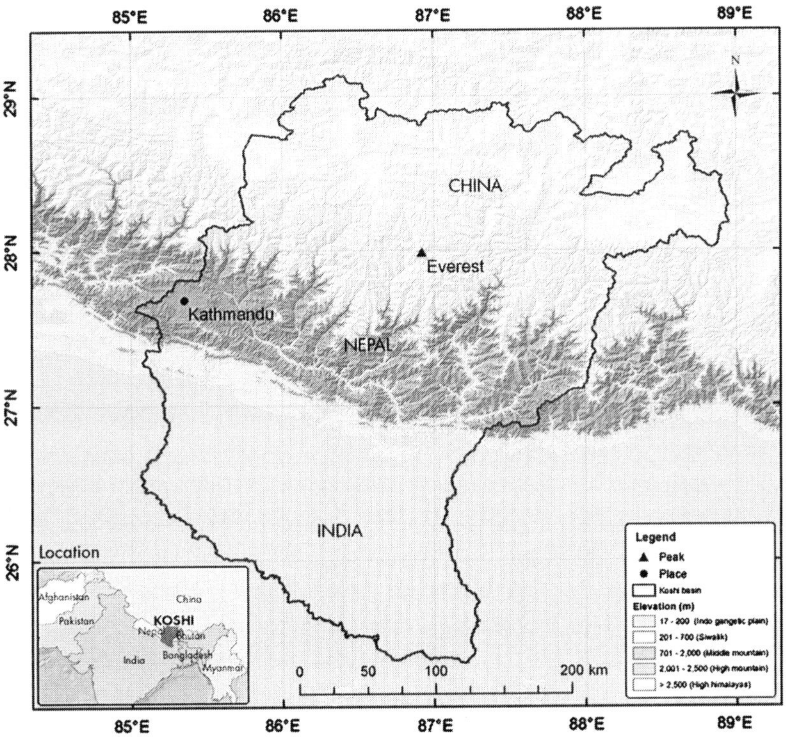

Figure 1. Location of the Koshi river basin.

The region includes five climatic zones determined mainly by elevation (Khanal *et al.*, **2007**) (Figure 1) and has nine distinct eco-regions. The vegetation is influenced by differences in soil, topography, and climate and varies greatly across the regions from deciduous forest in the tropical and subtropical region, through mixed forest in the temperate zones and alpine forest in the sub-alpine zone, to the rangelands and grassland of the high mountains. The extreme changes in elevation in the mountain area from high peaks to deep valleys means that there can be a great variation of vegetation within the different physiographic zones (Shrestha and Devkota, **2010**).

The climate in the basin varies from humid tropical in the south, through subtropical and temperate, to cold and arid in the north. The climate in the southern part of the basin is strongly influenced by the South Asian monsoon, whereas to the north the Tibetan plateau lies in a rain shadow area.

The annual precipitation ranges from 207 mm in the trans-Himalaya to more than 3000 mm in the eastern mountains and mid-mountains of Nepal (Neupane *et al.*, **2013**). Precipitation patterns in the basin are directly associated with the summer monsoon, with about 80% of the annual precipitation falling between June and September. Owing to the great variation in topography, the spatial and temporal complexity of rainfall is large within short distances. In the temperate region there is a clear correlation between land elevation and precipitation.

The Koshi river system has a high potential for investment in hydropower development as well as irrigation in the downstream areas. In addition, the basin contains important ecosystems and protected areas that support a high level of biodiversity and provide a range of ecosystem services that sustain livelihoods. However, the diverse topography, young geological formation, high degree of glaciation, and strong monsoon influence make the basin particularly prone to erosion and sedimentation and natural hazards such as glacial lake outburst floods (GLOF), flash and riverine floods, landslides, debris flows, and drought (Shrestha *et al.*, **2010**). Poor women and men are the most vulnerable and hardest hit by natural hazards, and it is thought that the hazards may increase in magnitude and frequency with climate change (Goswami *et al.*, **2006**). Increasing population, urbanization, natural resource degradation, and encroachment have also added pressures on the freshwater ecosystems of the Koshi basin. The livelihoods of a large part of the rural population, and especially the poor, depend on the Koshi basin's ecosystem services and arable land. Neupane *et al.* (**2013**) suggested climate change is likely to lead to an increase in water scarcity in the basin, which will ultimately influence the agro-based livelihoods of the basin population and threaten food security.

2.2 Data

Daily precipitation and temperature (both maximum and minimum) values for all available gauges within the Nepal portion of the Koshi basin were acquired from the Department of Hydrology and Meteorology (DHM) of Nepal. Data for two stations in Tibet Autonomous Region, China – Nielamu and Dingri – were provided by the Tibetan Meteorological Bureau. Data for the Indian part of the basin were extracted from the 1° × 1° India Meteorological Department (IMD) grid dataset (Rajeevan *et al.*, **2005**; Srivastava *et al.*, **2009**). In all, precipitation data was obtained from 73 stations and temperature data from 36 stations. In addition six IMD grid points were used. Following screening for data quality (see below), 50 stations were selected to provide daily precipitation data for the study; and 20 stations to provide daily temperature data (Figure 2).

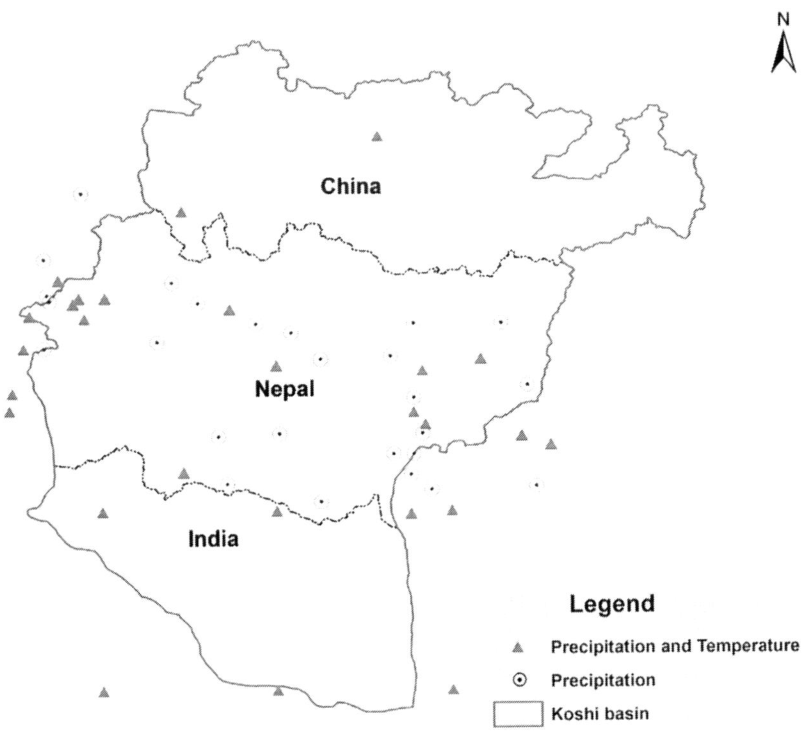

Figure 2. Location of the stations used in the study.

Studies of climate change require a long period of data in order to identify long-term trends rather than short-term fluctuations (WMO, **1996**). Temperature and precipitation data were obtained for a period of 35 years from 1975 to 2010. The aim was to encompass all the climatic zones in the Koshi basin. This could not be completely achieved due to the complex topography of the basin and the limited number of stations available, but spatial coverage was increased by including nearby stations from areas just beyond the basin boundary (Figure 2). The 50 stations that provided precipitation data lay at elevations from 33 to 4300 masl. Of the 20 stations that provided temperature data, six were located in the high hills and mountains above 2000 masl; six in the hill region at elevations between 1300 m and 1750 masl, and eight in the Indo-Gangetic plains at elevations below 91 m. To capture the different topography or climatology diversity of the basin we divided the stations into five sub-division High Himalaya (HH); High Mountain (HM); Middle Mountain (MM); Siwalik (S); Indo-Gangetic Plain (IG) (Tables 3 and 4, Figures 1, 5-10 and 12-14).

2.3 Quality control and homogenization

One of the most important steps in the study was to ensure the quality and completeness of the data. Stations with only a short period of data records or more than 5% of data missing were discarded. The precipitation and temperature data underwent several steps of screening using visual checks of data plots using MS Excel and RClimDex for erroneous outliers. The computer program RClimDex identifies erroneous temperature and precipitation data on a first run, such as daily precipitation amounts less than 0 mm or days with T_{max} less than T_{min}. Second step identifies outliers in daily maximum and minimum temperature, which have to be manually checked, validate, corrected by looking tempQC.cvs, prcpQC.cvs, and tepstdQC.cvs in a subdirectory called log. For temperature, they are defined as daily values outlying a user-defined threshold determined by mean ± SD. In our case, we choose four standard deviations as the thresholds for a finer quality control of the data. Both for precipitation and temperature, data plots are available for visual inspections to reveal more outliers as well as a variety of problems that cause changes in the seasonal cycle or variance of the data in a same subdirectory plot. For precipitation histograms of the data are created which reveal problems that show up when looking at whole data (You *et al.*, **2008**).

The next step was to test the homogeneity of the data series. A homogeneous climate time series is defined as one in which variations are caused only by variations in climate and not, for example, by changes in station location, observation procedures and practice, or instrumentation (Aguilar *et al.*, **2003**, **2005**). The purpose of homogeneity test is to remove the non-climate factors from the data as much as possible before the climate data can be reliably used for climate change study. The RHtest software package was applied to data for homogeneity testing because it can be used without reference series and freely available. RHtest can help to identify step changes in a time series by comparing the goodness of fit of a two phase regression model with that of a linear trend for the entire series. It is used to help identify sudden jumps (change points) in indices of temperature and precipitation data. As done by Caesar *et al.* (**2011**) we did not attempt to adjust the data as a result of the homogeneity tests and excluded the data that appear to be potentially suspect.

Homogeneity assessment and adjustment can be quite complex and it often requires close neighbour stations, detailed station history and a great amount of time (Vincent *et al.*, **2005**).

But this is often not possible due to the sparseness of the station network. In this study, we considered stations individually without using a reference series, as suitable reference series were unavailable due to lack of homogeneous series data, lack of proper metadata and very scattered spatial distribution of stations and use only *F* tests. *F* tests allows the time series being tested to have a linear trend throughout the whole period of data record (means no shift in the trend component), with the annual cycle, linear trend and lag-1 autocorrelation of the base series being estimated in tandem through iterative procedures, while accounting for all the identified mean shifts (Wang, **2008**). More details on RHtest is given in *RHtestsV3 User Manual*(Wang and Feng, **2010**) and in Whan *et al.* (**2013**), Caesar *et al.* (**2011**), Klein Tank *et al.* (2009), You *et al.* (**2008**).

Figure 3 shows an example of homogeneity testing of precipitation data from stations SI1115 and SI1016. The values for SI1115 (a) show no sudden jumps in the data series whereas those for SI1016 (b) show a sudden jump after 1990. Tests using the *F*-test showed that the jumps were significant. Stations that showed significant jumps (significant change points) of this type were rejected, and the data were not used.

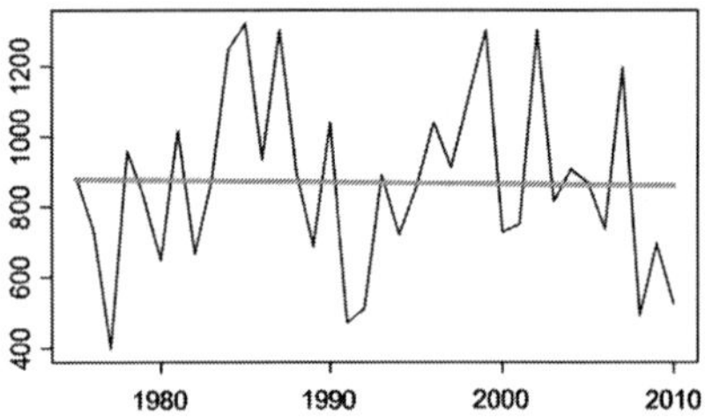

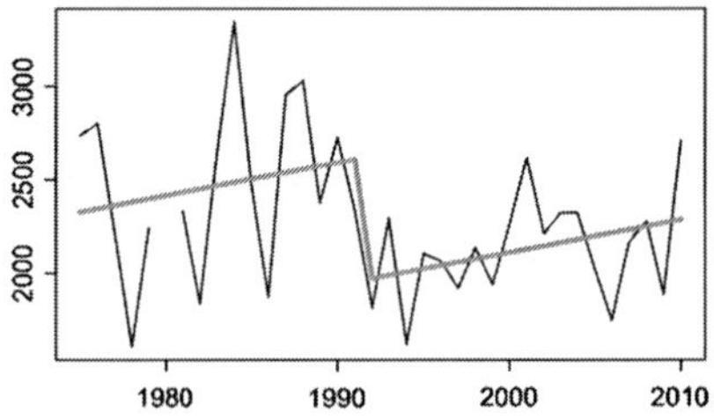

Figure 3. RClimDex RHtest results for data homogeneity. (a) Homogeneous precipitation data series with no significant break point; (b) non-homogeneous precipitation data series with significant break point.

After data quality control and homogeneity assessment; 20 stations were left for analysis of temperature indices (from maximum and minimum temperature) and 50 stations were left for analysis of precipitation indices (Tables 3 and 4). For the indices calculation in RClimdex, upper and lower threshold values were defined for daily maximum and minimum temperatures, and user define daily precipitation threshold for rainfall. An average temperatures of MAM (March, April, May) were taken for upper threshold of daily maximum temperature and an average temperatures of DJF (December, January, February) were taken for upper threshold of daily minimum temperature and an average of daily rainfalls from JJAS (June, July, August, September) were taken for user define daily precipitation threshold based on the historically observed data (1975–2010; Islam, **2009**).

2.4 Trend and indices analysis

The long-term trends were analysed using simple linear regression (Wang and Feng, **2010**), and significance was analysed using Student's *t*-test (Miller and Miller, **1988**). Daily temperature and precipitation data were used to calculate extreme climate change indices using the specially designed software RClimDex (1.0) (http://cccma.seos.uvic.ca/ETCCDI/ RClimDex/ rclimdex.r). RClimDex (1.0) is a freely available tool designed to provide a user friendly interface to compute indices of climate extremes and runs under the R programming environment (http://www.r-project.org/). RClimDex has become a standard tool for developing climate change indices and has been used by many authors, for example Baidya *et al.*

(**2008**), Islam (**2009**), Whan*et al.* (**2013**), Alexander *et al.* (**2006**), State and Service (**2008**), Manton *et al.* (**2001**), Klein Tank *et al.* (**2006**), and Donat *et al.* (**2013**). RClimDex performs in the R Console and includes a data quality control procedure before computing the indices (see above). It computes 27 core indices (11 for precipitation and 16 for temperature) recommended by the Commission for Climatology (CCl), World Climate Research Programme (WCRP) of the Climate Variability and Predictability Component (CLIVAR) project and Expert Team for Climate Change Detection and Indices (ETCCDI). Of the 27 indices, 23 are relevant for the region and were selected for the study (Table 1).

2.5 Comparison of observed data with 1° × 1° IMD grid dataset

Owing to limited availability or access of observed historical meteorological data from Indian part of Koshi basin, 1° × 1° IMD grid dataset were used for climate indices analysis (Tables 3 and 4). The study from King *et al.* (**2013**) indicates that in general gridded datasets are suitable for investigating extreme rainfall trends and variability, but suggest that the data should be first validated and correlated. This is particularly important as rest of the data are station based. For the validity of the 1° × 1° IMD grid dataset, the gridded data was compared with few available observed station data from Bihar, India (Figure 4). For this, grid point X85.5, Y25.5 was compared with Patna station, X85.5, Y26.5 was compared with Buchana station, X86.5, Y25.5 was compared with Bhagalpur station and X87.5, Y25.5 was compared with Sabour station.

The results show that almost all climate indices generated by IMD grid data follow the same trend and sign as observed station data but only few threshold indices (R10mm, R20mm) and SDII showed slightly difference (Table 2). Time series comparison of indices at Biratnagar and Patna with corresponding grid points do not show temporal divergence in the value (not shown due to lack of space) indicating that the gridded data represent the observation adequately. The correlation coefficient between gridded datasets and stations were more than 0.9 in minimum and maximum temperature but less than 0.5 regarding the rainfall. From this we concluded that, while there was an overall reduction in the number of stations contributing to the Indian gridded dataset, there were no significant changes in the network density in the study region until the end of the study period and therefore the IMD grid data captures the climatology of the lower basin.

Table 1. ETCCDI recommended climate indices used in the study.

ID	Indicator name	Definition	Unit
TXx	Max. T_{max}	Monthly maximum value of daily maximum temp	°C
TNx	Max. T_{min}	Monthly maximum value of daily minimum temp	°C
TXn	Min. T_{max}	Monthly minimum value of daily maximum temp	°C
TNn	Min. T_{min}	Monthly minimum value of daily minimum temp	°C
TN10p	Cool nights	Percentage of days when TN < 10th percentile	%
TX10p	Cool days	Percentage of days when TX < 10th percentile	%
TN90p	Warm nights	Percentage of days when TN > 90th percentile	%
TX90p	Warm days	Percentage of days when TX > 90th percentile	%
WSDI	Warm spell duration indicator	Annual count of days with at least 6 consecutive days when TX > 90th percentile	%
CSDI	Cold spell duration indicator	Annual count of days with at least 6 consecutive days when TN < 10th percentile	%
DTR	Diurnal temperature range	Monthly mean difference between TX and TN	°C
RX1day	Max. 1-day precipitation amount	Monthly maximum 1-day precipitation	mm
Rx5day	Max. 5-day precipitation amount	Monthly maximum consecutive 5-day precipitation	mm
SDII	Simple daily intensity index	Annual total precipitation divided by the number of wet days (defined as PRCP ≥ 1.0 mm) in the year	mm/day
R10	Number of heavy precipitation days	Annual count of days when PRCP ≥ 10 mm	Days
R20	Number of very heavy precipitation days	Annual count of days when PRCP ≥ 20 mm	Days
Rnn	Number of days above nn (50, 100, 150) mm	Annual count of days when PRCP ≥ nn mm. nn is user defined threshold	Days
CDD	Consecutive dry days	Maximum number of consecutive days with RR < 1 mm	Days
CWD	Consecutive wet days	Maximum number of consecutive days with RR ≥ 1 mm	Days
R95p	Very wet days	Annual total PRCP when RR > 95th percentile	mm
R99p	Extremely wet days	Annual total PRCP when RR > 99th percentile	mm
PRCPTOT	Annual total wet-day precipitation	Annual total PRCP in wet days (RR ≥ 1 mm)	mm

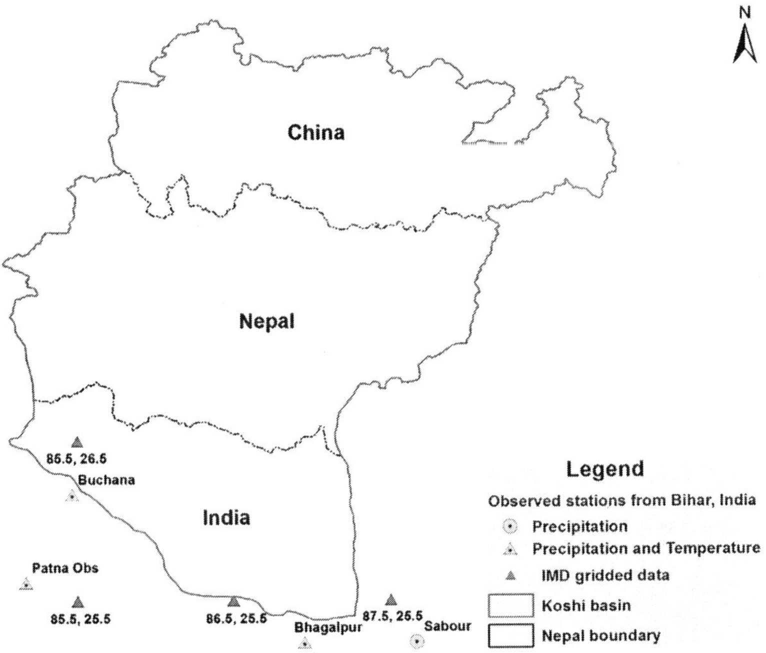

Figure 4. The stations used to compare observed station data with 1° × 1° IMD grid dataset in Koshi India part.

3 RESULTS AND DISCUSSION

3.1 Seasonal temperature trends

The long-term (1975–2010) trends in seasonal maximum and minimum temperatures are shown in Figures5 and 6. The maximum (Figure 5) and minimum (Figure 6) seasonal temperatures showed an increasing trend at the majority of stations and most of the values were statistically significant at a 95% significance level. Ji and Kang (**2014**) highlighted that changes in extreme temperature become increasingly pronounced from South to North China, with the most significant changes occurring on the Tibetan Plateau. The few decreasing trends were mostly not statistically significant except in the IGs, where the maximum seasonal temperature showed a decreasing trend in all seasons except the monsoon, although the decrease was only statistically significant in winter and pre-monsoon, which is consistent with the results reported by Ji *et al.* (**2015**). At most stations maximum seasonal temperatures had increased at a rate of more than

0.3 °C decade^{-1}. The increases in seasonal minimum temperature were generally higher than those in seasonal maximum temperature, which is consistent with the results reported by Alexander *et al.* (**2006**). There was no conspicuous difference in the change in minimum seasonal temperatures between the plains and the mountains, with increases observed at almost all stations. The rate of increase in minimum seasonal temperature was somewhat higher in winter than in other seasons, and more often statistically significant, which is consistent with the results reported by Donat *et al.* (**2013**). Overall, the results showed widespread significant changes in temperature trends consistent with warming, especially for those indices derived from daily minimum temperature, with stronger trends in more recent decades.

3.2 Temperature indices

The trends in temperature indices derived from the maximum and minimum daily temperatures are shown in Table 3 and those are arithmetic mean of the values of the 20 stations used in this study. The monthly maximum value of daily maximum temperatures (TXx) increased on average by 0.1 °C decade^{-1}. Six of the 20 stations showed a statistically significant increase and four (all in the IGs) a statistically significant decrease. The monthly maximum value of daily minimum temperatures (TNx) increased on average by 0.2 °C decade^{-1}; 14 of the 20 stations showed an increase (nine statistically significant). The monthly minimum value of minimum temperature (TNn) increased on average at a rate of 0.3 °C decade^{-1}; 18 of the 20 stations showed an increase (six statistically significant). This finding is consistent with previous study done by Liu *et al.* (**2006**) over the eastern and central Tibetan Plateau during 1961–2003. The monthly minimum value of daily maximum temperature (TXn) increased on average at a rate of 0.1 °C decade^{-1}; 12 of the 20 stations showed an increase (seven statistically significant), and eight, all in the IGs, a decrease (six statistically significant) (Figure 7). Baidya *et al.* (**2008**) suggested that the negative trend in TXn in the plains might be the result of the fog episodes in winter which have become more frequent over the past decade, sometimes lasting for more than a week or even a month, which reduced the maximum temperature significantly. A study done by Ji *et al.* (**2011**) shows that negative trend in TXn in the plains could be due to emission of anthropogenic particles such as sulphate and organic carbon which scatter the incoming solar radiation and reduction in the radiative forcing causing the decrease of temperature in IGs and another study by Ji *et al.* (**2015**) shows that negative trend in TXn in the plains could be due to increased haze occurrence and its dimming effect. Similarly, You *et al.* (**2008**) suggested that warming in the MM, HM and HH most probably due to the increase in anthropogenic greenhouse gases emission contributing and change in cloud cover.

The number of warm days, i.e. days in which the daily maximum temperature was above the 90th percentile (TX90p), increased at all 12 hill stations and 2 of the 8 stations in the plains, with the increase statistically significant at 13 stations; it decreased at the remaining 6 stations in the plains, although the decrease was not statistically significant (Figure 8). The number of cool days, i.e. days in which the daily maximum temperature was below the 10th percentile (TX10p), decreased at 10 of the 12 hill stations (4 statistically significant) and increased at all 8 of the stations in the plains (2 statistically significant) (Figure 9). The increase in cool days over the plains could be partly due to the smaller difference between maximum and minimum temperatures during fog episodes, with the afternoon maximum temperature close to the minimum temperature, indicating a cold day but relatively warmer night (Manandhar, **2006**). The number of warm nights, i.e. nights in which the daily minimum temperature was above the 90th percentile (TN90p), increased at all 20 stations (17 statistically significant) (Figure 10). The number of cool nights, i.e. nights in which the daily minimum temperature was below the 10th percentile (TN10p) decreased at 14 stations (7 statistically significant), and increased at 6 stations, all in the hills, (2 statistically significant).

The warm spell duration indicator (WSDI), i.e. the annual count of days with at least 6 consecutive days with a maximum temperature greater than the 95th percentile, increased on average by 15.7 days decade^{-1} (Table3). The WSDI increased at all 12 hill/mountain stations and 2 plains stations (13 statistically significant) and decreased at the remaining 6 plains stations, but the trend was not statistically significant. The cold spell duration indicator (CSDI), i.e. the annual count of days with at least 6 consecutive days with a minimum temperature less than the 10th percentile, decreased on average by 0.4 days decade^{-1}. The CSDI decreased at five hill/mountain stations (three statistically significant) and all eight plains stations (one significant), and increased at the remaining seven hill/mountain stations (none significant) (Table 3). Taken together, the results suggest that overall days are becoming warmer.

Table 2. Comparison of observed stations with IMD grids.

Indices	Units	Patna station	X85.5, Y25.5 (IMD grid)	Sabour station	X87.5, Y25.5 (IMD grid)	Bhagalpur station	X86.5, Y25.5 (IMD grid)	Buchana station	X85.5, Y26.5 (IMD grid)
TXx	°C	−0.005	−0.042			−0.015	−0.039	0.035	−0.026
TXn	°C	−0.04	−0.032			−0.094	−0.063	−0.114	−0.061
TNx	°C	0.009	0.001			0.028	0.029	0.021	0.027
TNn	°C	−0.015	−0.014			−0.023	0.003	0.032	0.003
TX10p	%	0.025	0.023						
TX90p	%	0.007	−0.07						
TN10p	%	−0.158	−0.214						
TN90p	%	0.24	0.151						
WSDI	%	0.148	0.007						
CSDI	%	−0.076	−0.049						
DTR	°C	−0.027	−0.022			−0.094	−0.041	−0.032	−0.04
Rx1day	mm	−0.545	−0.954	1.066	3.456	0.515	2.327	−4.39	1.11
Rx5day	mm	−1.207	−1.012	2.016	5.527	0.831	6.096	−5.262	−1.973
SDII	mm	−0.042	0.014	0.106	0.105	0.103	0.199	−0.623	−0.137
R10mm	Days	−0.079	0.014	0.048	0.481	−0.14	0.594	−0.308	0.282
R20mm	Days	−0.1	0.085	0.041	0.163	−0.055	0.507	−0.427	−0.009
R50mm	Days	−0.079	−0.024	0.082	0.147	0.026	0.154	−0.177	−0.098
CDD	Days	1.656	0.256	1.302	1.033	3.287	0.909	1.529	0.05
CWD	Days	0.003	0.015	0.121	0.128	−0.014	0.121	0.049	0.085
R95p	mm	−3.687	−2.26	8.277	14.052	3.461	15.289	−17.238	−5.613
R99p	mm	−1.295	−2.477	3.154	9.347	3.042	10.258	−11.606	0.347
PRCPTOT	mm	−6.136	−1.663	7.115	19.056	−1.066	26.147	−25.39	1.676

Start and end year are different for different stations: Patna (1975–2009), Sabour (1975–2005), Bhagalpur (1975–2009) and Buchana (1982–2007). But except Patna there are lots of no data in the remaining three stations.

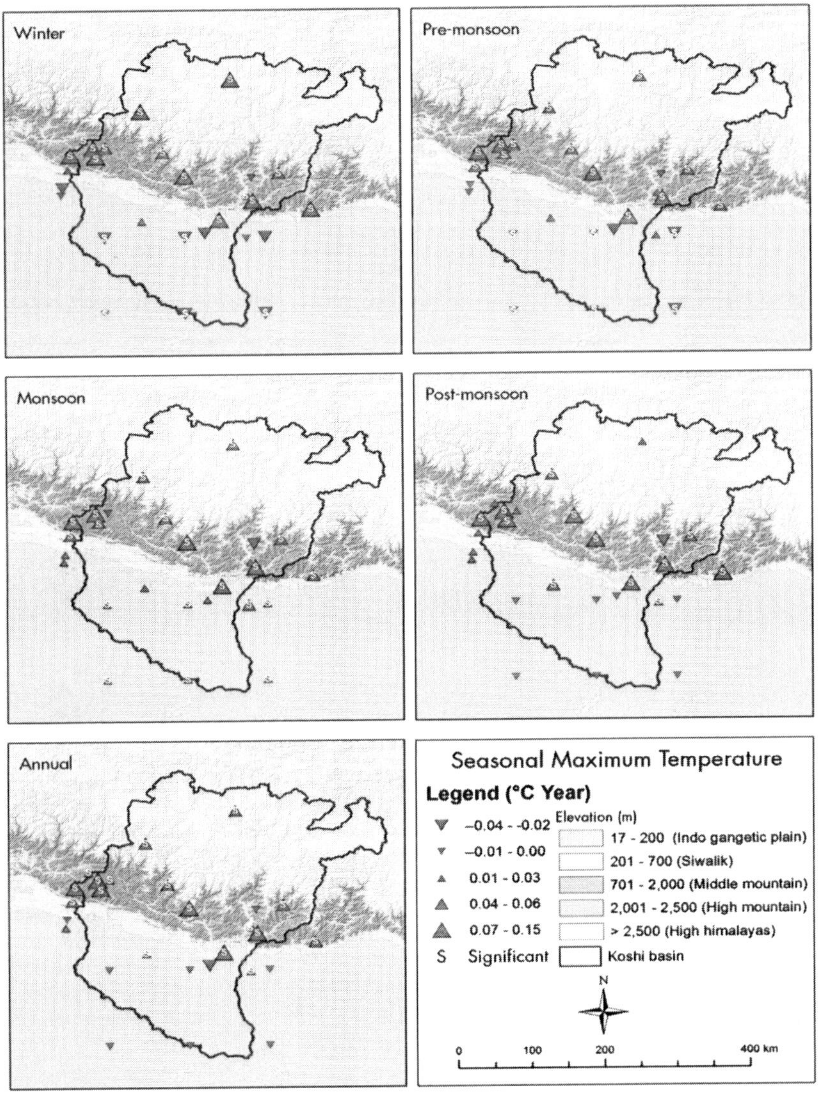

Figure 5. Trends in maximum seasonal temperature for 1975–2010: (a) winter (DJF); (b) pre-monsoon (MAM); (c) monsoon (JJAS); (d) post-monsoon (ON); (e) annual.

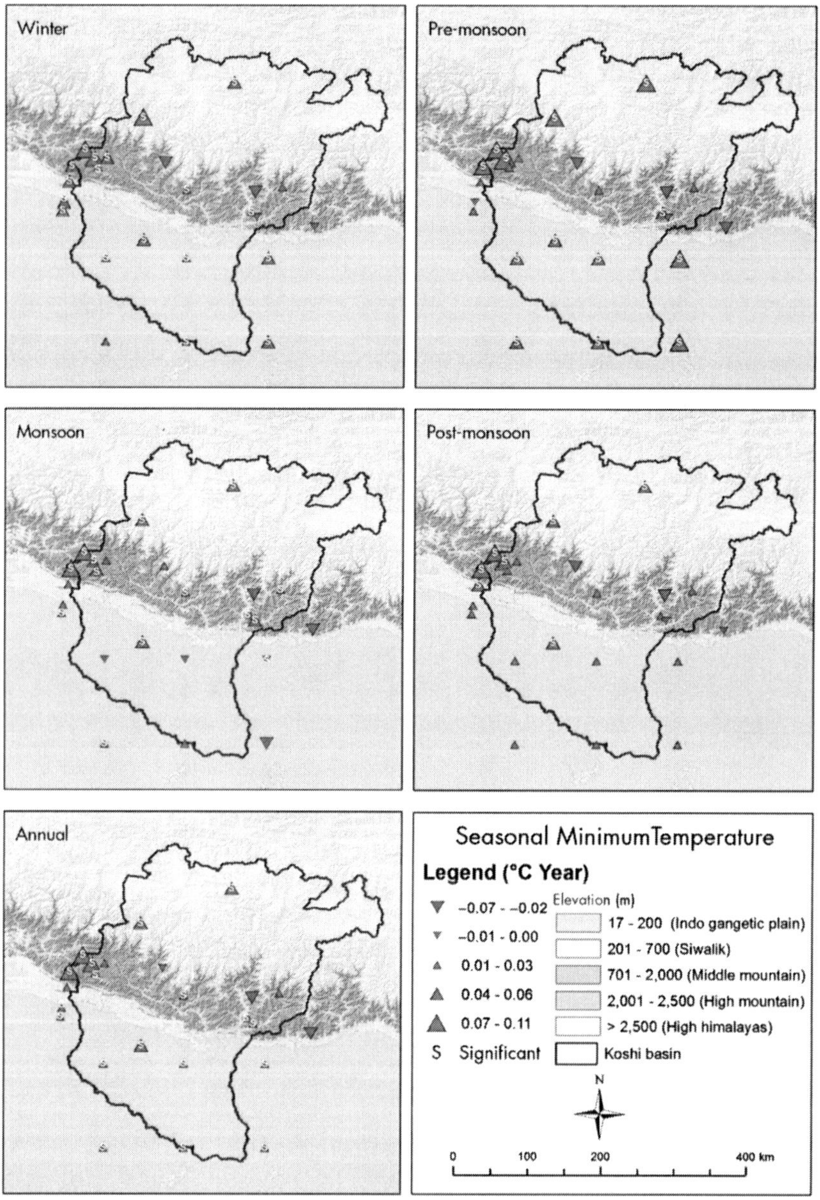

Figure 6. Trends in minimum seasonal temperature for 1975–2010: (a) winter (DJF); (b) pre-monsoon (MAM); (c) monsoon (JJAS); (d) post-monsoon (ON); (e) annual.

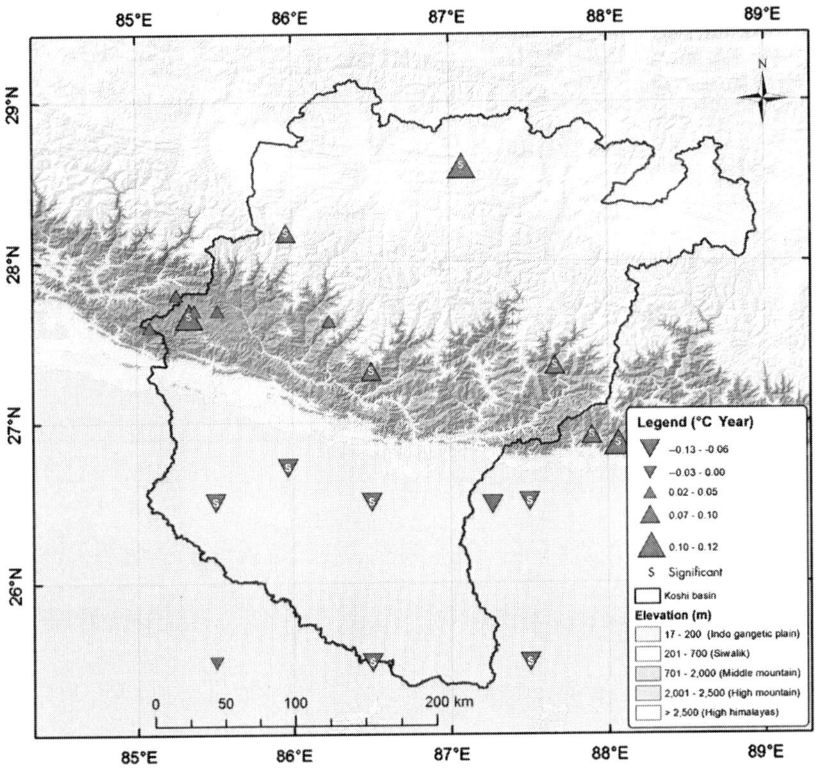

Figure 7. Trends in monthly minimum value of daily maximum temperature (TXn).

The diurnal temperature range (DTR) remained constant on average, but this was the result of opposing trends within the basin. The DTR increased at 10 of the 12 hill/mountain stations (all statistically significant), and decreased at all of the plains stations (6 statistically significant). The decrease in DTR over the IGs is probably due to the minimum temperature warming faster than the maximum temperature. This finding is consistent with previous studies in Nepal using daily temperature data from 1971 to 2006 (Baidya *et al.*, **2008**), and with previous global studies using daily global data (Alexander *et al.*, **2006**) and monthly global data (Jones *et al.*, **1999**), and regional studies using daily data (Yan *et al.*, **2002**; Klein Tank *et al.*, **2006**).

3.3 Rainfall indices

Figure 11 shows the average annual total wet day rainfall across the 50 rainfall stations in the Koshi basin from 1975 to 2010. There was considerable fluctuation, but no significant long-term trend, which is consistent with the findings of Klein Tank *et al.* (**2006**) and You *et al.* (**2008**). This study was mainly concerned with trends in precipitation extremes. Various rainfall-related indices (Table 4) were calculated for the 50 stations; the results are summarized in Table 4 and discussed in more detail below.

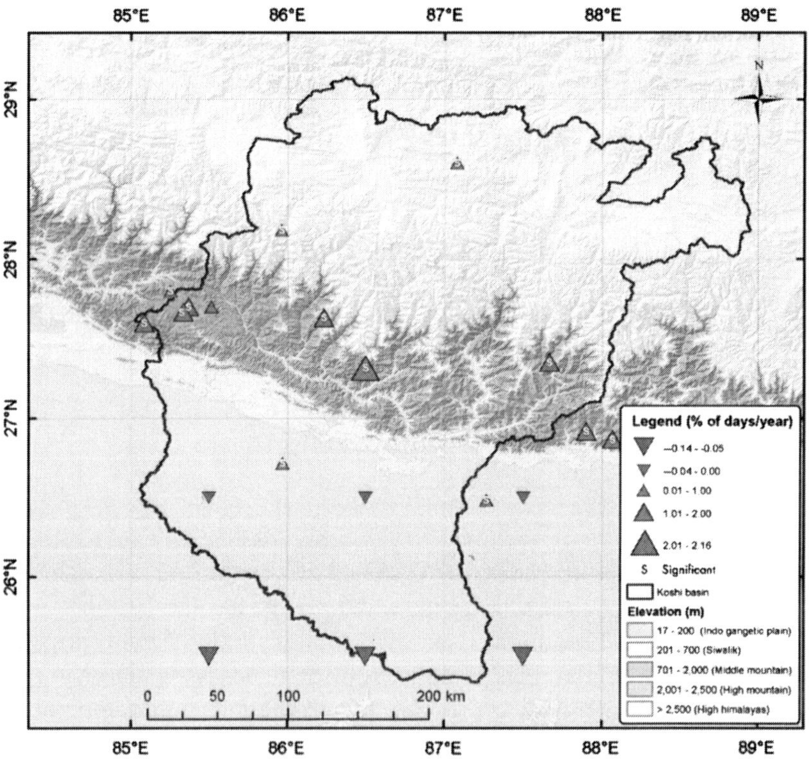

Figure 8. Trends in the percentage of days in which TX > 90th percentile (TX90p).

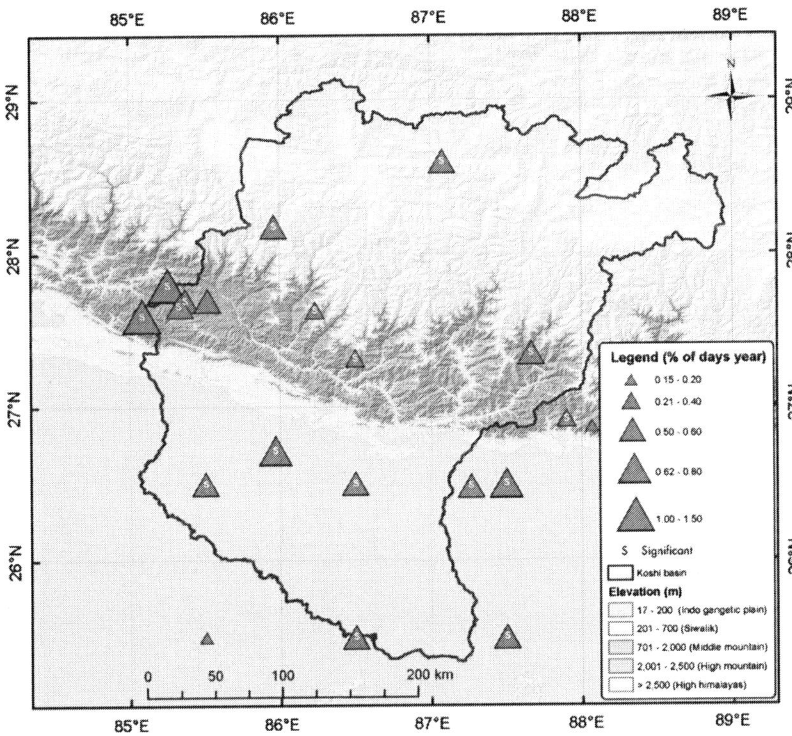

Figure 9. Trends in the percentage of days in which TX < 10th percentile (TX10p).

The monthly maximum 1-day precipitation (RX1day) decreased at 27 of the 50 stations (6 statistically significant), and increased at 23 stations (3 statistically significant), including all except one of the plains stations located at or below 100 masl (Figure 12). Similarly, the number of consecutive dry days (RR <1 mm, CDD) increased at 47 of the 50 stations (12 statistically significant). The number of consecutive wet days (RR ≥ 1 mm, CWD) increased at 33 of the 50 stations (5 statistically significant) and decreased at 17 (3 statistically significant). These findings are partially consistent with previous studies by Manton *et al.* (**2001**), which indicate a decrease in the monthly maximum 1-day precipitation using daily rainfall data from 1961 to 1998 in Southeast Asia; and by Singh *et al.* (**2014**) and Sivakumar and Stefanski (**2011**), which indicate an increase in the frequency of drought and increasing trend of CDD. The trend in CWD may be of more concern than the trend in CDD as it can give an idea of whether an area is experiencing more extremes in precipitation during the rainy season (Kruger, **2006**), particularly important where people are living on flood-prone areas and when most agricultural activity takes place. And also CDD largely reflects events during the dry season in a monsoonal climate and is therefore of limited importance.

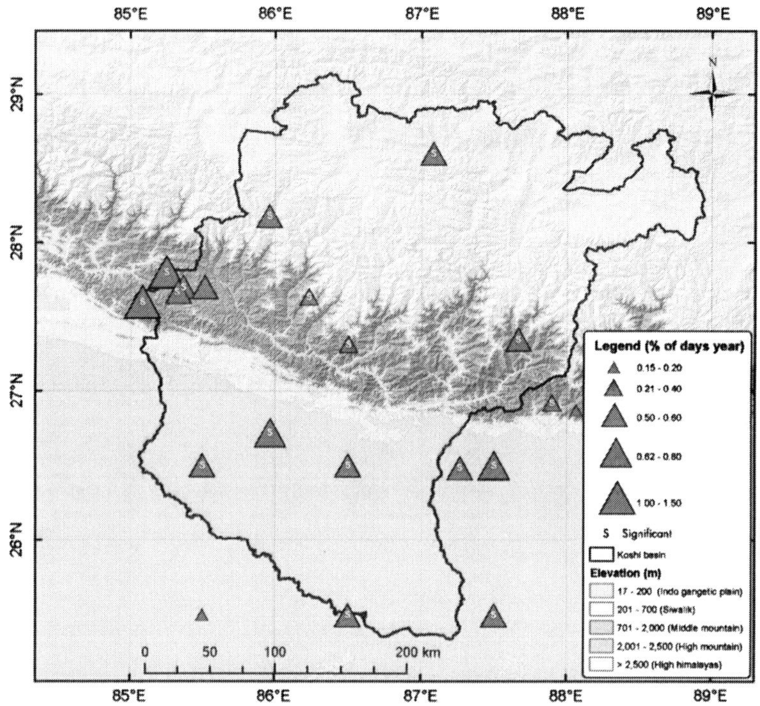

Figure 10. Trends in the percentage of days in which TN > 90th percentile (TN90p).

The monthly maximum consecutive 5-day precipitation (RX5day) increased on average by 6.1 mm decade^{-1}. RX5day increased at 30 stations (2 statistically significant) and decreased at 20 stations (2 statistically significant). The annual count of heavy precipitation days with precipitation ≥ 10 mm (R10), 20 mm (R20), or 50 mm (R50), increased at 32 (3 statistically significant), 31 (3 statistically significant), and 32 (8 statistically significant) of the 50 stations (64%) respectively. This is consistent with the results reported by Singh *et al.* (**2014**), Baidya *et al.* (**2008**), and Alexander *et al.* (**2006**).

Table 3. Trends in annual temperature indices over the Koshi basin (1975–2010).

StnIndex	Lat. (DD)	Long. (DD)	Elev. (m)	TXx	TXn	TNx	TNn	TX10p	TX90p	TN10p	TN90p	WSDI	CSDI	DTR
S55655 (China)	28.60	87.06	4300 (HH)	0.02	**0.12**	**0.02**	**0.08**	**−0.30**	**0.32**	**−0.50**	**0.53**	**0.51**	**−0.21**	−0.01
S55664 (China)	28.18	85.96	3310 (HH)	0.01	**0.08**	**0.02**	**0.08**	**−0.29**	**0.47**	**−0.50**	**0.53**	**0.61**	**−0.21**	0.01
SI905 (Nepal)	27.60	85.08	2314 (HM)	**0.07**	0.05	**0.12**	0.00	−0.11	**1.59**	−0.03	**1.00**	**3.52**	0.10	0.03
SI1043 (Nepal)	27.70	85.52	2163 (HM)	0.00	0.03	−0.01	**0.05**	0.00	0.36	0.12	0.50	0.02	0.07	0.00
SI1007 (Nepal)	27.80	85.25	2064 (HM)	**0.05**	0.04	**0.05**	**0.15**	0.11	**0.40**	**−0.20**	**1.50**	**0.56**	**−0.34**	**−0.06**
SI1103 (Nepal)	27.63	86.23	2003 (HM)	**0.05**	0.03	−0.01	−0.02	**−0.10**	**1.22**	**0.43**	**0.21**	**2.47**	0.38	**0.08**
SI1405 (Nepal)	27.35	87.67	1732 (MM)	**0.05**	**0.08**	−0.01	−0.01	−0.08	**1.55**	0.07	**0.50**	**3.33**	0.05	**0.05**
SI1206 (Nepal)	27.32	86.50	1720 (MM)	**0.09**	**0.08**	0.01	0.00	−0.07	**2.16**	0.14	**0.28**	**5.21**	0.13	**0.10**
SI1416 (Nepal)	26.87	88.07	1678 (MM)	0.03	**0.10**	−0.02	0.00	**−0.11**	**1.38**	0.47	0.19	**3.08**	0.88	**0.02**
SI1029 (Nepal)	27.67	85.33	1350 (MM)	−0.01	**0.12**	−0.01	**0.06**	−0.09	**1.33**	**−0.15**	**0.50**	**3.48**	−0.09	**0.02**
SI1030 (Nepal)	27.70	85.37	1336 (MM)	**0.07**	0.02	0.01	**0.05**	−0.04	**1.73**	−0.13	**0.59**	**4.12**	−0.06	**0.04**
SI1407 (Nepal)	26.92	87.90	1300 (MM)	0.01	**0.08**	0.01	0.00	−0.06	**1.51**	**0.24**	**0.33**	**3.95**	0.18	**0.08**
SI1111 (Nepal)	26.72	85.97	90 (IG)	0.00	**−0.13**	0.02	0.03	**0.25**	**0.43**	−0.03	**0.75**	**0.50**	**−0.18**	**−0.03**
X875Y265 (India)	26.50	87.50	79 (IG)	**−0.04**	**−0.06**	**0.04**	0.03	0.06	−0.05	−0.39	**0.62**	−0.15	−0.20	**−0.04**
X865Y265 (India)	26.50	86.50	77 (IG)	−0.03	**−0.07**	**0.04**	0.00	0.06	−0.04	**−0.42**	**0.58**	−0.04	−0.36	**−0.05**
SI1319 (Nepal)	26.48	87.27	72 (IG)	0.00	−0.08	−0.02	0.03	**0.22**	**0.66**	−0.03	**0.57**	**0.87**	−0.08	−0.01
X855Y265 (India)	26.50	85.50	63 (IG)	−0.03	**−0.06**	**0.03**	0.00	0.05	−0.02	**−0.37**	**0.53**	−0.08	−0.22	**−0.04**
X855Y255 (India)	25.50	85.50	50 (IG)	**−0.04**	−0.03	0.00	0.01	0.02	−0.07	−0.21	0.15	−0.01	−0.05	−0.02
X865Y255 (India)	25.50	86.50	39 (IG)	**−0.04**	**−0.06**	**0.03**	0.00	0.08	−0.07	**−0.43**	**0.50**	−0.16	−0.10	**−0.04**
X875Y255 (India)	25.50	87.50	33 (IG)	**−0.05**	**−0.07**	**0.04**	0.02	0.11	−0.14	−0.55	**0.55**	−0.30	−0.43	**−0.06**
Average				0.01	0.01	0.02	0.03	−0.01	0.74	−0.12	0.55	1.57	−0.04	0.00
Stations with + positive trend				11	12	14	18	10	14	6	20	14	7	10
Stations with negative trend				8	8	6	2	10	6	14	0	6	13	10
Stations with 0 trend				1	0	0	0	0	0	0	0	0	0	0

Bold indicates statistically significant at 5%.

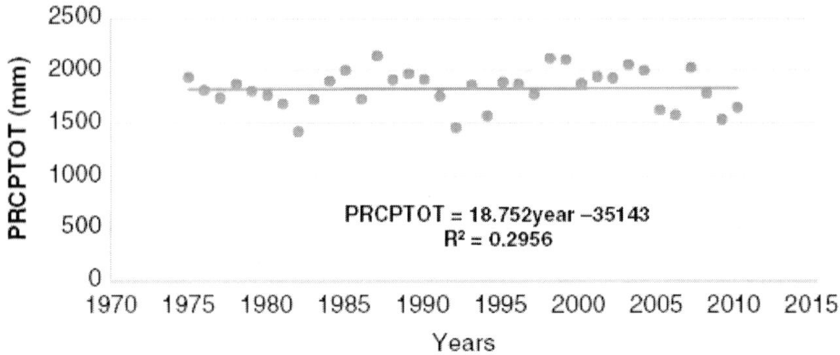

Figure 11. Average annual total wet day rainfall across the Koshi river basin.

The annual total precipitation on very wet days (days with rainfall >95th percentile; R95p) and that on extremely wet days (R99P) increased over Koshi basin which is consistent with the results of the study by Caesar *et al.* (**2011**). Overall, R95P increased at 30 stations (8 statistically significant) and decreased at 20 (4 statistically significant, Figure 13); while R99P increased at 24 stations (3 statistically significant) and decreased at 26 (3 statistically significant). Some stations showed a very high increase (up to 153 mm $decade^{-1}$ on the IGs) and others a marked decrease (up to −161 mm $decade^{-1}$ in the mid-mountains). The results indicate an overall increase in precipitation extremes.

The annual total wet day precipitation (PRCPTOT) increased on average by 24.7 mm $decade^{-1}$, but there was a mixed pattern of increasing and decreasing trends across the basin. PRCPTOT increased at 30 stations (4 statistically significant) and decreased at 20 (4 statistically significant). Some stations showed a marked increase (up to 262 mm $decade^{-1}$ in the plains) and some a marked decrease (up to −285 mm $decade^{-1}$ in the mid-mountains) (Figure 14). These results are consistent with previous studies by Baidya *et al.* (**2008**).The simple daily intensity index (SDII) also increased on average at a rate of 0.3 mm day^{-1} $decade^{-1}$. Taken together the results indicate that overall rainfall over the Koshi basin is increasing.

Table 4. Trends in annual rainfall indices over the Koshi basin.

Stn Index	Lat. (DD)	Long. (DD)	Elev. (m)	RX1day mm	RX5day mm	SDII mm/day	R10 mm	R20 mm	R50 mm	CDD days	CWD days	R95p (mm)	R99p (mm)	PRCP TOT (mm)
S55655 (China)	28.60	87.06	4300 (HH)	1.63	5.64	0.29	0.24	0.17	0.08	0.40	0.02	13.14	12.22	15.37
S55664 (China)	28.18	85.96	3310 (HH)	1.39	2.32	0.10	0.05	0.07	0.11	−0.19	0.81	10.11	12.16	20.64
SI1219 (Nepal)	27.50	86.58	2378 (HM)	0.15	0.45	**0.06**	0.09	0.11	0.01	0.36	0.07	4.84	0.22	3.51
SI1043 (Nepal)	27.70	85.52	2163 (HM)	0.36	1.09	−0.03	0.02	0.07	−0.02	0.85	0.09	−1.11	0.30	3.33
SI1204 (Nepal)	27.35	86.75	2143 (HM)	**−1.46**	−0.45	**−0.19**	**0.41**	−0.04	**−0.12**	0.13	**1.01**	**−9.52**	−3.12	−0.33
SI1007 (Nepal)	27.80	85.25	2064 (HM)	−0.45	−0.09	−0.06	−0.20	−0.17	−0.02	1.02	0.31	−0.63	−0.02	−5.28
SI1103 (Nepal)	27.63	86.23	2003 (HM)	**0.56**	1.19	**0.08**	0.25	0.32	0.07	0.27	0.13	**6.53**	**4.56**	**13.18**
SI1102 (Nepal)	27.67	86.05	1940 (MM)	**−0.83**	0.09	−0.03	−0.33	−0.06	0.04	0.70	−0.23	2.35	0.25	−4.94
SI1001 (Nepal)	28.28	85.38	1900 (MM)	−0.08	0.62	0.04	0.48	**0.27**	0.00	0.36	0.16	5.21	−0.90	9.26
SI1406 (Nepal)	27.20	87.93	1829 (MM)	0.37	0.38	0.04	0.27	**0.25**	0.05	0.10	0.39	2.16	1.64	7.76
SI1403 (Nepal)	27.55	87.78	1780 (MM)	0.07	0.39	**0.06**	0.13	0.14	**0.06**	0.41	0.10	**7.55**	3.17	6.42
SI1405 (Nepal)	27.35	87.67	1732 (MM)	0.75	1.23	−0.02	0.53	0.21	**0.00**	−2.65	0.08	1.04	−1.01	13.35
SI1206 (Nepal)	27.32	86.50	1720 (MM)	−0.07	0.56	0.04	0.01	0.05	0.03	0.55	0.07	2.28	−0.60	2.75
SI1416 (Nepal)	26.87	88.07	1678 (MM)	−0.29	−0.48	0.03	−0.22	−0.16	−0.05	**1.08**	−0.23	−4.55	−1.24	−13.19
SI1224 (Nepal)	27.55	86.38	1662 (MM)	**−0.91**	−0.71	0.01	0.01	−0.22	**−0.06**	0.30	−0.41	**−8.20**	−2.23	**−14.22**
SI1015 (Nepal)	27.68	85.20	1630 (MM)	0.07	−1.25	**−0.21**	**−1.06**	**−0.55**	**−0.18**	0.59	**−0.84**	**−8.49**	−1.84	**−28.51**
SI1301 (Nepal)	27.55	87.28	1497 (MM)	−1.20	−2.01	−0.02	0.19	0.28	**0.02**	−0.27	0.33	−4.40	−8.78	7.31
SI1022 (Nepal)	27.58	85.40	1400 (MM)	−0.61	−0.58	−0.01	**−0.23**	−0.10	**−0.08**	0.40	0.03	**−7.23**	−2.96	**−9.69**
SI1029 (Nepal)	27.67	85.33	1350 (MM)	−0.36	−0.64	−0.03	−0.17	−0.08	−0.03	**1.07**	−0.03	−4.07	−1.38	−3.98
SI1030 (Nepal)	27.70	85.37	1336 (MM)	0.35	0.88	0.03	−0.02	0.14	0.02	**1.02**	−0.03	2.51	0.10	2.75
SI1303 (Nepal)	27.28	87.33	1329 (MM)	−0.37	−0.78	0.01	0.22	0.12	**0.02**	0.78	0.03	−0.16	−0.39	3.93
SI1407 (Nepal)	26.92	87.90	1300 (MM)	0.11	−0.38	**0.18**	−0.19	−0.04	0.07	0.35	−0.13	5.09	0.50	−0.97
SI1027 (Nepal)	27.78	85.90	1220 (MM)	**−1.37**	**−2.33**	−0.12	0.27	0.02	−0.14	0.90	0.52	**−16.13**	**−6.09**	−6.54
SI1325 (Nepal)	27.37	87.15	1190 (MM)	−0.74	−0.83	**0.12**	0.03	0.11	**−0.02**	0.60	−0.08	−3.43	−4.02	−3.32

SI1213 (Nepal)	26.93	86.52	1175 (MM)	**−3.04**	**−4.18**	−0.04	−0.07	0.04	**−0.17**	0.91	−0.08	**−14.6**	**−5.93**	**−11.36**
SI1307 (Nepal)	26.98	87.35	1160 (MM)	−0.01	0.99	0.02	−0.05	−0.05	**0.02**	**1.06**	0.01	0.28	0.23	−2.28
SI1115 (Nepal)	27.45	85.82	1098 (MM)	−0.68	0.30	**−0.17**	−0.13	−0.06	−0.03	0.85	0.03	0.32	−0.48	−0.60
SI1038 (Nepal)	27.72	85.18	1085 (MM)	−0.20	0.03	−0.01	−0.17	−0.06	0.00	1.00	−0.12	−0.42	0.56	−3.64
SI919 (Nepal)	27.42	85.17	1030 (MM)	2.15	3.71	**0.17**	0.54	0.41	0.13	**1.25**	0.09	13.37	8.04	24.51
SI1004 (Nepal)	27.92	85.17	1003 (MM)	−0.12	−1.33	**0.13**	−0.42	−0.09	−0.05	1.14	**−0.41**	−1.04	−0.23	−11.91
SI1311 (Nepal)	26.82	87.28	444 (S)	−0.29	0.82	−0.01	−0.17	−0.24	**−0.02**	0.10	**−0.26**	−0.41	−1.37	−6.44
SI1305 (Nepal)	27.13	87.28	412 (S)	**−0.87**	−1.34	0.00	0.17	0.00	**−0.05**	**1.26**	0.07	−3.26	**−2.43**	0.56
SI1308 (Nepal)	26.93	87.33	365 (S)	0.29	1.30	0.08	0.07	0.00	**0.08**	**1.32**	0.02	5.07	2.23	4.16
SI1320 (Nepal)	26.70	87.27	200 (IG)	−0.74	−0.44	0.09	0.09	0.18	0.07	**1.38**	−0.12	1.96	−1.02	6.20
SI1316 (Nepal)	26.82	87.17	183 (IG)	−1.13	1.23	0.03	0.18	0.11	0.03	0.65	**0.15**	3.00	−1.62	4.25
SI1112 (Nepal)	26.92	86.17	165 (IG)	−0.84	1.48	**0.23**	**−0.35**	−0.09	0.05	**1.63**	0.05	4.22	−2.05	−1.48
SI1312 (Nepal)	26.62	87.38	152 (IG)	−1.12	1.14	−0.04	−0.12	0.09	0.09	1.01	0.03	1.04	−4.46	3.44
SI909 (Nepal)	27.17	84.98	130 (IG)	−1.23	−1.51	0.03	−0.08	0.07	0.05	0.39	**0.13**	−1.06	−3.34	0.55
SI1409 (Nepal)	26.63	87.98	122 (IG)	1.00	1.56	0.10	0.01	0.09	**0.26**	0.85	−0.13	**13.87**	1.63	**18.22**
SI911 (Nepal)	27.07	84.97	115 (IG)	−0.69	−0.06	0.05	0.06	0.07	0.07	**1.37**	**0.20**	0.70	−2.10	3.59
SI1216 (Nepal)	26.65	86.22	102 (IG)	0.63	0.41	−0.13	0.02	−0.08	−0.02	**1.27**	**0.20**	0.33	0.60	0.03
SI1223 (Nepal)	26.55	86.75	91 (IG)	0.89	2.94	0.01	0.09	0.04	**0.02**	1.10	−0.03	3.84	1.89	3.59
SI1111 (Nepal)	26.72	85.97	90 (IG)	**2.38**	4.80	**0.24**	0.09	0.13	0.11	**1.30**	0.03	**10.34**	4.81	12.25
X875Y265 (India)	26.50	87.50	79 (IG)	1.17	3.00	0.10	0.26	0.35	0.18	0.22	0.02	6.23	3.48	20.32
X865Y265 (India)	26.50	86.50	77 (IG)	0.39	1.77	−0.01	−0.37	−0.29	0.01	0.79	−0.06	2.44	0.58	−7.46
SI1319 (Nepal)	26.48	87.27	72 (IG)	0.10	0.89	0.01	0.01	0.07	**0.02**	1.16	−0.02	0.85	0.44	3.11
X855Y265 (India)	26.50	85.50	63 (IG)	1.11	−1.97	−0.14	0.28	−0.01	−0.10	0.05	0.09	−5.61	0.35	1.68
X855Y255 (India)	25.50	85.50	50 (IG)	−0.95	−1.01	0.01	0.01	0.09	0.04	0.26	0.02	−2.26	−2.48	−1.66
X865Y255 (India)	25.50	86.50	39 (IG)	2.33	**6.10**	**0.20**	**0.59**	**0.51**	0.15	0.91	0.12	**15.29**	**10.26**	**26.15**
X875Y255 (India)	25.50	87.50	33 (IG)	**3.46**	**5.53**	0.11	**0.48**	0.16	0.15	1.03	0.13	**14.05**	**9.35**	**19.06**
Average				0.02	0.61	0.03	0.04	0.05	0.02	0.67	0.05	1.27	0.35	2.47
Stations with + trend				23	30	31	32	31	32	47	33	30	24	30
Stations with − trend				27	20	18	18	19	18	3	17	20	26	20
Stations With 0 trend				0	0	1	0	0	0	0	0	0	0	0
%With + trend stations				46	60	62	64	62	64	94	66	60	48	60
%With-trend stations				54	40	36	36	38	36	6	34	40	52	40
%With no trend stations				0	0	2	0	0	0	0	0	0	0	0

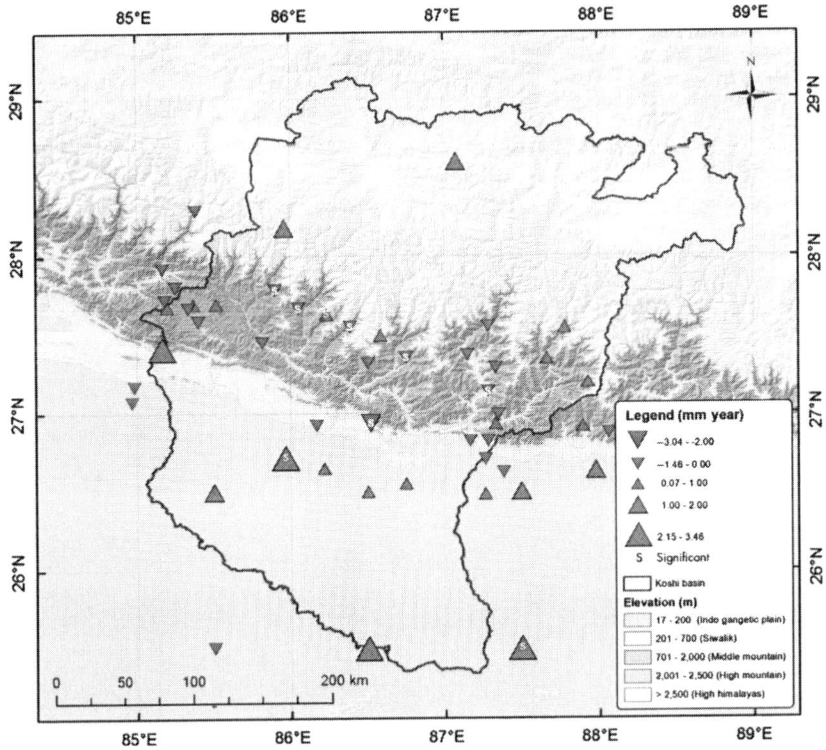

Figure 12. Trends in monthly maximum 1-day precipitation (Rx1day).

The changes in precipitation extremes for 1975 to 2010 are not as significant as the temperature extremes, but do indicate heavier rainfall over fewer days in many places, with an increase in drought periods, and marked variation between the plains areas and hills and mountains, as well as local variations, especially within the hill and mountain areas. The finding that the precipitation changes are less significant than those for temperature are consistent with the results of the study by King *et al.* (**2013**); Klein Tank *et al.* (**2006**) and others have found this result too.

4 CONCLUSION

The study looked at the long-term trends and patterns in temperature and precipitation extremes and average values over the Koshi basin from 1975 to 2010. There was a clear indication of a warming trend in maximum and minimum seasonal temperatures, but with some spatial variation. While the stations in the hills and mountains generally showed a significant warming, in the plains warming was absent or temperatures (especially maximum temperature) were decreasing. Warming was apparent in all seasons, but with the least change in the post-monsoon season. The most significant increment in minimum temperature was seen in winter.

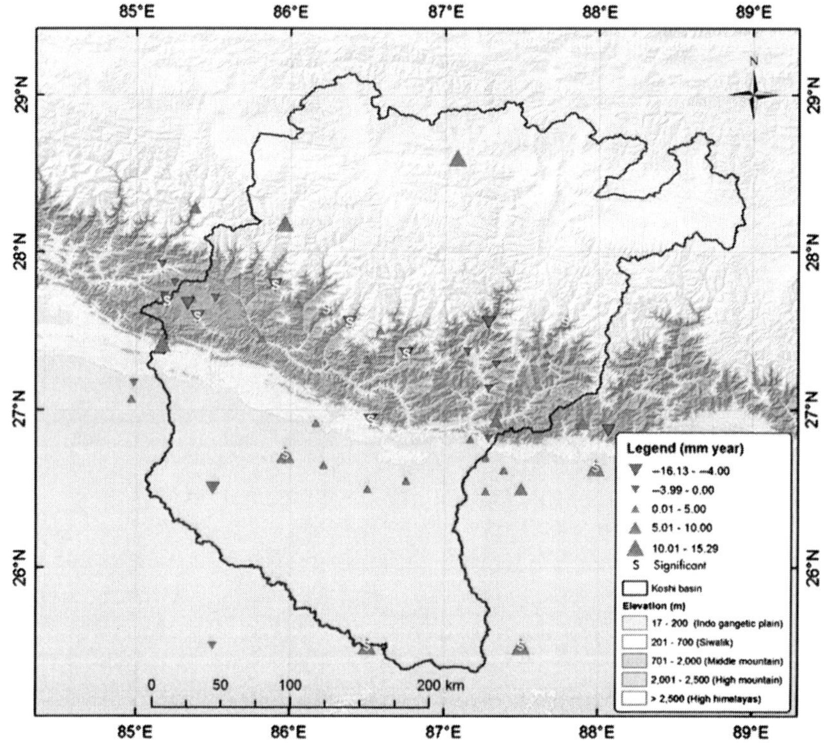

Figure 13. Trends in annual total PRCP when RR > 95th percentile (R95p).

The results also showed widespread changes in temperature extremes associated with warming, especially for those indices derived from daily minimum temperature; most of these results were statistically significant. More than 90% of stations showed an increase in monthly minimum value of minimum temperature, and 70% of stations a decrease in the annual occurrence of nights in which the daily minimum temperature was below the 10th percentile. In general, there were large coherent patterns in the basin with contrasting trends between the hills and mountains and the plains. The rising trends in maximum and minimum values of daily maximum and minimum temperatures clearly show that temperatures in the basin are increasing, although the statistical significance was not sufficient for a definitive analysis. In general, a much larger percentage of stations showed significant changes in minimum temperature than in maximum temperature, and the minimum temperature increased three times faster than the maximum temperature. All stations in the basin showed an increase in the number of warm nights (TN90p), most statistically significant, and 70% a decrease in the number of cool nights (TN10p). On average the number of cool days (TX10p) is decreasing and the number of warm days (TX90p) is increasing. The duration of warm spells is increasing (WSDI) at a rate of 15.7 days decade^{-1} while the duration of cold spells (CSDI) is slightly decreasing.

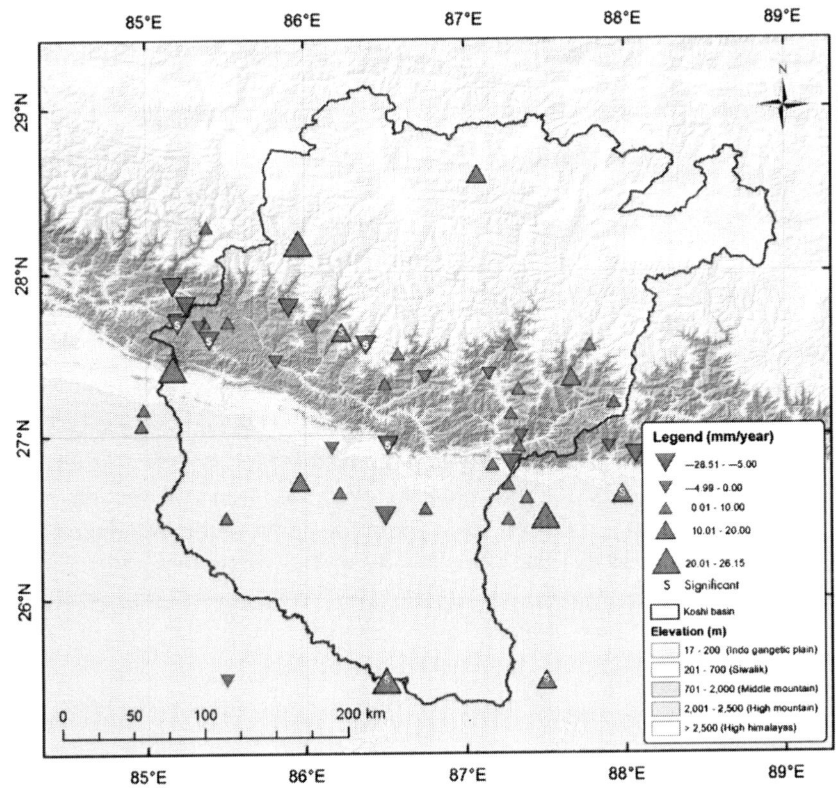

Figure 14. Trends in annual total precipitation in wet days with RR ≥ 1 mm (PRCPTOT).

The changes in the plains region were often different to those in the hills and mountains. The number of summer days, monthly maximum and minimum values of daily maximum temperature (TXx, TXn), percentage of days when the maximum value was >90th percentile (TX90p), percentage of days when the minimum value was <10th percentile (TN10p), warm and cold spell duration (WSDI and CSDI), and DTR all decreased at the stations in the IGs, while the number of cold days increased. This difference may in part be due to the increased duration of fog episodes in the plains in recent years. Overall the results of the study are consistent with those of previous studies, but provide special insights into the differences between the mountains and the plains.

Overall, the number of heavy precipitation days (R10) and number of extremely heavy precipitation days (R20) show increasing trend at 62 and 64% of weather stations, respectively and the number of very wet days increased at a faster rate than the number of extremely wet days. The SDII also increased slightly. The precipitation results indicated a tendency towards shorter and more intense spells of rainfall. Consecutive dry days (CDD) increased at a high rate (6.7 days decade^{-1}) at most stations, and much higher than the number of consecutive wet days (CWD) (0.5 days decade^{-1}). Overall, the annual total precipitation increased over the basin between the start and end of the period, as did the annual total wet day precipitation, but there was no consistent trend. Taken together, the results suggest that there is a mixed pattern of changes in precipitation extremes which support the idea that the Koshi basin is becoming generally wetter, but the changes were not statistically significant at all stations.

The changes in DTR, duration of warm and cold spells, total precipitation, SDII, and consecutive dry days, could have significant societal impacts. The Koshi basin is home to a large population; many live below the poverty line and their coping capacity, especially in marginal areas, is low. Thus, there is a strong need to understand the vulnerability of the basin's inhabitants and to mainstream good practices for adaptation. This study will help the relevant stakeholders to look more clearly at the frequency, magnitude, and duration of climate extremes in the water resources sector, and the potential socioeconomic impact, especially the impact on agriculture, health, and other climate-related livelihood activities.

ACKNOWLEDGEMENTS

This study was undertaken under the Himalayan Climate Change Adaptation Programme (HICAP). HICAP is implemented jointly by ICIMOD, CICERO, and GRID-Arendal in collaboration with local partners and is supported by the Governments of Norway and Sweden. The authors would like to express their sincere gratitude to the Department of Hydrology and Meteorology (DHM), Nepal, for providing the data and Mr. Gauri Shankar Dangol of ICIMOD for preparing the maps. The views and interpretations in this publication are those of the authors. They are not necessarily attributable to ICIMOD and do not imply the expression of any opinion by ICIMOD concerning the delimitation of frontiers or boundaries.

REFERENCES

1. Aguilar, E., Auer I, Brunet M, Peterson TC, Wieringa J. 2003. Guidelines of climate metadata and homogenization. WCDMP No. 53 – WMO/TD-No. 1186, Geneva, Switzerland.

2. Aguilar E, Peterson TC, Obando PR, Frutos R, Retana JA, Solera M, Soley J, García IG, Araujo RM, Santos AR, Valle VE,Brunet M, Aguilar L, Aívarez L, Bautista M, CastañÓn C, Herrera L, Ruano E, Sinay JJ, Sánchez E, Oviedo GIH, Obed F,Salgado JE, Vázquez JL, Baca M, Gutíerrez M, Centella C, Espinosa J, Martínez D, Olmedo B, Espinoza CEO, Núñez R,Haylock M, Benavides H, Mayorga R. 2005. Changes in precipitation and temperature extremes in Central America and northern South America, 1961–2003. *J. Geophys. Res.* **110**: D23107

3. Alexander LV, Zhang X, Peterson TC, Caesar J, Gleason B, Klein Tank AMG, Haylock M, Collins D, Trewin B,Rahimzadeh F, Tagipour A, Rupa Kumar K, Revadekar J, Griffiths G, Vincent L, Stephenson DB, Burn J, Aguilar E,Brunet M, Taylor M, New M, Zhai P, Rusticucci M, Vazquez-Aguirre JL. 2006. Global observed changes in daily climate extremes of temperature and precipitation. *J. Geophys. Res.* **111**: D05109

4. Baidya SK, Shrestha ML, Sheikh MM. 2008. Trends in daily climatic extremes of temperature and precipitation in Nepal. *J. Hydrol. Meteorol.* **5**(1): 38–50.

5. Caesar J, Alexande LV, Trewin B, Tse-ring K, Sorany L, Vuniyayawa V, Keosavang N, Shimana A, Htay MM,Karmacharya J, Jayasinghearachchi DA, Sakkamart J, Soares E, Hung LT, Thuong LT, Hue CT, Dung NTT, Hung PV,Cuong HD, Cuongo NM, Sirabaha S. 2011. Changes in temperature and precipitation extremes over the Indo-Pacific region from 1971 to 2005. *Int. J. Climatol.* **31**: 791–801.

6. Donat MG, Alexander LV, Yang H, Durre I, Vose R, Dunn RJH, Willett KM, Aguilar E, Brunet M, Caesar J, Hewitson B,Jack C, Klein Tank AMG, Kruger AC, Marengo J, Peterson TC, Renom M, Oria Rojas C, Rusticucci M, Salinger J, Elrayah AS, Sekele SS, Srivastava AK, Trewin B, Villarroel C, Vincent LA, Zhai P, Zhang X, Kitching S. 2013. Updated analyses of temperature and precipitation extreme indices since the beginning of the twentieth century: the HadEX2 dataset. *J. Geophys. Res.* **118**: 2098–2118 Frich P, Alexander LV, Della-Marta P, Gleason B, Haylock M, Klein Tank AMG, Peterson TC. 2002. Observed coherent changes in climate extremes during the second half of the twentieth century. *Clim. Res.* **19**: 193–212.

7. Goswami BN, Venugopal V, Sengupta D, Madhusoodanan MS, Xavier PK. 2006. Increasing trend of extreme rain events over India in a warming environment. *Science* **314**: 1442–1444.

8. Hingane LS, Rupa Kumar K, Ramana Murthy BV. 1985. Long-term trends of surface air temperature in India. *J. Climatol.* **5**: 521–528.

9. Islam N. 2009. Understanding the rainfall climatology and detection of extreme weather events in the SAARC region part II utilization of RCM data. SMRC-No. 29, SAARC Meteorological Research Centre (SMRC), Dhaka, 1–58.

10. Ji Z, Kang S. 2014. Evaluation of extreme climate events using a regional climate model for China. *Int. J. Climatol.* **35**:888–902

11. Ji ZM, Kang SC, Zhang D, Zhu C, Wu J, Xu Y. 2011. Simulation of carbonaceous aerosols over the Third Pole and adjacent regions: distribution, transportation, deposition, and climatic effects. *Clim. Dyn.* **36**: 1633–1647

12. Ji ZM, Kang SC, Cong ZY, Zhang QG, Yao YD. 2015. Simulation of carbonaceous aerosols over the Third Pole and adjacent regions: distribution, transportation, deposition, and climatic effects. *Clim. Dyn.* **45**(9): 2831–2846 Jones PD, New M, Parker DE, Martin S, Rigor IG.

1999. Surface air temperature and its variations over the last 150 years. *Rev. Geophys.* **37**: 173–199.

13. Khanal NR, Shrestha MS, Ghimire M (eds). 2007. Preparing for Flood Disaster. Mapping and Assessing Hazard in the Ratu Watershed, Nepal. ICIMOD: Kathmandu.

14. King AD, Alexander LV, Donat MG. 2013. The efficacy of using gridded data to examine extreme rainfall characteristics: a case study for Australia. *Int. J. Climatol.* **33**: 2376–2387

15. Klein Tank AMG, Peterson TC, Quadir DA, Dorji S, Zou X, Tang H, Santhosh K, Joshi UR, Jaswal AK, Kolli RK, Sikder AB,Deshpande NR, Revadekar JV, Yeleuova K, Vandasheva S, Faleyeva M, Gomboluudev P, Budhathoki KP, Hussain A,Afzaal M, Chandrapala L, Anvar H, Amanmurad D, Asanova VS, Jones PD, New MG, Spektorman T. 2006. Changes in daily temperature and precipitation extremes in central and south Asia. *J. Geophys. Res.* **111**: D16105

16. Klein Tank AMG, Zwiers FW, Zhang X. 2009. Guidelines on analysis of extremes in a changing climate in support of informed decisions for adaptation. WCDMP – No. 72-WMO/TD No. 1500, Geneva, Switzerland.

17. Kruger AC. 2006. Observed trends in daily precipitation indices in South Africa: 1910–2004. *Int. J. Climatol.* **26**(15):2275–85

18. Liu X, Yin Z-Y, Shao X, Qin N. 2006. Temporal trends and variability of daily maximum and minimum, extreme temperature events, and growing season length over the eastern and central Tibetan Plateau during 1961–2003. *J. Geophys. Res.* **111**: D19109

19. Manandhar KB. 2006. The fog episode in southern Terai plains of Nepal: some observations and concepts. *J. Hydrol. Meteorol.* **5**: 95–101.

20. Manton MJ, Della-Marta PM, Haylock MR, Hennessy KJ, Nicholls N, Chambers LE, Collins DA, Daw G, Finet A,Gunawan D, Inape K, Isobe H, Kestin TS, Lefale P, Leyu CH, Lwin T, Maitrepierre L, Ouprasitwong N, Page CM,Pahalad J, Plummer N, Salinger MJ, Suppiah R, Tran VL, Trewin B, Tibig I, Yee D. 2001. Trends in extreme daily rainfall and temperature in Southeast Asia and the South Pacific: 1961–1998. *Int. J. Climatol.* **21**: 269–284

21. Miller JC, Miller JN. 1988. Statistics for Analytical Chemistry. Ellis Horwood: New York, NY.

22. Neupane N, Murthy MSR, Rasul G, Wahid S, Shrestha AB, Uddin K. 2013. Integrated biophysical and socioeconomic model for adaptation to climate change for agriculture and water in the Koshi basin. In Handbook of Climate Change Adaptation. Springer-Verlag: Berlin, Heidelberg

23. Nicholls N, Gruza GV, Jouzel J, Karl TR, Ogallo LR, Parker DE. 1996. Observed climate variability and change. InClimate Change 1995: The Science of Climate Change, Houghton JT, Filho LGM, Callander BA, Harris N, Kattenberg A,Maskell K (eds). Cambridge University Press: Cambridge, UK, 133–192.

24. Rajeevan M, Jyoti B, Kale JD, Lal B. 2005. Development of a high resolution daily gridded rainfall data for Indian region. IMD Meterological Monograph. No: Climatology 22, Pune, India. 27 pp.

25. Sharma KP, Berrien Moore III, Charles JV. 2000. Anthropogenic climatic and hydrologic trends in the Kosi Basin, Himalaya. *Clim. Change* **47**: 141–65.

26. Shrestha AB, Devkota LP. 2010. Climate Change in the Eastern Himalayas: Observed Trends and Model Projections. ICIMOD: Kathmandu, 20.

27. Shrestha AB, Wake CP, Mayewski PA, Dibb JE. 1999. Maximum temperature trends in the Himalaya and its vicinity. An analysis based on temperature records from Nepal for the period 1971–94. *J. Clim.* **12**: 2775–2787.

28. Shrestha AB, Erikssona M, Mool P, Ghimire P, Mishra B, Khanal NR. 2010. Glacial lake outburst flood risk assessment of Sun Koshi basin, Nepal. *Geomat. Nat. Hazard. Risk* **1**: 157–169.

29. Singh D, Tsiang M, Rajaratnam B, Diffenbaugh NS. 2014. Observed changes in extreme wet and dry spells during the South Asian summer monsoon season. *Nat. Clim. Change* **4**: 456–461

30. Sivakumar MVK, Stefanski R. 2011. Chapter 2: Climate change and food security in South Asia. In Climate Change in South Asia, Lal R, Sivakumar

MVK, Faiz SMA, Mustafizur Rahman AHM, Islam KR (eds). Springer Netherlands:Dordrecht, The Netherlands, 13–30.

31. Srivastava AK, Rajeevan M, Kshirsagar SR. 2009. Development of a high resolution daily gridded temperature data set (1969–2005) for the Indian region. *Atmos. Sci. Lett.* **10**: 249–254

32. State T, Service M. 2008. Trends in Turkey Climate Extreme Indices from 1971 to 2004. BALWOIS Conferences publications, In *Third International Conference BALWOIS*, 27–31 May 2008, Ohrid, Republic of Macedonia, 1–9.

33. Vincent LA, Peterson TC, Barros VR, Marino MB, Rusticucci M, Carrasco G, Ramirez E, Alves LM, Ambrizzi T, Berlato MA, Grimm AM, Marengo JA, Molion L, Moncunill DF, Rebello E, Anunciacao YMT, Quintana J, Santos JL, Baez J,Coronel G, Garcia J, Trebejo I, Bidegain M, Haylock MR, Karoly D. 2005. Observed trends in indices of daily temperature extremes in South America 1960–2000. *J. Clim.* **18**: 5011–5023.

34. Wang XL. 2008. Accounting for autocorrelation in detecting mean-shift in climate data series using the penalized maximal t of f test. *J. Appl. Meteorol. Climatol.* **47**: 2423–2444.

35. Wang XL, Feng Y. 2010. *RHtestsV3_UserManual*. Climate Research Division Atmospheric Science and Technology Directorate Science and Technology Branch, Environment Canada Toronto Ontario, Canada. http:// cccma.seos.uvic.ca/ETCCDI/software.shtml (accessed 1 March 2014).

36. Whan K, Alexander LV, Imielska A, McGree S, Jones D, Ene E, Finaulahi S, Inape K, Jacklick L, Kumar R, Laurent V,Malala H, Malsale P, Pulehetoa-Mitiepo R, Ngemaes M, Peltier A, Porteous A, Seuseu S, Skilling E, Tahani L, Toorua U,Vaiimene M. 2013. Trends and variability of temperature extremes in the Tropical Western Pacific. *Int. J. Climatol.***34**(8): 2585–603

37. WMO. 1996. Some methods of climatological analysis Technical Note No. 81, WMO-No. 199. TP.103, Geneva, Switzerland.

38. Yan Z, Jones PD, Davies TD, Moberg A, Bergström H, Camuffo D, Cocheo C, Maugeri M, Demarée GR, Verhoeve T,Thoen E, Barriendos M, Rodríguez R, Martín-Vide J, Yang C. 2002. Trends of extreme

temperatures in Europe and China based on daily observations. *Clim. Change* **53**: 355–392

39. You Q, Kang S, Aguilar E, Yan Y. 2008. Changes in daily climate extremes in the eastern and central Tibetan Plateau during 1961–2005. *J. Geophys. Res.* **113**: D07101

INDEX